Patterns of Intuition

Gerhard Nierhaus
Editor

Patterns of Intuition

Musical Creativity in the Light of Algorithmic Composition

Editor
Gerhard Nierhaus
Institute of Electronic Music and Acoustics
University of Music and Performing
Arts Graz
Graz, Austria

ISBN 978-94-024-0092-2 ISBN 978-94-017-9561-6 (eBook)
DOI 10.1007/978-94-017-9561-6

Springer Dordrecht Heidelberg New York London

Softcover reprint of the hardcover 1st edition 2015

Printed on acid-free paper

Springer is part of Springer Science+Business Media (www.springer.com)

Acknowledgments

I wish to thank all institutions and people who contributed to the work for the project and the making of this book. In the first place, I wish to acknowledge the role of the Austrian Science Fund (FWF), whose funding made possible a project of this scale. I am grateful also to Springer Verlag for their professional and pleasant cooperation throughout the publication of this volume. I am particularly thankful to my colleagues Daniel Mayer and Hanns Holger Rutz, as well as the contributing researchers and, most of all, the composers who, over a large period of time, patiently, enthusiastically, and with substantial personal effort, exposed their practice for the benefit of an unusual exploration of their compositional approaches. I would like also to mention Tamara Friebel who was responsible for the translation of German texts and proofreading.

Contents

Introduction

Gerhard Nierhaus

'Give a sequence of six numbers by choosing randomly from 1, 2 or 3!'—Most people would respond to this task with number sequences such as the following: '1 2 3 2 3 1', '2 3 2 1 3 1', '3 3 2 1 1 3', '1 2 1 1 3 2', '1 3 1 2 2 3'. The character of the sequences given would not be essentially different if the task were slightly varied in sequence length or quantity of numbers to choose from.

One may now ask whether there are any commonalities to such sequences, and whether there are any latent rules at work during their forming, rules as yet unknown. In approaching an answer to this question, one might transfer the task to a computer program. Within the sequences produced by the program there will occasionally be ones such as '1 1 1 1 3 1' or '1 3 1 3 1 3', or ones like '2 2 2 2 2 2' or '1 2 3 1 2 3'. Such will however only appear exceptionally in the sequences generated by humans.[1] Notions (in themselves formally correct or incorrect) and strategies (used intentionally, or, in other cases, automatically or unconsciously) for a 'random selection' differ between individuals.

Hypothetically, and in order to gain some insight into the structure of those humanly generated sequences, the following "rule of thumb" can be applied to the making of such sequences: 'When forming a sequence, strive to use all numbers, and seek to avoid obvious patterns'. A next step would aim at a formally correct representation of this "rule",[2] in turn followed by implementing a software capable of producing such sequences. A user evaluation could then deliver clues as to the adequacy of the hypothesis that the structure of the human generated sequences can be approximated by the formalised rule of thumb.

[1] Whenever such sequences are found they will arguably stem from someone with a background in statistics, who has reflected on the task and probably possesses a good sense of humour.
[2] The criterion 'seek to avoid obvious patterns' already raises a number of tricky and interesting questions to the task.

G. Nierhaus (✉)
Institute of Electronic Music and Acoustics,
University of Music and Performing Arts Graz, Graz, Austria
e-mail: nierhaus@iem.at

G. Nierhaus (ed.), *Patterns of Intuition*,
DOI 10.1007/978-94-017-9561-6_1

Such and similar were the thoughts that—transferred to the domain of music—led me to initiate a project in which some aspects of composition are viewed from a different perspective by means of algorithmic composition. I envisaged a kind of musical analysis that begins with the composers' structural ideas and, by way of a dialogical process, makes the ideas visible on a more objectifiable plane.

The present book is a result of a three year research project (*Patterns of Intuition*, funded by the Austrian Science Fund (FWF), project number: AR-79), in which my colleagues Daniel Mayer, Hanns Holger Rutz and myself stood in a creative dialogue with numerous composers, seeking to trace important facets of their respective individual compositional approaches. In all this, the composers themselves chose a point of departure, where upon we focused on researching a specific aspect or compositional principle, proceeding thereafter in a dialogical manner with the artists. Generally, the procedure was thus:

Presentation of a compositional principle. Formalisation of the approach and implementation in the form of a computer program. Computer generation of musical material. Evaluation of the results by the composer. Modification of the strategy of formalisation with respect to the identified objections. Entry into new and further cycles of generation and evaluation until correlation between the computer generated results and the composers' aesthetic preferences is sufficiently high, or the limits of formalisation have been reached (which might be the case for various reasons). As indicated by the latter limits of formalisation, compositional decisions beyond this point are reached intuitively, and are thus outside the reach of a meta level representation.

The project Patterns of Intuition was therefore not aimed at addressing musical intuition as a whole in completely formalisable terms; rather, the project aim was to shed light on those particular aspects of intuitively made decisions that can be related back to implicit rules or constraints applied by the composer.

One example of such a process is the collaboration with composer Clemens Nachtmann. Nachtmann's work is led by his avoidance of tonal associations. In his case we wrote a program which works ex negativo so to speak: 'everything is possible, but!' Concentrating on chords, we at first formulated simple constraints to exclude pitch constellations with associations of tonality. We then presented diverse instances of generated chords to Nachtmann, which he then evaluated. After a number of cycles through which we followed his observations and critical comments we arrived at a dense web of constraining conditions, which in the end selected only 14 chords (from a vast number) that would ruffle no feathers if used within an 'orthodox' new music context.

In a case such as this, a traditional score analysis would not have been able to deliver a full description of all harmonic constraints underlying the forming of such chords, since it can only rely on exemplificatory materials. At the same time it is clear that the results of analyses coming from this and similar projects within POINT are not cast in stone—we are dealing with snapshots from a compositional process, often within the context of a single piece, during the course of which structures most often undergo further changes and transformations. Nachtmann himself has commented pointedly on this and other aspects of his project contribution.

The basic approach of the project, with its generative and evaluative cycles, obviously describes an idealised model. Clearly, within the framework of such processes there appear numerous side effects, which feed back into the results of the analyses. To give some examples: composers are generally unfamiliar with a situation in which they discuss their compositional work during its origination and, at same time, evaluate generated structures with respect to their own goals. Besides, the criteria for evaluation can change during the course of such a process, even those referring to their own work, so that it can seem appealing not only to analytically observe the results of what was computationally generated, but to introduce them into the ongoing creative process. In broad allusion to quantum mechanics one might say that the observation changes the outcome.

In each case of collaboration with the composers the approach taken was markedly different, and it did not follow the described cycles of generation and evaluation in every case. The same diversity was present in the individual compositional practices and aesthetic positions of the composers. There were a large range of different approaches, starting from Elisabeth Harnik's working with improvisational structures, through the attempt of an automatic classification of personal preferences in the case of Matthias Sköld, to Bart Vanhecke's and Peter Lackner's work with interval- and tone-rows.

Structural Overview

In the first part of the book—*Composers' Projects*—each chapter describes the collaboration with a composer. The chapters begin with a presentation of the composer's artistic background. This is followed by sections called *Artistic Approach* and *Exploring a Compositional Process*, concluding with a *Project Review*. Each *Artistic Approach* section features the composers' discussions in relation to the following topic areas: (1) Statement: A concise description of their personal aesthetic position and their compositional approach; (2) Personal aesthetics: This concerns details of individual practices; (3) Formalisation and intuition: The composer's views on the field of tension between formalisation and intuition; (4) Evaluation and self-reflection: How each composer appraises and conceives the results of her or his work; (5) Project expectations: Insights the composer hopes to gain through work on the project.

The section *Exploring a Compositional Process* describes the collaboration between composer and project team.

The section *Project Review* is dedicated to composers' discussions of the outcomes of the collaboration, considering especially whether it led them towards new insights on their own compositional process.

Regarding the chapter contents: Next to being a composer, **Elisabeth Harnik** is a well known piano improviser. In her project, she sought to understand some of the stylistic choices she makes in her chosen musical constellations. For this, we recorded improvisations and generated new ones using prefix- and suffix-trees within a variety of context based methods.

Clemens Nachtmann's aesthetics avoids tonal associations. Together with the composer we arrived at a system by which we computationally determined significant criteria matching Nachtmann's choice of chord materials and aesthetic practice and verified by Nachtmann in various stages of evaluations. The system combines a method of exclusion with complete enumeration of all solutions.

Eva Reiter took sounds she recorded from a range of machines as a point of inspiration for a string quartet. The research collaboration addressed potential correlations between the original audio files and the finished quartet on the level of the sounding structure.

A part of **Clemens Gadenstätter's** work is based on a complex system of intertwined metaphoric expressions. We aimed at modelling this network of relations by way of a generative grammar, and to compare possible derivations of the system with the solutions he arrived at himself. Aspects of weak synesthesia and metaphor theory are of further relevance to Gadenstätter's work. **Thomas Eder** contributed a linguistic perspective to the composer's research.

Interlocking musical patterns and polyrhythmic structures are among the characteristics of **Dimitri Papageorgiou's** compositional practice. In this project we formalised these techniques, after which Papageorgiou showed how these formalisations apply within the context of a number of his compositions.

Transitions between harmonic fields are important to **Katharina Klement's** work. The aim was to find specific principles for strategies of morphing to approximate her handcrafted transitions.

Orestis Toufektsis at times works with harmonic processes, shaping them more or less intuitively. Based on genetic algorithms, we developed a system which enabled Toufektsis to generate different chord sequences. He then evaluated these chord sequences as to their compositional adequacy under different criteria. Seeking to keep the evaluation criteria flexible and to provoke surprising solutions, a human fitness rating was used rather than an algorithmic fitness function.

Alexander Stankovski works with a technique he calls 'mirroring', a technique not dissimilar to a use of palindromes. Stankovski's technique involves a conscious variation of nesting mirrors within mirrors, and also applying the mirroring procedure to different musical parameters.

With **Mattias Sköld** we investigated whether machine learning might assist our understanding of what makes musical structures 'interesting' rather than 'unininteresting'.

Djuro Zivkovic often works with chord sequences created from combinations of difference tones. We implemented his approach in various ways and compared the results with Zivkovic's handwritten solutions.

Bart Vanhecke uses 54-tone interval-rows as a basis for compositional elaboration; central to the present project was the question whether "optimal" rows computed via brute force procedures would be of additional compositional value when compared to those rows already considered optimal by the composer.

Peter Lackner's practice features an innovative approach to the systematisation of 12-tone rows. In the second section of this chapter this systematisation is presented in terms of mathematical music theory by mathematician **Harald Fripertinger** in collaboration with the composer.

Interdisciplinary Contributions

The collaborations with the composers should also to be viewed within the context of different disciplines. Given the involvement of creative processes, this project can certainly be conceived as a form of artistic research, while at the same time the analytical focus situates it into a musicological context. Beyond this, the kinds of methods used also make this project an undertaking in algorithmic composition. The underlying methodology—namely the working through of cycles of generation and evaluation—characterises the project as experimental and last but not least, the project's results open up discourses which can be oriented according to a variety of different perspectives. In order to look "outside the box", so to speak, a number of outstanding researchers (who are, in part, active as artists also) were invited to respond to the project and its outcomes informed by their different perspectives; to offer independent contributions on the topic or, alternatively, more general views from their respective research fields.

Regarding the chapter contents: **Darla Crispin** reflects upon the contemporary status of composition both as an artistic and as an academic practice, as seen from the perspective of artistic research. She speculates on how some of the creative listening practices described by composers within the POINT project might help to revivify the relationship between sound, structure and meaning which lies at the heart of a healthy compositional tradition.

William Brooks situates the POINT project in the context of experimental practice in music, especially in the tradition of pragmatist aesthetics initiated by John Dewey.

Nicolas Donin offers an epistemological reflection on the way composers' discursive and self-critical skills are embedded in POINT and more generally in artistic research. Self-analysis as a tool for (scientific or artistic) research is both needed and challenging, as recent debates in psychology and phenomenology show.

Sandeep Bhagwati poses the question whether algorithmic composition might one day replace human music making. Starting with our fear of intelligence amplifiers, and delving into the presence and future of listening, he explores the aesthetic impact of computational musicking on our understanding of what music is and what it could be.

Guerino Mazzola analyses part I of Pierre Boulez's *Structures pour deux pianos* and proposes a resynthesis by a computational approach.

David Cope describes how the use of computers in the composing process is a natural outgrowth and continuation of how composers have been using algorithms for composing since the very beginning of recorded time.

Postscript

This project considered a wide range of compositional approaches.[3] Had we limited ourselves to the work of a single or only a few composers, the analyses could, of course, have been brought to a deeper level. However, I placed more importance on the integration of composers from a very diverse range of aesthetic positions and individual practices into the project. Many questions and issues have had to remain unanswered; yet they have opened up space for further intriguing discourse. I hope that the projects presented in this book inspire future work in this direction.

Gerhard Nierhaus, Graz, 29th August 2014

[3] From initially 16 collaborations with different composers we selected 12 which were documented in this book.

Composers' Projects

Elisabeth Harnik/Improvisational Re-assemblies

Elisabeth Harnik, Hanns Holger Rutz and Gerhard Nierhaus

Elisabeth Harnik was born in Graz, Austria, and received her first musical education at the age of five.[1] At the age of 10 she started playing the piano, an instrument that became a constant companion during her musical development. After finishing school she initially studied piano at the Music University of Graz. During her student time she turned at first to jazz and jazz-singing, working with Ward Swingle (Swingle Singers) and continued her education with Ines Reiger, Sheile Jordan, and Jay Clayton in the field of vocal improvisation. Harnik received further important impulses as a pianist by studying the repertoire of contemporary music, participating at the Vienna days of contemporary piano music and she continued to work as an improvisation musician. Harnik did not find until her intrinsic approach of the instrument with free improvisation until meeting the French double bass player Joëlle Léandre, whose musical journey from classical music to improvisation she shared. In the following years she worked as a pianist in various areas of improvisational music and participated, amongst others, in the classes of Peter Kowald, Lauren Newton or David Moss. As a pianist, Harnik looks for the challenge to dissolve or disperse the long-established norms and apparently fixed boundaries of the instrument, where she considers it her task and challenge to permanently re-invent her playing and her instrument.

[1] Biographical introduction and texts from the composer translated from the German by Tamara Friebel.

E. Harnik
Institute for Composition, Music Theory, Music History and Conducting,
University of Music and Performing Arts Graz, Graz, Austria
e-mail: elisabeth.harnik@kug.ac.at

H.H. Rutz · G. Nierhaus (✉)
Institute of Electronic Music and Acoustics,
University of Music and Performing Arts Graz, Graz, Austria
e-mail: nierhaus@iem.at

H.H. Rutz
e-mail: rutz@iem.at

G. Nierhaus (ed.), *Patterns of Intuition*,
DOI 10.1007/978-94-017-9561-6_2

However, in her artistic desire to "create" she was looking for an additional means of expression, and this is where her first compositions emerged. An encounter with the Swiss composer Beat Furrer during her participation in Haubenstock-Ramati's *Amerika* conducted by Furrer a few years earlier is still alive in her memory. Harnik received essential further impulses and stimuli for the artistic development from the visits of a "Deep Listening Workshop" with the American composer and accordion player Pauline Oliveros.

After these events, Harnik studied composition at the Music University of Graz with Beat Furrer. Soon after finishing this study, composing quickly became a second essential aspect of Harnik's artistic activities, alongside her practice as a free improvisation pianist. Harnik performed as a piano soloist and in ensembles with prominent representatives of improvisational music at national and international festivals; her composition activities also lead to commissions and performances of her works by well-known soloists and ensembles.

Despite the predominant separation of composed and improvised music in the present performance climate, there are more and more overlaps between both disciplines at festivals for contemporary music or improvised music emerging. In some of her works Harnik relies on a strategy where one influences the other, balancing a connection which uses economical and practical means between improvisation and composition, moving from a confrontation to a synthesis, nevertheless both fields of activities remain in the majority of Harnik's oeuvre rather disjoint. When it comes to composing it is the fascination to move freely along the time-axis as well as the possibility to work meticulously on details of the realisation of sound and form. Improvisation is more about its enforced linear time lapse, but on the other hand she sees it as a "going backward into the future"—with the presentiment of approaching a future which is still open, that has to be shaped artistically as it emerges.

In Harnik's compositional work, she rarely starts at the beginning of a piece; she likes to move erratically along the time line, where structures of a later section often feedback to previous parts. With respect to structures, she likes to work with complex rhythmical and melodic patterns, which are combined and selected in different ways. The musical progressions are notated with utmost precision, which in their frequent complexity open the sought-after "new".

In the compositions of Harnik there is often a refreshing friction and/or tension between self-imposed rules and their modifications, even a breaking of the rules caused by intuitive decisions. The rules open an area of discourse, which gets evaluated and processed by the musical intuition as well as having the effect of completely re-forming the composition.

In her current work, the search for methods to give a composition more flexibility and elasticity, without losing the precision of conventional notation is an important focus of her artistic exploration. In a recent piece, *grafting* (*veredeln, aufsetzten, anreichern...*)[2] she translates methods from other working practices into her composition, for example, the role of how an improvisation orchestra uses signs and hints to initiate their play. These practices widen her scope, leading to modifications

[2] "veredeln, aufsetzten, anreichern" roughly translates to: "refine, setup, accumulate".

and changes within the compositional work. The processes act as a "medium" in order to be able to implement a flexible zone in the conventional musical score writing.

Re-framing II (inside the frame is what we're leaving out) for string quartet is composed using an "elastic form". The sequence of the form is set. Within the sections, however, are options for the individual players. The performer can alternate between different types of notational reading. Depending on the selected type of reading, the shape of the time and rhythm-melodic patterns are affected. Through this process, the time frame is reinterpreted multiple times, to bring flexibility within the established structure of the work.

Artistic Approach

Statement

> The nautilus is a nomad which explores the oceans on its vast journeys. It collects particles of each investigated place to build its shell, becoming a sort of collection of its explorations. Every year the shell forms and adds a new chamber. The old chamber is sealed and the animal moves into the new chamber...

I see parallels in my artistic work as a composer and improviser to the journey of the nautilus. In both disciplines of composition and improvisation there is a drive for me to obtain something "new" within a particular framework of conditions and thus to extend the boundaries.

As a professional pianist and improviser, my hands have acquired a rich repertoire of gestures. This is further refined, extended or also revised by regular frequent practice and reflexion. It can be described like a ritual: from a state of alert curiosity, in which some decisions are consciously left up in the air, I let myself be guided by the expectation of what will come. I have an attentive anticipation of the possible outcome, but one which can still remain foreign or strange to me. It is like while playing, something can spontaneously occur which is new to the previous context. Hand and ear "localise" the incident and almost "anticipate" the foreign element. I then take this new engagement on with a readiness to take a risk and follow it up. When composing I also choose certain working methods, which make me follow up particular musical incidents spontaneously. Mostly, I do not know which result will come from it, but that is what constitutes the excitement in both disciplines. They are only differing ways to obtain a sought-after "new".

I consider composing and improvising as a kind of interplay between the calculated and the inconceivable: a reflexion about a developed sound vocabulary—be it via preconceived of spontaneous interventions—and a tracing of an unconscious inner structure.

Personal Aesthetics

Whether I write a piece in the conventional sense or I play an improvisation, both are highly complex creative processes. I like to put improvisation and composition as counterparts to each another, and the discussion often ends up being a kind of power struggle or trial of strength where either the one or the other loses. For me however both composition and improvisation represent a complex interplay of activities, which assigns meaning to musical material—I appreciate both disciplines because I can reach something with both different creative methods.

The possibility to move freely along the time-line when writing, to later exchange what's already written with new findings and insight—to let this influence future sections back in the beginning—leads to a completely different approach compared to the linear time structure of an improvisation. On the contrary the challenge of improvisation lies precisely in the brilliance of the moment since no posteriori correction is possible. The role of listening is crucial, which transfers and takes me into a state of subtle presence. Everything that is heard—the carrier of information and relation—is composed or made up of sudden, imminent direct sensory perceptions and sensations, or of a pensive leaning towards old experiences and intuitive presumptions.

In my work as an improviser I meet musicians from all different musical backgrounds. My personal aesthetic is based on a repertoire, which I have collected over many years in my improvisation and composition practice. It is affected by my cultural heritage and education and also by international and intercultural collaborations with performers of various musical genres. Contemporary music, jazz, electronic music, rock music and Indian music have crucially influenced my handling of aesthetic preferences. Improvised music is an artistic area that is influenced by different approaches and positions.

I would call my aesthetic as an improviser "integrative" rather than anything else. It is impossible to deny my central-European heritage—nevertheless I observe, especially in my practice as an improvisation artist, that by the exchange with musicians of other cultures and different genres I am repeatedly encouraged to consider the often unconsciously adopted concepts of western avant-garde art and music. This implicates that I allow a pluralistic point of view in the aesthetic of my improvisation, but of course, there are always boundaries.

Improvisation occurs often as a collaborative act. In my opinion this requires one to be open to "foreign" aesthetics and to be ready to leave behind your own preferences. I would go even further and say that in a group improvisation the group sound, respectively the form of the moment takes primacy over the aesthetic of the individual members. In a group improvisation the various kinds of information processing change. Separated and sequential linear sound vocabulary—with or without a preconceived system—is combined with non-linear, presently sounding, imagined or remembered information.

When composing conventionally or in a solo improvisation, the dimension of the collective nuance is of course missing, which is so eminently important in a group improvisation. I alone am the "author/originator/creator" of my actions. Nevertheless I often manage also to take on a multi-perspective when composing or playing solos, which allows a plurality of discourses to happen simultaneously, whose individual layers can arbitrarily interrupt each other or respectively pass into fore- or background.

Formalisation and Intuition

Each composition and improvisation carries within a certain interrelation between "interpretation" as formalisation and "spontaneity" as intuition. It is therefore interesting as a composer and improviser to gain within this respective framework something "new".

In recent times, when I compose with pen and paper, I work increasingly with patterns, which I formulate as a form of basic configuration of sounds, which react, to different filter processes. For the filter processes, which blend in and out the sound and motion patterns I use mostly rigid rule-based systems like cellular automata.[3] The almost automatic execution of the rules allows me to react intuitively to the emerging body of sound. Unexpected musical situations often arise for me, which can significantly change the course of a composition, or sound qualities detach themselves from the initially formulated pattern, sound qualities which were not yet determined at the beginning of the composition process. It is an integrative process in which forgetting the rules of a system play an important role since otherwise no change, no transformation is possible. The moment of the sudden "neglect or oblivion" in order to follow up an intuitive idea appears in my work method often as an "insertion", which is incorporated retroactively in the composition—sometimes also retrospectively. Therein, the driving engine is the improvising of solutions, which do justice to the system of rules as well as to the intuition.

The skill of improvising appears however, in the ability to anticipate the sum of all processed information without a comprehensive formal plan or design. Sound after sound, silence after silence is added where the respective form of the moment adapts itself to the actuality. Music itself is considered a field, which is open to all sides, which wants to be worked on artistically. In the flow of an improvisation an overemphasis of intellectual reflexion can detract from the spontaneous action and reaction. Derek Bailey uses the following image: you can approach the unknown with a method or a compass, but with a map you would never get there.[4]

POINT: Our project focuses on your artistic work as a solo improviser, what are the most important components for you in a solo improvisation?

---

[3] A cellular automaton consists of a number of cells, which may assume a certain number of states. The temporal development of the system is represented in an n-dimensional cell space, where the cells change their states accordingly to their states and the states of the neighbouring cells.

[4] Translated from the German "Man kann sich dem Unbekannten mit einer Methode und einem Kompass nähern, aber mit einer Landkarte würde man niemals dorthin gelangen".

Harnik: As a composer, when scoring music, I have all the time I would need to finish a composition. As an improviser I create the sound in the moment. In doing so I put myself into a meditative state to follow intuitively an internal structure, whereas the role of a composer and interpreter is merged in the process. The mental and corporal preparation as an improviser/performer for a concert is very important. The performance where creation of music is in "real time" leads to it becoming an event.

The stimulating challenge of a solo improvisation lies in the possibility to deal consciously with one's own personal use of material. Without external intervention I immerse myself in an inner dialogue and am thus able to further explore my performance. Apart from the technical and conceptual exploration of the instrument, solo improvisation is based on the integration of certain elements in real time, with the option of bringing new material into the "game". This spontaneous handling of the material is only possible because the patterns of movement are automated to an extent, freeing up one's concentration to execute and perform new gestures. The particular instrument I play on is also a factor here because instruments can be very different in their build and can "disturb", for instance, the application of "known" material. If an instrument does not react like one expects then this possible irritation holds the potential for a spontaneous finding of solutions.

Moreover, in the course of an improvisation I can react to instantaneous situations in two different kinds of ways, which can be called, according to Lydia Goehr[5] "Improvisation Extempore" and "Improvisation Impromptu". The "Improvisation Extempore" denotes a familiar concept of every day music, namely to make music out of the moment and to develop it. The "Improvisation Impromptu" approaches the example of daily life as originated from a fracture, a problem, where an emergence necessitates an immediate (re)action. We have to react right away, without developing the reaction. In order to create room in a solo improvisation for the "Improvisation Impromptu" I often provoke unforeseen disturbances by risky preparations or materials, which are never fully controllable like mechanical toys, falling objects and similar things.

When improvising I also work very strongly with a knowledge and memory from the body of the instrument. Clusters, chords, and tonal sequences—both in intention and execution—are coupled to basic positions of my hands like "narrow hand", "somewhat open hand" and "far open hand". I also possess a repertoire of movement patterns of the hand along the keyboard, from conventional techniques of playing to self-developed performance techniques.

From my own playing a catalogue of typical basic material can be isolated which is subject to permanent selection and extension: diverse gestures at the keyboard such as melodic micro-segments, chord pattern, cluster forms, rhythmical cells as well as extended techniques, for example the use of mobile and fixed preparation of the interior of the piano, and more common materials from a combination of play on the keyboard and the inside of the piano, glissando effects, percussive play on the instrument body, linear processes of development, sound types, texture types, etc.

---

[5] Professor of Philosophy at Columbia University, New York.

All this basic material has a common allowance for ambiguity, where changes and adaptations must be possible if necessary. It is also advantageous if these ambiguities can be combined with versatility or if they are not too precisely defined in the area of application. I prefer the use of my bare hands, for instance, when playing in the interior of the piano, compared to using beaters and drumsticks, since quick changes in the sound production are easier done with the hands.

From the viewpoint of an "observer without commentary" I follow the sound formations and refine them, guide them into a certain direction or also reject them in some cases. Altogether one can observe that the sound colour potential of the material and its possible structural development takes primacy over the pitch organisation. Of course the pitch and temporal organisation of the musical events also play a significant role. During an improvisation however, the interval constellations are for me considerably more important than the selection of actual pitches. On the temporal level I work mostly intuitively, with a free combination of aperiodic material and rhythmical micro-segments where an instantaneous forming and sensing plays an important role.

Evaluation and Self-reflection

I do not "think" but at the same time it feels like "knowledge" as my eyes are mostly closed; it is a kind of "no-mind" state. If I think very deliberately about what to play next, I only manage with great difficulty to get into this state of "flow", yet this does not mean that there are no conscious decisions during an improvisation. Conscious moments serve me an "in-between stop" and I don't put too much emphasis on them since I want to be always ready to give up the conscious "control" in the right moment. It seems that I rely on my "bodily memory" and simultaneously move into the role of a "non-commentary" observer, which subtly directs the play.

Project Expectation

As a composer and improviser I am in a permanent dialogue with my own repertoire and the associated possibilities of structuring time. This way of dedicated awareness of the material constantly accompanies my artistic process. From participating in this project I expect a deepening of this debate. First of all I hope to unravel some unconscious processes and the implied knowledge of these processes. Amongst other things I am thus interested in the criteria by which I recognise and ascertain spontaneous discoveries or lucky coincidences, which may open new paths because these form mostly in conjunction with intuitive forces, the basis for artistic decisions. Yet the formation of such criteria can also imply wrong ways and dead ends. These imperfections and mistakes found at the edge between solving and finding problems are important for development.

I think that the analysis of my piano improvisation can also bring out this aspect of "failure", which in return is a possibility to better understand my own methods.

How far it is possible to address the aspect of "embodiment" I cannot estimate. The connection between "hand" and "head" is crucial in my performance practice. As a "composer-performer" I become one with the sound and with the instrument. The basic impulse for every movement are my hands—their size for instance, or the way in which they cooperate, etc. This has a strong influence on my improvisation. This project is, in any case, a new way of reflexion. It contains a new perspective to study and analyse the "pathways of my hands".

Exploring a Compositional Process

POINT: We decided to focus on Harnik's improvisational work for our research. In order to gather some empirical data, we arranged a session in which she would play a number of small "snippets", improvising with a strict constraint such as using only chords of a given number of voices. We recognised Harnik's objection that this situation was highly unusual, however we still considered it useful for some initial observations. Figure 1 shows the relative frequency of frame intervals occurring within the total body of these improvisations. In contrast and reflecting the internal interval structure, Fig. 2 shows histograms of the neighbouring intervals occurring within chords of given sizes.

With respect to the frame intervals, the major seventh is particularly prominent, whereas minor seventh and major sixth are seldom. There are only few instances where octaves occur. With respect to the layered intervals the fourth and the tritone are prominent, except for the series of chords of four voices, where the major third is very frequent.

Harnik: It is of course clear that within my normal improvisation process, such sequences of constrained chords are unlikely to occur. Harmonic consonances arise, though, due to diverse conditions, such as the physicality of my hands, movement patterns that have developed in the course of my improvisational activity, and also arise due to the transformation of melodic phrases. Nevertheless, these analyses show very clearly my harmonic preferences and motivate me to consciously break the patterns.

Would it also be possible to create new musical structures from my improvised material? I have indeed seen some interesting approaches to regenerating Bach preludes from existing preludes during our meetings. Such an approach would also be exciting for me, as it might be able to produce something like a mirror of my improvisational preferences.

POINT: There are various possibilities to generate musical structures using a corpus of existing data, such as using context based methods operating on prefix- or suffix-trees. A particularly interesting method is the *context snake* [3, pp. 112–117], an algorithm that moves along a context tree, effectively providing variable length Markov chains. The next section will introduce this concept and the possible configurations.

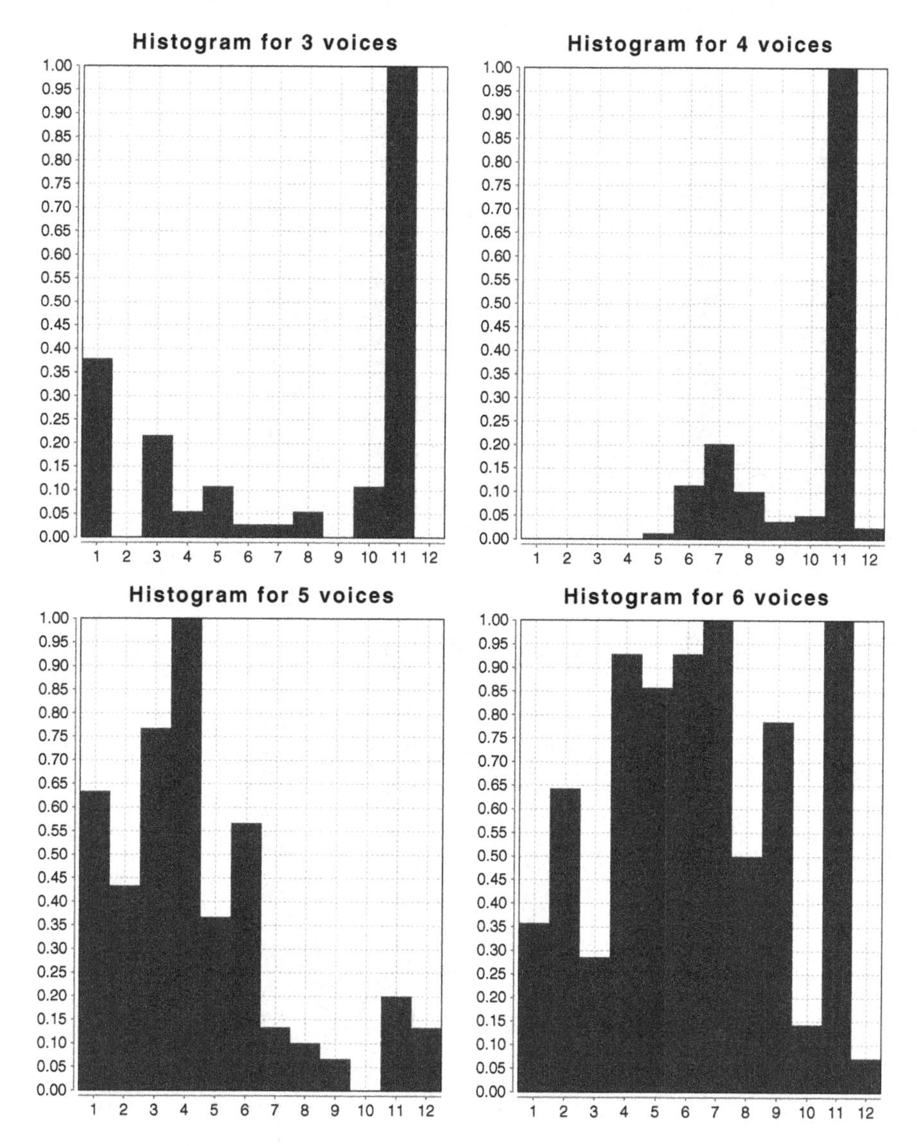

Fig. 1 Frame intervals in the chord-only improvisations, for a given number of voices. Intervals greater than an octave are wrapped

Since we have access to the data produced by Harnik's play and more data can be produced on demand, we decided to train a computer algorithm so that it could somehow reproduce the improvisations, thereby revealing certain aspects that are modelled convincingly, and others that are not well captured. This would engage Harnik in a dialogue and help to explicate the aspects of the play that are only intuitively and implicitly known.

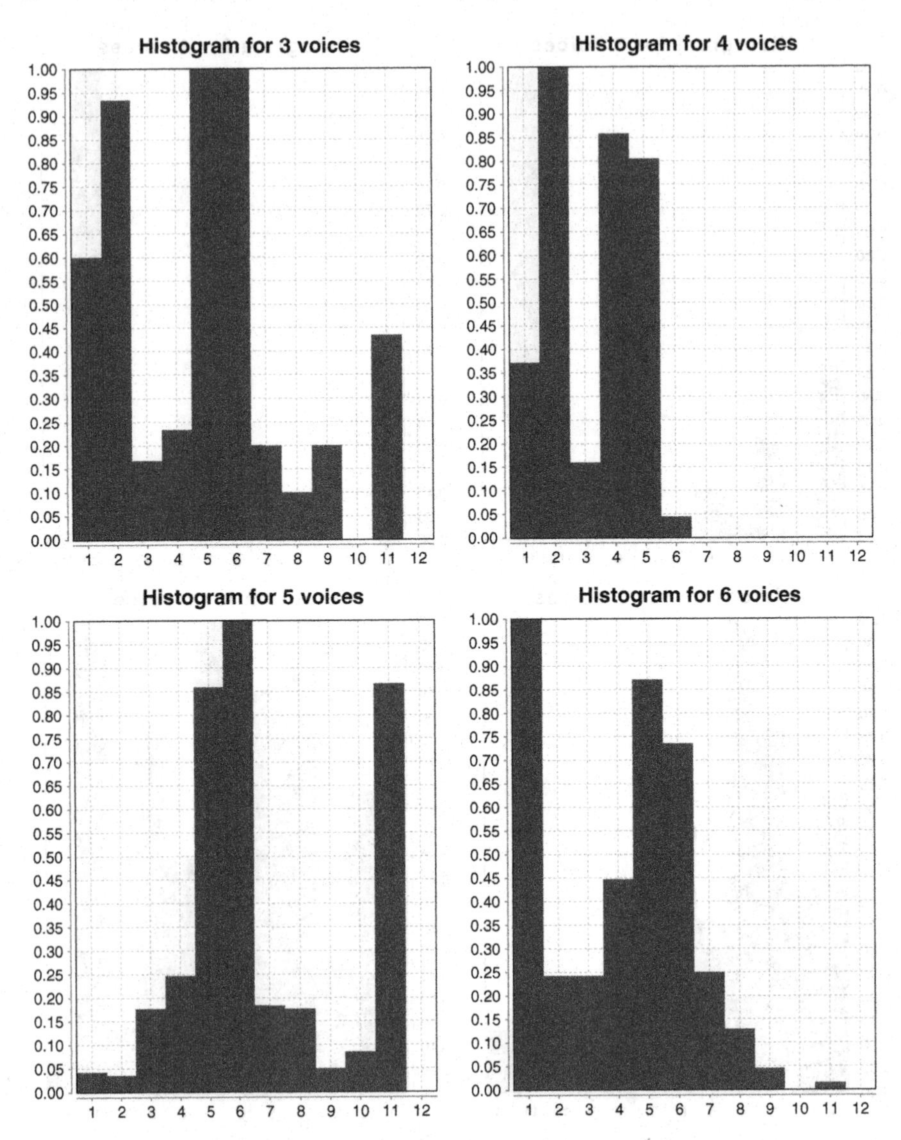

Fig. 2 Layered intervals in the chord-only improvisations, for a given number of voices

A classical approach of modelling a sequence of events—such as pitches played on the piano or letters forming words of text—is to create a table of probabilities that describe the chances of getting from a particular event or state to another event or state. The table of probabilities may be the result of analysing an actual body of events (the corpus). Using chance operations, new chains can then be formed which resemble the original corpus with respect to the statistical properties of event frequency and transition frequency. These chains are called Markov chains, because

Table 1 First-order Markov transition table for intervals in a free improvisation

	0	1	2	3	4	5	6	7	8	9	10	11
0	0.05	**0.22**	0.17	0.08	0.03	0.06	0.07	0.04	0.04	0.05	0.05	0.13
1	0.05	0.13	**0.17**	0.05	0.08	0.06	0.10	0.11	0.07	0.07	0.05	0.06
2	0.05	0.13	**0.13**	0.07	0.08	0.11	0.09	0.08	0.09	0.06	0.04	0.04
3	0.02	0.11	**0.20**	0.08	0.08	0.08	0.03	0.08	0.08	0.04	0.06	0.14
4	0.04	0.10	**0.21**	0.07	0.09	0.06	0.10	0.10	0.04	0.07	0.06	0.05
5	0.07	0.12	0.12	0.06	0.06	0.11	0.10	0.08	0.05	0.06	0.04	**0.14**
6	0.05	0.15	**0.16**	0.05	0.07	0.10	0.05	0.10	0.06	0.08	0.05	0.09
7	0.04	**0.14**	0.13	0.10	0.10	0.07	0.08	0.09	0.08	0.03	0.05	0.10
8	0.02	**0.16**	0.12	0.13	0.07	0.05	0.12	0.11	0.04	0.09	0.02	0.07
9	0.04	0.11	0.13	0.08	0.06	0.06	0.09	0.08	0.06	0.05	0.09	**0.17**
10	0.06	0.11	**0.14**	0.08	0.09	0.10	0.07	0.06	0.04	0.09	0.04	0.12
11	0.04	0.13	0.10	0.09	**0.14**	0.06	0.04	0.07	0.10	0.07	0.06	0.09

Each cell shows the probability of a transition from the row index to the column index. The sum of each row is 100 %. The largest probability of each row is shown in bold-face

they have been invented by Russian mathematician Andrey Andreyevich Markov at the beginning of the 20th century.[6]

As an example, Table 1 shows a transition matrix created from looking at the succession of intervals in the recording of one of Harnik's improvisations. The intervals are shown as the number of semitones modulus octaves. Looking at the first row, the probability that a pitch repetition (unison) is followed by another pitch repetition is 5 %, whereas the likelihood that a unison is followed by a minor second is 22 %. Using this table and a random number generator, one could now generate new sequences of pitches that reflect these probabilities.

The problem with this approach is that the generative process is not sensitive to rules or probabilities that involve a longer back trace than just the preceding element. For instance, the corpus might contain transitions $A \rightarrow B$ and $B \rightarrow C$, but no subsequence $A \rightarrow B \rightarrow C$ exists. A first-order Markov process that only looks at the last element to produce the successor may come up with this result. One can use higher-order Markov chains to avoid this problem. In a second-order process, transition probabilities are given for pairs of preceding elements. On the other hand, the higher the order, i.e. the more the transition rules are constrained by looking at the longer backtrace of the sequence, the less likely one finds alternative transitions. The effect is that the original corpus will be more or less recreated without variation. At the same time, patterns that clearly reflect low-order Markov processes are concealed in such higher order representations.

To navigate between these two extremes—context-insensitivity at low orders and lack of variability at high orders—Kohonen has proposed the use of variable-length Markov chains [2]. His generative algorithm tries to use long contexts (high orders)

[6] For an overview of Markov chains, see for example [1, Chap. 11] and [3, Chap. 3].

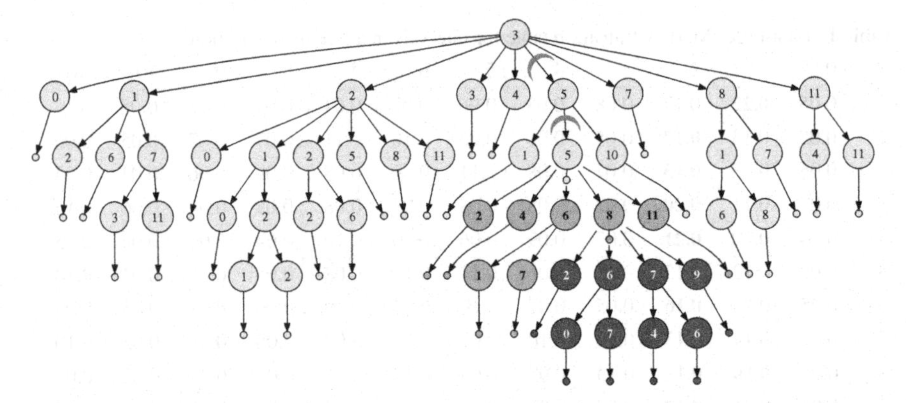

Fig. 3 Snake motion through a context trees of intervals. The initial tree, starting with element 3 and shown in *light gray*, successively expanded trees in medium and *dark gray*

but is restricted by a *depth parameter*, ensuring that the exploration stops before the maximum context length is reached, thus guaranteeing a choice in the successive elements of the generated sequence. A particular rendering of a variable-length Markov algorithm is the context snake. It builds a tree structure of the overall context. The "body" of the snake is the current context, a subsequence within the corpus. The tree structure allows us to find the successive elements of the current context. When there is zero or only one possible successor, the algorithm may either backtrack and move the snake's "head" towards other sub-trees, or it may truncate the context, forgetting older elements and shrinking the snake's "tail". Efficient search structures are available for the implementation such as suffix trees [5].

Figure 3 shows a traversal through such a suffix tree. The data used is a subset of the interval transitions used for Table 1.[7] The snake was initialised with only one element, 3. At this shallowest context depth, there are nine possible transitions: 0, 1, 2, 3, 4, 5, 7, 8, 11 (for simplicity, the edges are all drawn the same, although the transition probabilities differ). If, using a random number generator, 5 was selected as the successive element and appended to the snake's body, the context depth becomes 2, and now there are three alternative successors: 1, 5, 10. If 5 was selected again, the context depth or snake length becomes 3, but now the critical point has been reached where only one possible successor (6) exists. The algorithm could backtrack and try 1 or 10 instead of 5. Since these also do not provide longer context, the tail element 3 is removed and appended to the generated sequence. A new context tree starting with 5, 5 is found and the new set of successor elements becomes 2, 4, 6, 8, 11. The procedure is repeated as before, until the desired length of the generated output is reached.

Two aspects determine the quality of the generated sequences. Firstly, the size and exhaustiveness of the corpus—the larger the corpus, the more it reflects the knowledge embodied in Harnik's play, the more exhaustively it covers all the possible

[7] We used a smaller corpus to make the figure more readable.

ways of conceiving such improvisations. The second aspect is the type of element represented by the context trees. In the previous examples, we have used the intervals between successive notes. It did not make a difference between an upward and a downward interval, so one would probably want to preserve the interval direction. Instead of intervals, one could use the absolute pitches, or one could model entirely different parameters such as the dynamics of the notes, their durations, etc. A particular problem is posed by the request to model multiple parameters at once, such as pitch and duration. This will be discussed later in the chapter.

To begin with, we tried to regenerate plain chord sequences, using a given number of voices. Examples of the input material are shown in Fig. 4. To model the generation of new chord sequences, an example corpus was first converted from raw MIDI notes to chord objects. In order to keep the dimensionality of the vectors small and the amount of alternatives high, we used multiple context snakes whose outcomes were combined: the first snake generated was fed by vectors formed from the pitch class taken from the lowest and highest note of each chord. For example, looking again at Fig. 4, the first chord would produce frame pitch classes (G, G) or numerically (7, 7), the second chord would produce (Ab, C) or (8, 0). A second snake used tuples of the registers (octaves) in which the lowest and highest pitches of each chord occur. Using MIDI conventions, the first two chords of the previous example would yield tuples (3, 5) and (2, 5). If chords of mixed size should be modelled, another snake would just generate the chord sizes.

To model the interval structure between the frame intervals, we maintained a nested dictionary from frame interval size to chord size to chord intervals. After determining the lowest and highest pitch of a generated chord, using the pitch class and octave snakes, we looked into this dictionary for the thus given frame interval and chord size. If no entry was found, we looked at the next smaller or greater interval and chord sizes, until a body of chords was found. A random chord is then picked, and its intervals are used in a random layering. Example generations are shown in Fig. 5.

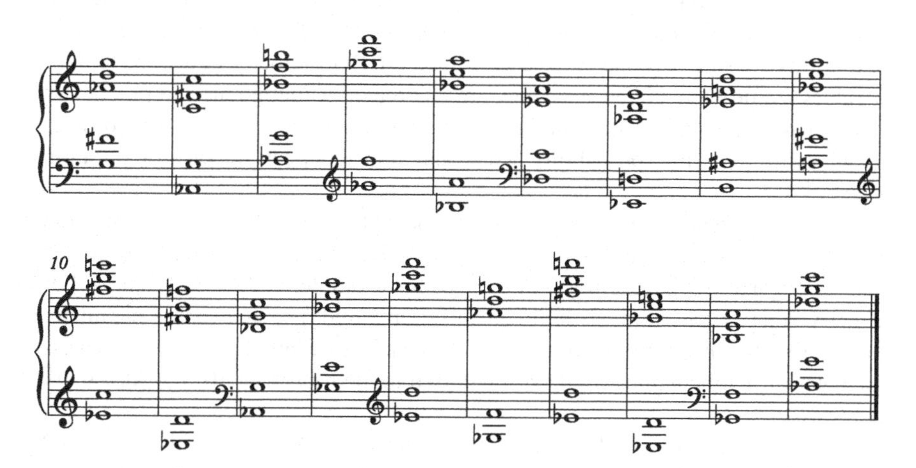

Fig. 4 Example chord sequence played by Harnik (cutout from recording No. 46)

Fig. 5 Regeneration from recording No. 46

POINT: What do you think about the chords from our regeneration?

Harnik: Apart from the chords that are not possible to play due to their position for the hands, the regenerations are convincing. The consciously preferred interval combinations are reflected in the regenerations very well. The chords in bars 10 and 11 I would rather have played as 6-part chords. The combination of fourth and tritone, respectively, in the chord of bar 13 is also a very unlikely scenario.

I would also have formed the sequence of chords differently. Chords in a row are usually intuitively grouped during playing. Pedal points both in treble and bass would not normally be part of my repertoire. It would be more likely to have a single pedal point either in the treble or in the bass, but in this case I would have placed the flow of these chords only under certain conditions, deliberately and with effects that would follow.

POINT: In the next step, we regenerated freely improvised material. In order to handle the articulation of horizontal sequences, the entry delays—the time that elapses between two successive notes—needed to be modelled, and also the dynamic contour was a desirable property to be accounted for. Both velocity values and temporal values are problematic because they are theoretically continuous and practically represented using fine grained digital resolution, such that in a MIDI recording. So only with very low probability we would find identical velocity or duration values.

To produce meaningful corpora, we reduced the resolution of velocity and temporal values using a coarseness parameter. The velocity is linearly quantised from its original MIDI resolution of 127 to, for example, $127/6 = 21$ steps. For the entry delay, we used logarithmic quantisation based on a coarseness parameter that specifies the number of steps per "time octave". For example, with a coarseness parameter of 2, time values would be quantised to the nearest of 10, 14, 20, 28, 40, 56, 80 ms, etc.

Again, in order to keep the tree branching factors in the corpus high, we used separate snakes to model the pitches and to model the entry delays. With the entry

Fig. 6 Cutout from recording No. 48

delays being formed both from melodic progressions and chords, chord structures automatically appeared depending on the entry delays (if a chord appeared in the corpus, the entry delays for all but one note were nearly zero).

Besides making a selection from recordings of Harnik's free improvisations, the initial note and the seed of the pseudo-random number generator—used when a tree has multiple branches—influenced the development of the generated material. Figure 6 shows an excerpt from a recording of Harnik's play, and in contrast Fig. 7 shows material regenerated using the context snake method.

POINT: What do you think about the regeneration from recording No. 48?

Harnik: The interval structure and also the rhythmic flow of the regeneration are convincing. It is striking however, that in my recording the interval of the initially played fifth is then reflected back in further bars of the piece. The interval "floats" permanently as a thought, without manifesting itself. This aspect is only captured in the beginning of the regeneration.

POINT: Figure 8 shows a different excerpt from a recording (No. 9) of Harnik's play. We ran another regeneration, combining this recording with the previously shown one (No. 48). An example from the regeneration is depicted in Fig. 9. In contrast to the previous example, we used a separate modelling of horizontal and

Fig. 7 Cutout from regeneration of recording No. 48

Fig. 8 Cutout from recording No. 9

Fig. 9 Cutout from regeneration from recordings No. 9 and No. 48

vertical structures here, alternating between them in the regeneration. Velocity and slight timing differences between the different notes of a chord are also incorporated, although not visible in the score.

For this alternative modelling, we partitioned the corpora into horizontal and vertical segments, modelling chords and melodic sequences separately. For example, the algorithm would start with a melodic fragment, choosing a number of notes according to the statistical distribution of sequence length. Next, a chord sequence would be generated as described above, incorporating the last melodic pitch. The results however sounded unnatural, probably because of the artificial division between purely horizontal and vertical segments.

In the discussion with Harnik, we concluded that horizontal and vertical structures can be understood as two renderings of the same underlying harmonic rules; melodic sequences thus can be seen as "horizontalised" chords, or chords as "collapsed" horizontal sequences. The technically simpler approach of the first regeneration, which disregarded any distinction between horizontal or vertical segments, was thus better suited.

POINT: What do you think about the regeneration from recordings No. 9 and No. 48?

Harnik: In this regeneration the flow of the rhythm is more successful than the interval structure. The beginning of the original recording No. 9 has an open-melodic character to it. From bar 4 the interval of the major second is spontaneously lit up and developed in the following sequence and at the end returns so that there is again an open melodic quality like at the beginning. The major second was thereby altered, for example, shifted chromatically or reduced to a minor second. The method of repeating the two tones only happens once. The regeneration also stresses an emphasis on single intervals, remaining involved with the repetition of sound. The choice of the six-part chord as a starting point for revealing the process I would definitely not have made. I would also not have played the repeating notes within such a quick gesture.

Overall, the regenerations are quite convincing. I observe that the subsequently conventionally notated originals and regenerations seem very strange to me. In fact, I don't have this kind of notation in mind when I improvise on the piano. A closer match would be a sort of fingering notation that better honours the cooperation of both hands. This dimension does not open itself up, and makes reconstruction of my own playing very difficult. This insight confirms my assumption that the physical memory and the movements influence the process greatly.

The tempo also plays an important role. The slower that I improvise, the more "analysis" is possible in real time. I can later on refer exactly to what I have played. My recordings would definitely be different if I reduced the speed. The examples given were deliberately performed at fast tempo, in order to investigate the unconscious flow of playing. I note that in addition to the conscious factoring out of diverse sound material for personal or aesthetic reasons, also the physical conditions of the body can effect the tonal expression. When playing at an instrument, complex patterns in movement occur. Fundamental in piano playing is the cooperation of the hands. I find it exciting to see patterns of solutions emerge from my play and how, in turn, they can be perturbed.

POINT: Harnik also pointed out that she was sceptical of the staccato character which can be observed in the acoustic rendering of the regenerated sequences. This resulted from the note durations being fixed, while the entry delays are modelled from the original corpus. Using the same logarithmic quantisation for the note durations, we were able to produce a new snake that used both quantised durations and entry delays as a combined vector. The result is a much more natural sounding articulation of the play.

Ideally, one would use a single vector that combines pitches, dynamics and durations, because these parameters are certainly not independent—for example, there might be a bias for low pitches to sound longer on average than the high pitches—and generating sequences from "zipping" the output of individual snakes can lead to unnatural situations.

To alleviate the shrinking connectivity of the trees, one can try to shrink the feature space. As an example, we tried to use interval steps instead of absolute pitches, and acceleration instead of velocity, but we traded one problem for another. Figure 10 illustrates the different effect achieved on the overall form level. From

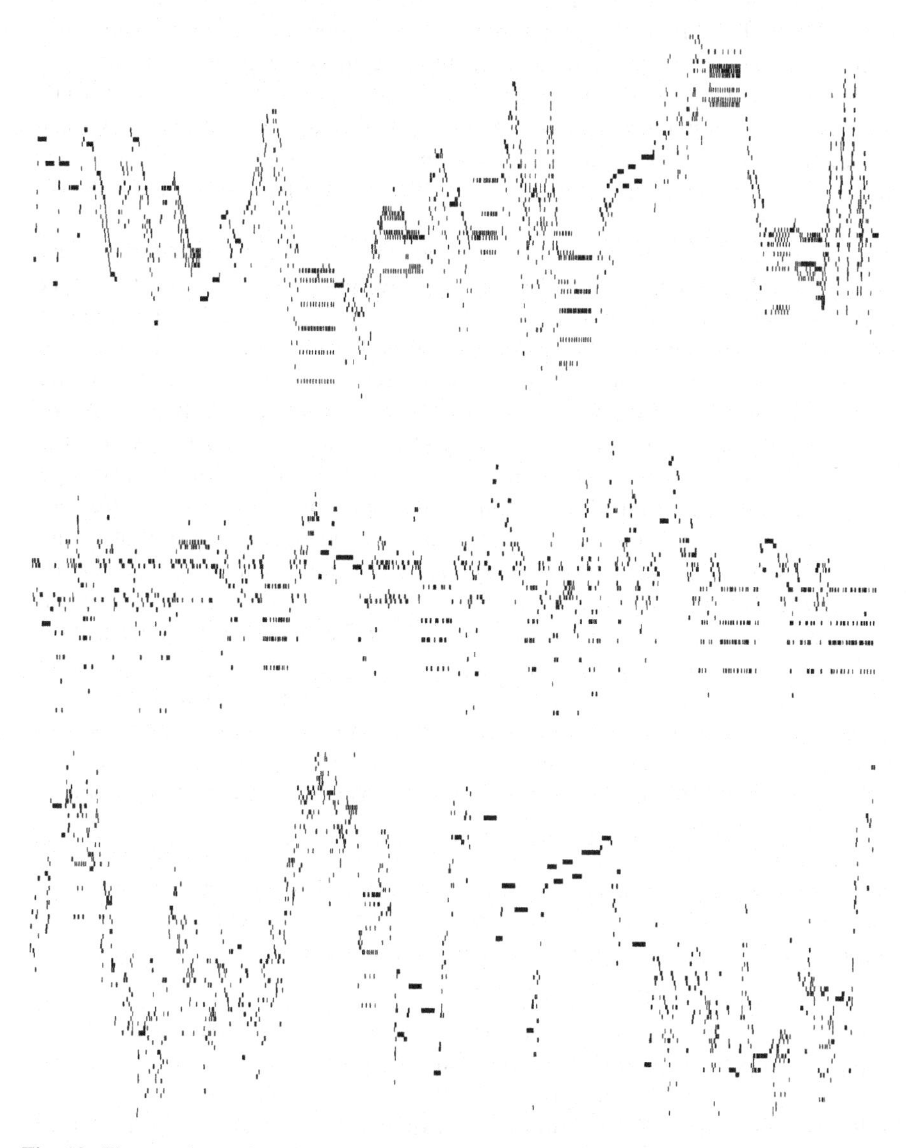

Fig. 10 The *top* shows the original recording No. 42, the *middle* shows a regeneration using a context snake for absolute pitches, the *bottom* shows a regeneration using a snake for relative pitches (interval steps)

the original recording No. 42, shown at the top, two sequences were synthetically generated. In the middle, one sees the sequence produced from absolute pitches. Clearly, we can identify sections that share structure with the original recording, however the algorithm does not pay attention to the overall form and development, it lacks the slow motions across the keyboard which are seen in Harnik's play. These motions would require much larger contexts. We imagine that a future approach could try to model these "low-frequency" motions by decimating the material and modelling multiple "rhythms" on different scales. The bottom of the figure shows the attempt to allow slower oscillations to occur by using interval steps instead of absolute pitches. In the global picture, much slower motions appear now. On the other hand, the clear correlates in micro gestures are lost. As will be discussed further down, not all keys are "equal", for instance the layout of the keyboard with white and black keys is relevant, something that is captured by the absolute pitch snake, but lost with the interval step model.

A more refined model should also take specific styles of playing and movement into account. For example, often the highest played pitch has a significantly longer duration. In fast pattern repetitions, often there is one note that is replaced or dropped or added in each iteration. When playing localised "blocks", these blocks are often connected by the lowest or highest note of the preceding block that inverts its function in the succeeding block. In general, many oscillatory forms such as A–B–A–B–A′–B′ can be found. There is in general a tendency of isolating specific elements from the play and elaborating them, going from "coarse to fine". Harnik described her strategy as an "enacted multitasking" or a "simulated multi-mind". She explains that her thought moves across different sound layers simultaneously. She is capable of adding these layers or "switching them off" at will, depending on the situation. A set of such "coordinates" of the sound space often define the initial situation in an improvisation.

To give an example, a "bothering" or irritating element is introduced. Also "mistakes" during the play function as a great trigger for changing the situation. In general, playing in an ensemble instead of solo, or having a prepared piano, makes it easier for such unforeseen elements to appear. Another important factor is the presence and dynamics of the audience. Further sources for a productive "irritation" might be a specific tuning of the piano (microtonality) or a noise emerging from the audience (a glass toppling over...).

We recorded video footage from Harnik playing on a grand piano. Figure 11 summarises some of the characteristic hand positions. Hands can be open or closed, there can be a small or large gap between them, or they are operating in "parallel", where usually the left hand is positioned above the right hand.

Overall, we could make out the following distinct forms:

- Glissando-forms, dragging the thumb across the white keys; this often used to separate repetitions of a particular gesture, where the repetition would change a particular aspect such as tempo or strength.
- Playing very fast and dense textures in the extreme high register of the piano, where one perceives rather an overall glassy granular texture instead of individual pitches.

Fig. 11 Stills from a video recording of Harnik playing. Different motion patterns can be seen, as discussed in the text

- Clusters: a number of cluster techniques are available, for example using the hand flat or laterally, employing the whole arm (usually the right arm), or gliding with the back of the hand across the keys.
- A preference of the left hand for the black keys and the right hand for white keys can be observed when they play parallel.

- A circular motion between the two hands.
- Incorporating the sustain pedal; both hands are free to stimulate the overtones.
- "Mute" keys; one hand holds a number of keys silently in the bass register, then the second hand adds accentuating sound material, often using staccato, thereby making the open strings of the mute keys reverberate. This way, different overtones can be heard.

When discussing her techniques, Harnik said that the initial impulse comes from a bodily memory, for example whether the hands open or close. Subsequently, there are a limited number of possible movement patterns, which are constrained by the structure and various positions and motions from her hands.

Harnik: David Sudnow describes in his book "Ways of the Hand: The Organization of Improvised Conduct" [4] how he learnt to improvise jazz at the piano ("Improvisation Extempore"!) After initial unsuccessful attempts to mimic the sound, a breakthrough only came by practicing scales, phrasing and chord sequences: "As I reached for chords, and reaching for chords in the song context involves reaching for patterns of chords, for characteristic sequences, I was gaining a sense of their location by going to them, experiencing a rate of movement and distance required at varying tempos, and developing, thereby, an embodied way of accomplishing distances. What 'there' means is how it is to go from place to place as an accomplishment. The symmetricality of the body, and that sort of extensional 'self-consciousness' that enables you to use a toothbrush without monitoring the course of the gesture and without smacking yourself in the face, entails a 'system' with elaborate distancing capabilities." [4, p. 12].

Instrumentalists from all performative practices can certainly confirm this experience. It does not matter what type of pattern you play, the feedback system between perception and motor skill applies. Any form of music, whether composed or improvised formulates "patterns". One can indeed "objectively" classify and transcribe these patterns, but the dynamical basis of the sensorimotor processes that is in a state of oneness with the instrument cannot be described.

Before a sound is created on the instrument, there is also a preparation in the body. Even before the sound is physically shaped, it is already modelled in the imagination in this state. In improvisation this ability is used more than in interpretation. Any type of movement is coupled to a more or less predictable sound result. The different basic positions of the hand are more open in their usage. They mark a "place" and can be interpreted differently depending on the situation. Remembered or imagined sound material can certainly trigger movement impulses. Conversely, however, a given impulsive movement can trigger sound and its processing. These decisions, out of necessity, often happen very quickly. Certain compositional methods, in which a real-time analysis is not possible, are eliminated from the outset. Position of the hands and types of movement define these loose sound "folds" that can be expanded depending on the situation. Active listening acts as an inner compass.

There is also another aspect concerning the hand gestures. They can, for example, not only be used for a particular expression to reinforce, or to strengthen. They can also be deliberately "played with". A gestural preparation can trigger a certain tonal expectation from the audience. It is possible to acknowledge this as a performer, or not to. The surprising "breaking" of expectations gives the audience a possibility to witness for themselves the flash of the moment. In addition, it keeps the creation alive. Due to irritation on the plane of movement possibilities, I am challenged as a performer to improvise a new solution. A variety of pianistic gestures—historic to contemporary—are available to draw from. Here I would emphasise again the cultural conditioning. Since my classical piano lessons, I have expanded my repertoire of gestures—partly from other genres but also by experimental development of my own movement patterns. It is interesting to me that I often assign gestures and playing techniques that come to me from other instruments. In this sense I observe a tendency of multi-instrumental gestures, which I bring into my repertoire.

POINT: What additional components come to fruition in an improvisation, for example, in response to the audience, preparations. . .? How about all the aspects that evade modelling, aspects that cannot be notated as pitches, durations and dynamics?

Harnik: Exclusive performance on the piano keys comes into my performance practice very rarely. In my improvisations it is mostly a combination of the piano keys and playing techniques in the interior of the piano. I use various types of preparations, including moveable objects. Furthermore, I also use sounds and noises that can be generated on the surface of the instrument or on the body itself. It is also possible to include the sound material that may be evoked by a loose piano stool or the creaking of a stage floor. Furthermore, the quality and acoustics of the concert room together with the presence of the audience is part of the creation in real-time.

Project Review by Elisabeth Harnik

The interaction of imagined sound and its realisation on the instrument is a central aspect of any improvisation for me. How strongly will physical memory and automated motion patterns control the improvisation? Would other sound solutions be provoked through the changes in movement or does the realisation on the instrument follow the sound imagination? The work and results of this project have delivered me an interesting incentive to observe these queries from a new perspective. It was very enlightening to me, especially through the diverse generated music examples in this project, to trace my musical patterns, which arise from an embodied knowledge, in an innovative way. That the regenerations arose only out of my well-rehearsed material, offered by evaluating the results, a musical counterpart, which allowed me, to a certain degree, to perceive my improvisational structures from an external perspective, thus to reflect in a new and unusual way. In the time of this project a greater incentive was built to deliberately break the collections of movements in my future improvisation practice.

I also found with respect to reflecting on my work, the question of an adequate notational representation of the improvisation processes very stimulating. Notation is not normally the goal of an improvisation. Until now, I have not found a method to do justice to transcribe my improvisations. I can well imagine that the future of my compositional work will reflect on transcribed or regenerated sound material, to extract from it, or to use it as a basis for various methods of composition. As an improvisational musician, it is more and more obvious to me that what especially interests me is how the "tonal language" emerges, regardless of the art or type of productions used.

References

1. Grinstead CM, Snell JL (1998) Introduction to probability. American Mathematical Society, Providence
2. Kohonen T (1989) A self-learning musical grammar, or 'associative memory of the second kind'. In: International joint conference on neural networks IJCNN. IEEE, pp 1–5
3. Nierhaus G (2009) Algorithmic composition: paradigms of automated music generation. Springer, New York
4. Sudnow D (1978) Ways of the hand: the organization of improvised conduct. Harvard University Press, Cambridge
5. Ukkonen E (1995) On-line construction of suffix trees. Algorithmica 14(3):249–260

Clemens Nachtmann/Forbidding Harmonies

Clemens Nachtmann, Daniel Mayer and Gerhard Nierhaus

Clemens Nachtmann was born in to a family of glassmakers, a frequently practiced profession in the region of Oberpfalz in northern Bavaria until the 1970s.[1] Among glassmakers, it has always been a cultural tradition to play music together and for family celebrations and similar festivities people play together mostly pieces of a folk-musical nature, which can also be enhanced by their own compositions.

At seven years of age Nachtmann started to learn the clarinet, an instrument that fascinated him due to some records he discovered from his parents, which he listened to over and over again. Shortly after, he received additional lessons in piano and classes of harmony and counterpoint in music theory at a local music school, which provided Nachtmann a solid grounding for his first trials of composition. A particular delight for him was also the studying and reading along of scores listening to music. Further on, Nachtmann also became interested in music theory and philosophical literature, where he found in the reading of Adorno, Marcuse and Marx the generosity and the extensiveness of thinking, which he heavily missed in general at home and in his home-region, where he was living back then.

Adding to his philosophical and musical interests came an engagement with painting. His final decision towards music, however, Nachtmann owes his music teacher at the local secondary school, who completely inspired him for the cause by his

[1] Biographical introduction and texts from the composer translated from the German by Tamara Friebel.

C. Nachtmann
Institute for Composition, Music Theory, Music History and Conducting,
University of Music and Performing Arts Graz, Graz, Austria
e-mail: clemens.nachtmann@kug.ac.at

D. Mayer · G. Nierhaus (✉)
Institute of Electronic Music and Acoustics,
University of Music and Performing Arts Graz, Graz, Austria
e-mail: nierhaus@iem.at

D. Mayer
e-mail: mayer@iem.at

G. Nierhaus (ed.), *Patterns of Intuition*,
DOI 10.1007/978-94-017-9561-6_3

challenging and dedicated tuition. His clarinet teacher first encouraged his earliest compositions. Due to concert visits and keen radio listening Nachtmann discovered the music from Bach to Mozart, Beethoven, Schubert, Wagner, Mahler and Strauss as far as Schönberg and Berg—included excursions into ancient music. The compositions, which emerged from this time on, were at first and foremost stylistic copies, which nevertheless developed primarily an understanding of musical forms via the work with basic and extended tonal harmony and metric. Exciting also for Nachtmann was in particular his discovery of unresolved dissonant sounds, of free atonality in general and the possibility of serial-like order of sounds. A similar thing occurred to him with serial composing, which seemed to him at first suspect, yet in the 1990s provided him with crucial impulses for his further thinking and composing.

The experience to stand rather alone with his musical views in the context to the preferences of his teachers meant that at the end of the day Nachtmann pursued his path quite persistently, even in the time as he was already formally studying composition, leading to some arguments with his composition teachers. After finishing school Nachtmann moved to Munich in order to study political sciences and shortly later also composition at the University of Music under Wilhelm Killmayer. The dominating refusal of modern music in the composition class as well as the social situation in Munich in the 80s made him pack up his tent and go to West-Berlin so as to continue there his studies of composition and political science. The entrance exam to the music university, where he failed due to the opinion of the jury, gave him contact to the composer Gösta Neuwirth, with whom he would ten years later, this time after he had passed the entrance exam into the composition class of Friedrich Goldmann, study counterpoint and later also music theory and from whom he would also receive essential stimuli for further composition work. In the in-between years Nachtmann moved in the circles of the left scene of West-Berlin, continued his theoretical and philosophical studies and completed his study of political science with Johannes Agnoli, to whom he owes much for his thinking and with whom he remained on cordial terms until his death in 2003.

Ever since this time Nachtmann began to publish regularly texts about philosophical, theoretical-sociological and aesthetical topics, a practice, which he carries on today. During this period no further pieces were made except for a few sketches. Nevertheless in the time of his "composition crisis" he educated himself further by reading, visits of operas and concerts as well as listening to and playing music at home.

In the late 90s he recommenced composing, starting to engage himself with the works and the methods of serial music, he listened to and studied music primarily of Helmut Lachenmann, Mattias Spahlinger and Nicolaus A. Huber, which fascinated him. During his composition study at the Berlin University of the Arts he valued especially the rare, but all the more fruitful reviews of his compositions with Gösta Neuwirth and Mathias Spahlinger—from these Nachtmann was given pivotal stimuli for this further development as a composer. He also received essential impetus through the discourse with his teacher Friedrich Goldmann and via the work with electronic music.

Through Beat Furrer, whose music he liked especially and whom he then personally met at the beginning of the 2000s, came the opportunity to change place and teach after finishing his composition studies in Berlin. Nachtmann graduated in composition and music theory with a thesis about an orchestral piece by Mathias Spahlinger,[2] and then went to Graz and completed the post-graduate course of composition with Furrer at the University of Music and Performing Arts Graz.

Here he finished his piece for ensemble *en dehors*, which he had already begun in Berlin and which allowed him to open up his composing and to develop technical methods, which have the capacity of evolving and extending, being so sufficiently determined yet general enough that they allow one to produce music with the most diverse different appearances.

Nachtmann is fascinated by music especially with respect to its characteristic of gaining knowledge: that in music and often in its smallest details something about the world can open up, that sounds can manifest a form of knowledge, as an experience, which constitutes a central motivation for the compositional work of Nachtmann. The question, in which way music can behave critically, is posed in each of his pieces—to charge them with political meaning or to deploy them for a political purpose he thinks little of: "Music can only be critical as autonomous (music), which is critical to itself, its material, its behaviour, to its own definition and thus also to the audience and its habits of listening and expectations."[3]

Artistic Approach

Statement

Concerning the practice of composing, I would like to quote my former teacher Wilhelm Killmayer who, when asked how he approaches composition, said: "I write down what comes to mind and what I believe is also interesting to others". In the same manner, what is interesting to me, first-hand, could be seen from a purely egotistical viewpoint: it is my drive to express things, which I haven't yet heard or heard in a particular way. To repeat or reproduce what already is there, I find utterly boring, and I would gladly leave it to those conformists in the cultural industry in the various genres, but to which unfortunately also belongs some parts of the new music establishment.

What also interests me, is to create with every piece, taking into account the risk of failure, something new, something even for me never fully predictable and to invite others to share this adventure, which is formed from all of that, by listening and reading. Due to this impulse I feel committed to new music, which constitutes itself by the rejection of all outlived musical conventions, expressions and formulas

[2] It concerns the composition *morendo* for Orchestra from the year 1974.

[3] From a discussion with the composer in 2012.

as atonal music in a broad sense. If, according to Adorno, the basic question of all music is how a whole can exist without doing violence to the individual parts [2, p. 62] then is new music, which refuses all outlived categories of completeness, the one, which attempts to answer this question with every new piece afresh and distinctively? This music is for me one of the most fascinating and "impossible to complete" experiments, which humans have engaged in forever and it is, despite the arduousness it sometimes causes, a continuing enjoyment for me to also participate in.

Personal Aesthetics

I do not have a "personal aesthetic" and believe that each is a contradiction in itself. Due to the fact that each aesthetic is directed towards forming general, i.e. generalizable categories, which allow for assessing pieces of art—its distinguishing feature is therefore that it is im- or over-personal. "Personal" on the contrary is all meaning, every point of view, in matters of the art shortly: taste. A "personal aesthetic" is thus an absurdity, although I have to admit that I encounter this absurdity quite often: a basically petty taste, which is blown right up to a pseudo-philosophical belief system. This I try to avoid by all means.

I naturally do have certain preferences and dislikes—and if I give into them unashamedly then I express them also clearly. But that, which I permit in my compositions and what I categorically exclude is not owed to a random whim, is no arbitrary specification but the result of reflection. All subjectivity enters into the reflection and morphs it decisively—at least it should, as reflection is otherwise only rationalisation in a psychoanalytic sense: legitimation of something, which one does in any case and doesn't want to challenge anyway. Composing is for me—and this aspect has become more important in time—a way of aesthetic research about "sounding" material, features thus as an artistic practice at the same time a so-to-speak "scientific"[4] character. In this context I hold the so-called serial composition in high regard, because it brought an irrefutable experience, which is at the basis of all new music, to self-consciousness. Falling back to it is only possible at the penalty of regression.

The continuing experience, which no composer, who is ready for an un-restrained self-reflection can elude, is that fundamentally composing is in a nominalist state since the downfall of tonality, i.e. music can no longer be imposed by any tonal-relations from a simple interval to a sophisticated texture, any rhythmical model, any scheme, any types, any form-development, but is condemned to freedom to some extent. In appropriate self-conception the so-called serial composing has drawn the ultimate theoretical and compositional consequences from the erosion of tonality executed for instance by Schönberg, Ives and Scriabin: by having claimed this process concerns utterly all constitutive, form-giving dimensions of music: not only

[4] Scientific in the sense of the impulse, which is at the bottom of all science and yet perishes frequently enough in the institutionalised science: a naive joy of discovery, which is at the same time self-critical since never satisfied with what is uncovered.

the simultaneous and successive relations of pitches, but also the duration, dynamic, articulation and timbre. So-called serial composing: because the essence of it is just not the twelve-tonal or serial scaling of various sound-parameters but the realisation of their form-constituting qualities, their specific cooperation, their reciprocal substitutability, and thus all together their ability to be composed. Only by serial composing it became possible to recognize the form-constituting potentials of those qualities of a musical texture which earlier, already suggested by the occidental notation, were considered secondary, as form-constituting phenomena and to include them into the compositional process. Thereby a development was started, which is in principle impossible to complete, and in logic of which the initially quite across-the-board perceived sound-parameters were further nuanced—thus, accidentals and embellishments like e.g. vibrati or trills were already recognised as form-giving moments by Stockhausen in his orchestral pieces *Punkte* and *Gruppen*, respectively, or by Helmut Lachenmann the analytically fragmented and freshly recombined processes of the sound-production at an instrument.

The historic "state of the musical material" after about 1900 is one where there is no more state, which, like for instance harmonic tonality, one could name a-priori, i.e. fixed beyond an individual piece; it is self-evident in new music that it no longer accepts any classification of music, form, idiom, or sound as self-evident—and thus its own form of existence is also no longer a given fact, but has turned into a problem, for which there are most different, each other complementing or diametrically opposing solutions. Therefore a composer has at first to prepare his/her material for each piece, to build a system of categories, within she/he can move freely. He/she has to decide all the pre-compositional set-up under whose premise the sounds can unfold their intrinsic life and drive. This means that the analysis of music, which previously was a downstream process of insight of a finished piece, became now an integral part also of composing itself.

Each newly created piece of new music thus ventures to playfully research afresh the unlimited possibilities of musical material and to cast the "research results" into a shape, which aims to answer the question "What is music?" in an individual and distinctive way. It is the idiosyncratic and peculiar dialectic of new music, which bids farewell to all systems and schemata apriori claiming generality and all "naturally" or "elementary" ideologised musical designations, that only new music makes it possible advance to the real and actual, i.e. elementary and irreducible designations of music and musical material. Hence, the musical relationship in new music became in a way easier than in the traditional music—and at the same time also more complicated, since for the composer all elementary relations run away to the incalculable and unlimited.

Each of my pieces features a "simple" connecting idea, a "theme", which is investigated in the course of the time, which the piece takes. Connecting means: to connect individual incidences, which may be very heterogenic and often indeed are. One of the central criteria of assessing if music turned out well is the richness of relations; the question that concerns me again and again—how a plenitude of relations, forms, textures can be derived from a simple starting material, how a very homogeneous material can yield very heterogeneous occurrences—I came already quite close to this through and since the work on *en dehors* and I remain still active in following it up.

The sextet *schnitte* of 2009, for example, develops in the concept and in the composition itself from the (in- and out-going) compressing and expanding movement of the accordion used in the piece; here, the original idea forms also the beginning of the movement. In *en dehors* the initial idea is even more abstract: it is about narrow and spacious structure as well as the processes, which mediate in between, a theme, which is developed in all musical facets and which correlates with a composed love-poem by Ernst Jandl. There are very often very elementary and seemingly abstract issues, which stimulate and tease me to develop music from them, and that since these are very concrete and tangible especially because of their abstract character and simultaneously they carry traits of self-absorbed play—everybody can imagine something with a trill, which is the theme of the piano piece *Bebung*, and at the same time a trill is abstract enough to be no "theme" or "motive" in the classical sense yet it is also sufficiently specific to be developed into the most divers directions: defined as a fast, flickering change between pitches in the interval of a (major or minor) second or as a prolonged changing note there are a whole set of parameters, which can be modified simultaneously, gradually, in the same and in opposite directions, respectively. However, it can also happen that the central idea is not yet apparent in the initial concept—it must then be established by the investigation of the material and its analysis using technical procedures.

I denote ideas according to their structure with notes or graphically or in a mixture of both. Graphical notations play an important role in the initial work, especially for longer, extended formal progressions. I can extract from them more direct and easily the "envelope" and the directionality, the energetics of the piece than from a less clear but more detailed score. With the notation it is similar as with the connecting ideas: the abstract is at the same time that, which is tangible. The first step is always to collect as many ideas as possible, i.e. to write them down because the writing itself catalyses again new ideas—and then to reorganise them once more on paper, such that maybe already the first references become visible and that out of it the first developments, progressions, etc. become deducible.

Always then, when I want to open up a piece, it is about spanning a coordinate network from the initial ideas, which is formed from the central elements and processes of the piece. To open up means: the initial ideas are given in different grades of clarity, the ones of clear shapes, which immediately allow notation in a score up to polyphonic textures or bigger, formal developments, ones standing in front of my eye very concretely and vaguely at the same time, which I therefore at first note verbally or graphically, and which I have to technically reproduce as such in all details and in the interplay with the more concise ideas: technique is the medium to catch up with and reel in something, which is in front of me, the means to take me quasi to the height of my own imagination. Artistic spontaneity, which makes itself absolute and thinks it is able to dispense with technique is no such thing, but only a reproduction of the own previous knowledge and thus becomes narrow-mindedness: one needs technique to precisely continue beyond that, to divest oneself, to realise in the first place which potential lies in the own ideas, preferences and intuitions.

Technical procedures have thus for me at first and foremost a negative quality in the sense of "negating"; the composing subject shall be negated in a specific sense, be challenged, be overcome, to be more precise: the own inevitable subjective limitation, which presents itself in practised, familiar and thus obvious preferences, opinions and partialities, which generally become in turn legitimated and rationalised by already acquired technical abilities. Precisely therefore I put a lot of emphasis on testing the from-each-piece-abstracted formalisation of a technical procedure always afresh and varying on the specific material and to reflect it critically: formalised procedures are likely to mislead someone to convenience and automatically fixated response-reactions and to an external, manipulative handling of the musical material—and by this it gets the negative, critical potential wrecked, which is inherent in them. Their negative and thus productive potential is that they serve to remove from my own abilities and ideas all what seems familiar and homelike to me, to alienate them from me and to show me what there is in "my own" ideas and what I would not likely discover, if I would only trust what I already know and can do. Technical processes serve well to return the "foreign" to the seemingly familiar and to overcome what Adorno refers to as the "private property character of experience"[5] with all its pettiness and narrow-mindedness in favour of a generosity and largeness of one's own thinking. The whole exercise bears facets of a self-analysis in the psychoanalytical sense: something own—a thought, an idea and with it the known technical abilities—is considered precisely, patiently, and fondly, yet also relentlessly and insistently, is "turned upside down and back up again" with all means of the technique until it starts to talk, develops its own life, to which I can relate to in return. The major point of interest in this matter is again the dialectic manner: the alienation caused by the technical procedure brings me in the same moment close to what I had in mind in a piece at first and it helps to open it up in all its details.

The "if to say so" positive effect of the technical procedure is that it allows me to build for the individual case of a certain piece a network of coordinates, a scaffolding in which I can then move freely when composing and which I can again leave if the piece is finished. That is always again a "va-banque-game" and it doesn't always succeed, sometimes not right away: because the establishing of the network as already mentioned shall not follow a formula, shall not be forcefully placed on an individual case—despite all the restraint, which is of course always part of the game; yet it should also not be a network, which ruptures at the first severe test.

Formalisation and Intuition

When considering where a composer one obtains one's ideas, I would like to distinguish two terms in the first place: "stimulus" and "intuition". Stimuli are all incidences

[5] It is not a coincidence that this emerges for Adorno's thinking as the central figure of thought for the first time in *Philosophie der neuen Musik* in a passage which refers to Schönberg's composing, where an essential element of contemporary music is pointedly the ability of the composer "always and always again, with each new attempt to throw away and deny what had previously been possessed." [3, p. 111] (quote translated from the German by Clemens Nachtmann).

and occurrences, things, situations, which provoke an idea either spontaneously or sometimes also just in memory. The stimulating things are of a diverse nature and almost a bit indifferent: it may be due to the unavoidable neurotic damage, the "déformation professionelle" professional deformation of a composer that she/he can make use musically and in composition virtually everything, which surrounds him/her and what he/she is subject to.

On the other hand intuition, the thing itself, which stands at the beginning when composing a new piece, can be of a very diverse nature: an individual musical event, a short formal passage, sometimes even the sketch of a whole piece or at least of a section from it.

At first, the shape of the initial idea influences of course the concrete process of the work: Extent, selection and direction of the technical procedures, to which an initial idea or a string of such ideas, respectively is subjected. In doing so the technical procedures are abstract, i.e. if taken individual are similar from piece to piece or even identical—and at the same time they are not similar, because in a way they are placed in an irreproducible, individually-concrete force-field by the initial idea and the therein utilised fantasy, that is the imagination what might come from it, and thus given a status, a relevance and a meaning: they lead to other things, yield results which were not obtained previously etc. Here I always like to use the image of various magnets, which put per se similar iron turnings into very different constellations. Such thought- and work-processes can therefore undoubtedly be formalised, but the formalisation should be made at the same time sufficiently determined and sufficiently abstract, thus in such a way to not only fit in an individual case yet also not like an all-fitting, indifferent algorithm: This is what I look for every time anew in my still non-computer aided procedures. The pool of technical possibilities expands and restricts itself with every new piece: expands, because new things are added or known things are developed in a new direction, respectively, restricts, because some procedures lose their value, become uninteresting or also absorbed into others. However, with the techniques also the understanding expands of the already discussed elemental structures of the musical textures, which in return may play a role in a new piece. The technical procedures, which I utilise, are of course not fully invented by me but the result of self-observation, listening and reading of other music and reading of theoretical literature; the essays of Pierre Boulez, in particular the ones from the 50s and 60s [4–7], have strongly influenced me therein, maybe even more than his music; this is true similarly for the considerations of Stockhausen about musical time from the 50s, which I still consider fundamental. Much of the technical procedures I have adopted, others I have rejected and finally some I have indeed myself invented. Intuition is thus for me not an antithesis to technique, to rationality and not at all a possession or asset of the subject, which it would need to defend against the alleged impertinence of the technique, on the contrary: "Unconscious knowledge not entirely subject to mechanisms of control explodes in inspiration and bursts through the wall of conventionalised judgements 'fitting reality'." as Adorno elaborates in such emotional words in his *Against Epistemology: A Metacritique* [1, p. 46]. Intuitions themselves are thus communicated by settled or deposited knowledge, which they break through in the same time, as they assert themselves to the subject as abrupt, sudden, non-commanding, ego-foreign ideas. I consider this passage of Adorno one of his most important, as it demonstrates his ability to enlighten rationally the non-rational thing (in this case the intuition), yet without rationalising it, i.e. to go beyond the measure of a terminology, yet without lifting it to an abstract antithesis to rationality like Henri Bergson, against whom Adorno polemicises here.

From a psychoanalytical perspective, intuitions represent the unconscious, which shows up again and again in the process of composition: naturally at the beginning is the idea—invasion might be even a more appropriate term—and then during composing the "non-transparent" push to arrange a passage in one specific way and not another despite maybe conflicting "rational" concerns. Composers like to talk in commando-jargon of the composition "strategy", which they chose and from the formal "decisions" they take, as if composing would be a chain full of logically consequential, one from another deducible conscious steps, which are always transparent in their premises and consequences. This overlooks or ignores that many decisions are not made in the full of consciousness but are noticeable dictated by a dark, unknown, internal force and for which the thematic connection with what was already composed is not or not sufficiently evident in the spot. The pre-composition exercises situated right at the beginning of composing, the formation of a coordinate network of possible relations have in this context also the character of improving confidence: at first I create a space in which I can move freely and unencumbered of all pre-considerations, with the trust that what will then happen in the preliminary claimed and unlocked terrain and thus which further ideas will approach, will all make a connection, even if it is not quite clear in the first moment. That means that in the beginning there is a maximum control such that I can be rid of the neurotic necessity of control and am not forced at every detail to make again basic decisions. The enormous technical effort which I make in every piece serves at the end of the day only to bring me to a "point of no return" from which on composing happens as a process almost devoid of a subject and that I am almost only an executing organ of a process, which it initiated by a subject, but which now develops its own dynamics. With technical means, objective and atmospheric conditions shall be established as optimally as possible such that things don't end after the first randomly occurring intuitions, but that they unravel like on a roll. All the technical procedures shall take from me my inhibitions and doubts, to entrust myself to the musical material without reserves and to enable me to be a match for what then incalculable comes at me. Insofar they have not only a constructive-musical but also a psychological sense: the one of a warm-up of what I henceforth want to let come at me.

Composing thus depends on whether one can offer oneself to the music, which one just lets develop, so insistently and as long as it becomes audible, where it wants to by itself. Therefore it is not about that the composer comes to the fore by perpetual decisions and takes himself/herself more important than he/she really is, but it is about that he puts condition in place under which sound can develop a life of their own and also to reveal it.

Adorno has once expressed this experience related to philosophical knowledge as follows: "Subjectiveness is present in knowledge in form of its negation.

A finding of insight or knowledge summons all our experiences only to demolish our experience."[6]

Such an approach connects the philosopher with the composer and makes an objective i.e. an object-saturated thinking precisely commensurable with composing: "I have an immensely strong experience that we do very little. A composer, for instance, does almost nothing. [...] I really need all my spontaneity, in order to do nothing, to see what there actually is. I have to try my hardest to do what 'I'm not doing' in order that something will 'be done'. It is the maximum effort in order to get a minimal effect."[7]

To make real, with exact and rigid plans what can't be planned and what is unknown is consistently my compositional utopia. Hermann Broch has formulated this relationship once in his novel "Ophelia" in the following way: "It is often the case as if you have set up your whole life to make yourself surprises, to act astonished and appalled by something, which you have caused yourself. And finally you believe it. Somehow like that must it always work if you invent yourself a story or compose music: you develop a plan and then let you be surprised by what comes" [8, p. 35].

Evaluation and Self-reflection

In the pre-composition phase of composing it is not yet the piece, which takes the centre, but it is about probing and exhausting the respective selected material as completely as possible with respect to its intrinsic potential. Conceptual procedures, which are done at the desk, are as equally important as the practical trying out of sounds—either by myself if I have the instrument available or with friends that are musicians. This experimentation goes along with the intellectual considerations and often continues into the actual composition process if it is about unusual and selected sounds.

It is the logic of the matter that in this exploration there are many results entirely or for the current piece useless, which however in return may be a stimulus for a new piece or which provides new insights, which can generally be useful for further composing. I take the decision about what is useful and what not, however not by means of the own fixed preferences or of a given and immovable plan, but on the basis of a general criterion, which is consistently the same in all phases and on all levels of the working process: tonal echoes or references, in particular the ones, which arise despite opposite intentions, are to be excluded systematically, in the individual

[6] Translation by Clemens Nachtmann from the German quote: "Die Subjektivität steckt in der Erkenntnis in Form ihrer eigenen Negation. Eine Erkenntnis bietet alle unsere Erfahrungen auf, nur um unsere Erfahrung zu vernichten." [10, 520f].

[7] Translation by Clemens Nachtmann from the German quote: "Ich mache so ungeheuer stark eine Erfahrung: wir machen nur ganz wenig. Ein Komponist z.B. tut fast überhaupt nichts [...] Ich brauche eigentlich meine ganze Spontaneität, um nichts zu tun, sondern nur zu sehen, was eigentlich ist. Ich muß mich mehr anstrengen, das zu tun, was ich nicht tue, sondern was getan wird. Die maximale Anstrengung zur Erreichung eines minimalen Effekts." [10, 520f].

musical parameters, in their cooperation on a local level (in a part of a form) or in the entire formal progression (like augmentation dramaturgy with climax, "sonata form" etc.). Otherwise I often let myself be inspired to a certain progression of form or time by the very originally unpredicted possibilities.

The boundary between "pre-composition" and the actual composing is indeed not distinct, unique and determinable in a general way. It means a conceptual distinction, which doesn't signify two clearly separated phases in time; it may well be and it also often that in the middle of a composition, i.e. in the stable deciding and elaborating of formal processes, more generally disposed, the "pre-compositional" work has to be taken afresh at a place for which one only realises during the elaboration of the composition that deep fundamental problems are posed. At the beginning one doesn't right away have an overview, but the more precise the first work steps are the less frequently one then holds up such fundamental questions during the course of composing. The critical, dialectic point of all these "pre-compositional" work steps is that they are designed in such a way to become superfluous and to be productively forgotten in the context of the production.

Project Approach: Avoiding Tonal Associations

POINT: What are the actual themes in your compositional work?

Nachtmann: Before the "actual composing" begins, it is generally always about building me a sound-framework and a time-framework from the analysis of the initial ideas. Depending on which sounds are in the centre of the piece, the techniques, the form of notation, the diagrams and tables (pitches or noises, tone forms like trills etc., articulations like al pont., pizz., dynamic grades) differ. The basic criteria according to which these frameworks or networks become constructed always remains the same: it is about gaining an abstract pool of elements, which then can be arranged in different formations and which is subject to various derivations in order to extend the material pool. Let's say for example that a piece shall work with distinct pitches: then, there is at first a pool of available pitches which correspond to the original idea that emerges from it. Depending if the original idea was "melodic" or "harmonic" in nature, the pool is then arranged into various simultaneous and sequential formations: chords and scales in general, in each case also into specific forms. This includes sequential: non-octave-changing scales, melodies with characteristic reflexions, definition of various ranges of scales/melodies/chords etc.; simultaneous: specific registration of chords and the play with octave-fixed and mobile tones and possibilities of passages between basic formations, like a melodic line, which freezes each tone to a chord, until they are freed again. In all formations permutative processes play a role like retrograde, rotation, palindromes, the dropping of certain tones, separation into certain subgroups, which are then subjected to the same or similar procedures etc.

Formations are thus the result of "permutative" processes of constant basis elements—derivations on the contrary extend the pool by introducing procedures like inversion, retrograde or transposition based on a certain formation: the original material is thus extended or multiplied, respectively, either by the same or by a higher number of new tones. Both with the formations as well as in particular with the extensions, temporal and formal considerations are already part of the game.

Altogether a variable network of points in a sound space is produced, where each single point, when taken as a tone is thought within a certain variation span (microtonal or larger intervals), but can on the other hand also stand for grades of brightness of noises. It is thus about elementary relations, which have always already represented formal processes abstractly or on a very small scale—formal processes which directly determine the composition or also just structure it in the background. This analytic, experimental and also playful examination of the material is altogether itself already like a dialectic variational procedure placed ahead of the piece: nothing remains like it was at the beginning, the "initial idea" as premise of the resulted of the technical processes become the premise of new works.

All these constructive procedures have, as already mentioned, at first and foremost a polemic and negative sense: tonal constellation and such, which act tonally or even only resemble tonality, shall be excluded. And this is true in a general sense: for tonal sound relations as well as for tonal relationships of durations, i.e. metric arrays/arrangements with quantitative equal yet qualitative unequally weighted units and the corresponding rhythms. Thus, it shall at the onset be avoided that any musical element is fixed as the "first" one, as an overall principle, from which everything else is derived; the aim is an equality of the sound qualities, in which these properties are not levelled out against each other—except if such a levelling out is itself the "theme" of the piece—but are able to stand qualitative specifically for each other. Thus, in place of a fundamental principle there is a permanent re-configuration of a self-variable sound material.

Project Expectations

For our research project we agreed to generate chords with preferably no tonal associations by developing exclusion criteria. What interests me in this context first of all, is the above mentioned "effect of alienation": what happens when inclusion- and exclusion criteria, which I use for my own, manual, at-home-at-my-desk-drafted composition procedures, are discussed and then transformed and revised using computer algorithms? Are my own criteria sufficient for what I have in mind or still way too imprecise and sketchy, or do they lead if applied consequently even to the opposite of what is intended?

Despite observing all the criteria are there chords formed, which still sound tonal? To discover the reason for this is the central question when evaluating the produced chords. Are there results of the computer-based calculation of chords thus indifferent to taste, preferences and intentions I would never have invented without the computer? Are the results maybe in their concrete form useless, but will inspire new thoughts and research? I believe that these and similarly located questions make me push further the self-reflection of my work—an indispensible must for each composer as I see it. To be able to observe and formulate how one's own imagination "works" is for me not only important for composing but also for teaching, which is consistently about finding the appropriate terms and images for the organisation of music in a discussion with other people, for the nature and texture of which there is no tradition, no model and non-defined types of forms.

Exploring a Compositional Process

POINT: We were aware that Clemens Nachtmann was sceptical about closed systematic approaches because they tend to establish norms in which he, as a creative individual with a historical awareness, is quite naturally opposed to. Hence we decided for a formalised strategy where rejection was the central element. After discussions about tonality and atonality it turned out that Nachtmann would be interested in searching and investigating chords that avoid a number of criteria.

POINT: How are you going to use chords in your compositional process? And how dense are the chords you are using?

Nachtmann: The organising intervals in my chord formations are generally the tritone (as the bisection of the octave) and then the octave itself. Very often I take as a starting point sound aggregates with a density of six tones, arranged over a framing-interval of two and a half octaves. Thereby the lowest tone serves as an axis-tone/centre-tone by means of which a chord is mirrored in two ways: either the interval-direction of the other tones is inverted, which produces an identical interval in the opposite direction and thus a new tone, or the interval size is inverted, that is a complementary interval is generated and the tone remains the same. Spread out horizontally a sound progression of six chords thus unfolds correspondingly to the density of the voices. In this way I have already gained a linear horizontal dimension.

Yet all decision criteria are variable, the density as well as the manner of arrangement (as chords of scales within each case variable framing-interval and interior-intervals). The density of voices can be reduced down to three parts or extended to up to nine parts, the framing-interval can be scaled down to one and a half octaves or even half an octave or it can be replaced by another framing-interval. The interior-intervals can be changed by further methods etc.

POINT: What kind of chords would you never use?

Nachtmann: I strictly avoid any chords which evoke or even suggest a tonal system—tonality in an extensive meaning of any tone system with hierarchic structure which can be fixed beyond the single work of music. This comprises harmonic tonality—a system in which all tones refer to a fundamental tone and the chord established on it—as well as modality: a system based on the hierarchy between perfect and imperfect intervals.

POINT: So we would have to search for chords that should:

1. consist of an arbitrary but fixed number of pitches
2. fit an arbitrary but fixed frame interval
3. fulfil a number of further constraints, most of them related to the intention to avoid tonal associations.

(1) and (2) can be clearly defined, a search for all chords of this kind is basically equivalent to the search for ordered partitions of integers. (3) is less clear, as "tonal association" is a vague term that might be interpreted based on knowledge that is specific for a certain culture (western music theory) as well as individual preferences. This is the point where ongoing discussions lead to refined restrictions more suited for the composer's demands concerning the harmonic material.

A computational implementation was established by a backtracking algorithm capable of full enumeration of (1) and (2), but at each step additionally taking into account constraints of type (3).

To give an easy example that can be carried through by hand: a minor sixth consists of eight halftone steps. The integer 8 can (ordering considered) be summed up by three integers in 21 different ways, each corresponding to an interval vector of three elements. The chord itself consists of four pitches and might be based on an arbitrary pitch, see Fig. 1.

First approximation: when e.g. avoiding "tonal" major thirds (partitions including 4) there remain 12 chords with, on average, a higher degree of atonality (Fig. 2).

Of course this is a rough simplification. The frame interval, being the inversion of a major third, has a tonal tendency itself, especially the first inversion of a major triad, which is fully contained in all above chords with Eb. As not every chord containing a major third will necessarily sound tonal, this means an important part of the work

$$8 = (1+1+6) = (1+2+5) = (1+3+4) = (1+4+3) = (1+5+2) = (1+6+1) = (2+1+5)$$

$$8 = (2+2+4) = (2+3+3) = (2+4+2) = (2+5+1) = (3+1+4) = (3+2+3) = (3+3+2)$$

$$8 = (3+4+1) = (4+1+3) = (4+2+2) = (4+3+1) = (5+1+2) = (5+2+1) = (6+1+1)$$

Fig. 1 Ordered partitions of number 8 into 3 integers; corresponding partitions of a minor sixth into 3 intervals

$$8 = (1 + 1 + 6) = (1 + 2 + 5) = (1 + 5 + 2) = (1 + 6 + 1) = (2 + 1 + 5) = (2 + 3 + 3)$$

$$8 = (2 + 5 + 1) = (3 + 2 + 3) = (3 + 3 + 2) = (5 + 1 + 2) = (5 + 2 + 1) = (6 + 1 + 1)$$

Fig. 2 Ordered partitions of Fig. 1 without number 4 as partial sum resp. without major third

was developing a catalogue of criteria for (3), those concerning tonality being the crucial ones.

When enumerating larger frame intervals the number of possible chords rapidly increases. Hereby many chords with large partial intervals were occurring which didn't fit Nachtmann's needs. So we restricted to a maximal interior interval of a major seventh, the first additional constraint:

(3a) Define minimal and maximal interior intervals (finally minor second—but this one only if beneath a major second—and major seventh).

POINT: Let's regard what you are going to exclude in building your chords, are there interval relations that you don't want to have?

Nachtmann: Within chords I exclude octaves and fifths for two different reasons. The fifth is the most problematic interval for me as even in elaborated non-tonal contexts it sounds archaic which is deeply connected with its position in the overtone series: it is the first interval after the prime or octave including two different pitches and as such it establishes—as an opposite to all the other following intervals, especially to its "inversion", the fourth—an unquestionable stability and founding effect which is also present in major and minor chords, the stability of which is due to the frame interval of the perfect fifth. So for me the fifth is the very archetype of tonality and therefore I'm either excluding it strictly or covering it carefully in case I can't avoid it anyway.

On the contrary I like octaves very much but I'm excluding them from the chords because the octave and the half of it, the tritone, are usually structuring the chords in the background; and moreover I'd like to have the freedom to transfer one or more tones of the chord in different octave registers when working on the chord material.

POINT: So we can formulate two further specifications of criterion (3):

(3b) Avoiding chords with any octave relation between two pitches.
(3c) Avoiding chords with any fifth (or fifth transposed by octave) between two pitches.

POINT: What are the most important chords that you usually exclude?

Fig. 3 Basic forms of 3-tone chords to be excluded as interior chords (3d1–3d6)

Nachtmann: I normally exclude the major and minor chord in any position as well as most of the seventh chords, including most of the incomplete versions and most of its inversions.

POINT: In summary we come to criterion (3d):

(3d) Exclude certain "tonal" 3-tone chords as interior chords.

In detail the following three-tone chords were to be sorted out literally (interval vector of basic form, see Fig. 3, in parenthesis):

(3d1) major triad (4 3) + inversions
(3d2) minor triad (3 4) + inversions
(3d3) diminished major chord (4 2) + inversions
(3d4) doubly diminished major chord (2 4) + inversions (can be regarded as inversion of seventh without fifth, transposition neglected)
(3d5) neutral seventh chord (7 3) + (some) inversions
(3d6) two consecutive major seconds (2 2)

POINT: The last chord above marches to a different drum, why is that?

Nachtmann: I consider the chords we are talking about as the basic and preliminary chord material of a hypothetical composition—and in this material I'd like to avoid a certain interval to be privileged, as it would be with two second intervals which tend to establish a cluster.

POINT: Whereas excluding 3-tone chords was quite a straight task, things are not so clear when there are larger chords to exclude. What if a major triad is contained in, let's say, a six-tone chord as first, third and sixth pitch? Other pitches might shadow its tonal appearance, but it strongly depends on the specific case, so excluding this way was not an option for a formalised strategy. Instead we decided to regard tonally-favoured 4-tone-chords as candidates for exclusion: a number of seventh chords and other four-tone chords containing triads (Fig. 4).

Fig. 4 Basic forms of 4-tone chords to be excluded as interior chords (3e1–3e14)

(3e1) diminished seventh chord (3 3 3) + inversions
(3e2) major minor (dominant) seventh chord (4 3 3) + inversions
(3e3) half-diminshed seventh chord (Tristan) (3 3 4) + inversions
(3e4) major major seventh chord (4 3 4) + inversions
(3e5) augmented major seventh chord (4 4 3) + inversions
(3e6) minor major seventh chord (3 4 4) + inversions
(3e7) added major sixth (4 3 2) + inversions
(3e8) major triad with added fourth (4 1 2) + (most) inversions
(3e9) major triad with augmented fourth (4 2 1) + (most) inversions
(3e10) major triad with major second (2 2 3) + inversions
(3e11) major triad with minor second (1 3 3) + (most) inversions
(3e12) major triad with minor third (3 1 3) + inversions
(3e13) neutral seventh chord with augmented fourth (6 1 3) + (most) inversions
(3e14) neutral seventh chord with minor second (1 6 3) + (most) inversions

Some criteria are overlapping, e.g. excluding a major triad in its basic form is excluding a major seventh in basic form, but not all of its inversions, so they are excluded explicitly. Some chord categories to be excluded are implicitly excluded by others completely, e.g. minor seventh (3 4 3) as inversion of added major sixth.

All in all about 400 chords belonging to the above categories were excluded. Defining them was a process that took some time and discussions. For some subcategories "some" or "most" inversions were excluded, meaning that reciprocally most or some were not excluded as Nachtmann judged them to be sufficiently far away from obvious tonal associations, especially when embedded into a larger chord.

With criteria defined, search was carried through for frame intervals between 13 and 35 halftone steps (one and three octaves without octaves itself) and chords between four and eight tones. For each of those constellations between a handful and some dozens of chords were found. With the final version of criteria Nachtmann regarded most of them as suitable harmonic material for his further compositional use.

This is a typical example for the reduction achieved: for a frame interval of 30 halftone steps (2 octaves plus tritone) there exist 3,654 chords with 5 tones. Many of them are very unbalanced with huge interior intervals, thus not usable for Nachtmann's needs. A restriction to a maximal interior interval of a major seventh reduces the number to 600, additionally excluding any octave or fifth relations to 87. With excluding interior chords as defined above we end up with these 14 chords, see Fig. 5.

POINT: How would you regard these chords, are there any further tonal residua you would avoid?

Nachtmann: In this sequence two chords could be especially prone to evoke a tonal context: No. 5 and No. 6, where (always read from bottom to top) the first, second, and fourth tone form a minor and major chord in second inversion respectively. In case of chord No. 5 this effect is almost entirely avoided by the fact that the impression of an incomplete chord of fourths (with perfect fourths) predominates. In contrast the major-character of chord No. 6 stands out immediately: the upper F# appears as a dissonant neighbouring-tone to F and the Bb is too weak to

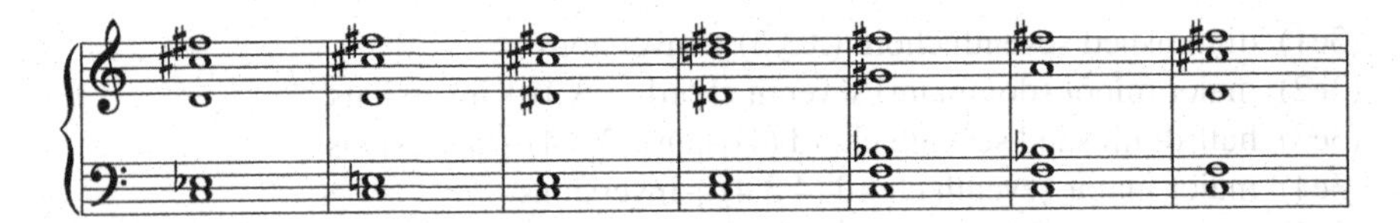

Fig. 5 5-tone chords with a frame interval of two octaves and a tritone, fulfilling all of Nachtmann's explicitly formulated exclusion criteria

disrupt the tonal effect. Something similar occurs with the chords No. 8 and No. 12: in both one perceives the tones one, two, three, and five (C-F-Eb-Gb) as incomplete seventh-ninth-chord, compared to which the C# respectively D# seem only as a mere "disturbing tone". More indecisive however is the effect of chord No. 9, in which there is a D instead of the C# in chord No. 8, so disturbance turns out somewhat stronger.

It is also remarkable how different chords No.1 and No. 4 sound, in which the lowest interval is a minor or a major third, respectively. In this exposed position it could possibly cause the effect of a triad: No. 1 and No. 3 which contain the same tones sound virtually not at all tonal, but No. 2 and No. 4 on the contrary are very tonal. What could be the reasons for that? All four chords contain a seventh from the second to the third tone and from the third to the fourth tone. In No.1 the tension of two major seventh counteracts a tonal gravitation and in the same way also the upper fourth is strongly dissonant with the two lower tones. In the second chord (No. 2) follows the (bottom) major third a minor seventh so that the lowest three tones seem like a spread of two major seconds and creates a mild, intrinsically resting sound effect, which makes the C# a side-tone of the C and degrades the upper F# as disturbing tone.

In No. 3 however a major seventh follows the major third, such that the first three tones form a major-/minor-third sound, which atonal-dissonant effect is underlined by the following C#, since also C, E, and C# constitute such a sound in inversion: the upper fourth dissonates further to this dissonant structure of thirds. Notably this effect is almost turned into the opposite in chord No. 4 by replacing a single tone: although here like in chord No. 1, the lower third is followed by two majors and thus in fact tension-charged seventh, the D melts into the lower third similarly well as in chord No. 2 and lets the D# lower into the status of a side-tone, against which the F# can not compete with, especially since it itself forms a third with the D.

I favour most the chords No. 13, which is the closest to me in its edginess, No. 10 and No. 11 with their rather mild dissonant effects and No. 14, in which the major second in the middle stops quite effectively, the tendency of the lower major sixth to combine with the F# to the remainder of a ninth-chord.

In Fig. 5 it can be seen that quite often pitch classes of a chord are situated within one tritone. In one of the early discussions there was also the idea to generate chords from a pool of a cluster with an ambitus smaller or equal a tritone. This would lead to suitable atonal chords too, nevertheless a whole category of chords would not have been found that way, the explicit definition of exclusion criteria turned out to be laborious by differentiation, but finally fruitful.

POINT: Can you imagine, that we could find a single chord according to the tonal-narrowing exclusion criteria, that could be used, without any qualms, as the "true" chord of new music?

Nachtmann: It would indeed be a sensation if one could do that: to have the "one" true chord, which one could throw into the face of all the neo-tonal and neo-modal composers and with which one could immediately put them to the wrong without ifs and buts, since the one and only new-music-chord would at the same time be an irrefutable and quasi scientific proof that the stuff, which they put together and compose, is not new music but antiquated scraps from the past!

However, besides these undeniable advantages and all jokes aside: I would not rejoice over this discovery since I would get rather bored if I had only one chord at disposal because that would mean that I would be stuck without alternatives. Moreover, the one and only atonal chord would immediately seem like the philosopher's stone, would become a sound garnished by a metaphysical grandeur in the sense of Scriabin's "mystical chord", Hauer's "twelve-tone-play", Stockhausen's "super-formula" or other ideological schisms. And since I am strictly against first, fundamental and fixed principles in composition and philosophy, I would become immediately suspicious and would try to counterwork it. Maybe in this way would emerge a potential for my first piece leaning towards tonality, how knows?!

POINT: What are your other compositional steps that you use, how do you deal with a chord once it is found?

Nachtmann: I mentioned already previously that my composition work initially develops based on verbal, graphical, visual or as a notated score of primary ideas in a framework of sounds and times, in which I can then freely move. In this coordinate network it becomes apparent, which relations exist between at the beginning of composition which are often still scattered and "unconnected" ideas and hence obtained musical elements with further possible relation can be consequentially deduced. In other words, the coordinate network already contains in itself basic rudimental formal processes, which can then affect the composition form, at the micro and macro levels.

Chords, as generated by us, are at first scattered sound-elements and the subsequent step is accordingly to derive possible relations based on them. This can mean many things: I can e.g. take a six-note-chord and mirror it around an axis-tone in the above described manner, from which I obtain a sound-progression, which already contains horizontal lines. Or I can take a chord and reduce the frame-interval from 2 and a half octaves to a simple tritone and take the notes of the chord which fit within as the basis of a linear progression. Or I can put two or more chords in relation and consider which tone they share and which differ; and new relations arise consistently or invariably depending on whether I use chords only in the original form or also transposed.

Similar results, even though with a lower density, I obtain if I chose for example only three tones of a chord and transpose them subsequently with respect to each of the six tones of the chord. In turn, I can then privilege the common and different tones in whatever manner: to fix them within an octave, for instance, so that they form a static partial chord or accentuated corner- and middle-tones of a linear progression. On the other hand I can either emphasise the "privileged" tones in the composition: by means of a certain duration, via a tone-form, a volume, a particular way of playing, articulation or tone colour—or I can put them into the background. The governing idea of all these work steps is: the vertical- and diagonal-harmonic and the horizontal-linear dimension shall be organised according to the same criteria and in this way also then structure the composition, when this is not directly audible in the sounding results.

Necessarily related to the elaboration of possible formal relations of sound qualities is of course the question, in which temporal relation, i.e. proportions these relations are realised. The elaboration/composition/construction of a temporal framework goes thus always hand in parallel with the work on the sound qualities. The conceptual organisation of the musical durations is owed at first mostly also to a primary idea and depends in its concrete characteristic majorly on the respective sound material and the intrinsic times, which it can form as well as on the used instruments in a composition.

The criteria of the sequences of durations are in general the same: they are generated or constructed so that they can be related to all imaginable musical parameters. Sound quality, i.e. pitch/noise, tone form, the way of playing, articulation, dynamic, tone colour, etc., each is, as already said, form giving and thus takes part in the articulation of the musical time. On the other hand these parameters can be projected onto several levels of musical time-organisation: a sequence of numbers, which are themselves already ordered according to additive or multiplicative criteria or axis around a mean value, can represent any related duration, for example an eighth note or perhaps also a quintuplet sixteenth note, the beat of a bar, a sequence of individual time-lags or a chronological unit like a second.

A particular sequence of durations can thus be musically used either literally as duration, as what is traditionally called "bar" or "metre", or also as in itself subdivided, underlying background duration, or even more elaborated like phrase, theme or a whole formal section. In doing so each duration can in itself be again flexible be divided either in measured of freely-executed durations, in the same way as the superordinate sequence or in another way, or flexibly arranged to larger arrays. In this manner I can change from the details over smaller and larger sections to the entire form. The conceptual distinction of rhythmical and bar-like durations is an attempt to overcome the tonal rhythmicity tied to pulses and beats without scarifying its achievements: the multiple levels and meanings of musical time. It is here possible to layer several such duration-sequences on top of each other and a "summation-rhythm" can be deduced or the duration-sequence can itself be understood as a sum, from which various single parts can be inferred.

By using several of these simultaneously on-going duration-sequences, highly complex textures can be derived from very simple basic elements. The question when and in what frequency the individual voices feature simultaneous impulses plays then often an important role for formal decisions: is it related in turn on the level of pitches most closely to the question if and how, for instance, common tones of sound complexes are emphasised or hidden. These duration-sequences have little in common with a conventional beat as they either consist of predominantly of irregular basic durations, which moreover do not constitute a usual bar, like e.g. $4 + 3 + 4 = 11$, or a traditional unit like a four-four bar is divided into a sequence $5 + 6 + 5 = 16$. The preference of sequences of irregular and diverse durations follows here from the consideration to a-priori avoid regularly repeating, periodic patterns on any micro- and macro-formal level. Concerning this duration conception I mostly owe to the music and writing of Pierre Boulez, Nicolaus A. Huber, Mathias Spahlinger and Elliott Carter [5, 11, 12].

Applied to the duration-sequences are also procedures of variations, which either rearrange or extend the material: to the former belong rotation techniques, regular or irregular augmentation/diminution, specific division or subsumption—dissection in pause and sound, inversion true to scale or approximately, "breakdown/segmentation of the rhythm by itself" etc. Extensions are possible, for instance, through multiplication of individual durations to groups, in proportional, regular or irregular proportions, where augmentation/diminution procedures can be applied to these groups, or through the extension of the configuration of durations by itself as configuration of pause, by nesting of configurations once as sound-, once as pause-series.

Of the highest interest in each piece is the question, how the intrinsic time in a multi-voice progression can be reproduced from piece-crucial sound-qualities: A ricochet of a string instrument is divided in precise rhythm or overall duration onto several instruments or specifies the sequence of basic durations of a section. For example an arpeggio is reproduced from precise cue-gaps or from a purposely shaken/blurred cue, sequences of cues of several instruments are chosen in order to de facto yield a trill etc. At the end of the work are then multi-stage, ambiguous, in-itself nested proportions of time, which normally approximate the formal development in full and in many details already very close to the end result.

In order that the description of the composition technique and the individual working steps are not lost in sophistry, it seems necessary to point out once more that I always pursue one objective. It is via technical means that I aim to keep alive my primary ideas in the freshness they presented themselves in the first moment, to unlock them in all their details and at the same time to transcend them and the isolation, in which they once appeared. The pre-composition working steps are for this reason a permanent confrontation of technical procedures, which generally lead very far away from the original idea and the composition idea itself.

The pre-compositional working-steps can be best described in a general and reasonably comprehensible way. It is much more complicated with the "actual" composing—basically I can only say that on the one hand it appears very matter-of-fact and almost banal, because it is only about elaborating what was already until then configured. On the other hand however, this elaboration is still so full of surprises, which are different from piece to piece and entail such diverse forms of reaction and patterns of conduct that it is virtually impossible for me to discuss it in general.

Already the determination of when the point is reached at which one can finally behave, to paraphrase Hegel, "in freedom to the object", without constantly having to resort to all the pre-sketches with all their schemata and tables, is always also an intuitive-arbitrary decision in the sense of the above discussed, which is, although motivated by facts, not fully covered. The irrevocable feeling that the preparations have been pursued long enough and that now the moment is really getting started, occurs when the formal structure as a whole and in its most important parts is proportionated to itself and partly also already in itself with respect to the "basic durations" and that at the same time a temporal structure of the sound qualities has become so clear that they yield a form and a dramaturgy, respectively, even if it remains still vague in some sections. The "actual" composing always iterates from the details to the entire form and back in a constant change of perspective: in turn on a higher level it is about the temporal arrangement and the proportions of sounds.

The general criteria for the specification of a musical formal development are again negative at first: those formal conventions should be avoided, which have sunken to emblems, tokens, and bad habits, shortly clichés. That doesn't mean that every time entirely "new" forms have to be "invented": it is quite possible to let transcendent traditional formal patterns quasi "from a distance" by using them in the first place in a broken, overgrown and ambiguous shape/conformation.

The "in the first place" is crucial here: because this method is the opposite of those rather well liked techniques in the area of new music, which present certain forms at first fairly affirmative like a quote, only to then dissolve, deconstruct, to spoof it, etc., a seemingly sceptical procedure, which is in truth completely uncritical because it leaves the entirely worn out tonal symbolism untouched, in which the intact, closed forms stand for the ideal world and the opening and respectively disintegration of forms symbolises the negative, broken world.

The material of new music, however, lies beyond this symbolism, since it is just not "in the first place" disposed for an a-priori granted form and thus the formal characters of construction and deconstruction hold an equality and value, which does not aim for a cliché-like symbolism.

Also in return it is very possible that the "invention" of new form leads back quasi on a detour to known forms, which however became something new, thanks to the detour: novel and newly conceptualised constructions, which "newly invent" a traditional form on a detour, which is then no longer the conveyed one.[8]

General questions of the form-formation in the course of the composing are: Should correlation and connectivity be established or right avoided? Is the lack of connectivity plausible or does it appear random and arbitrary and why is that so? Is

[8] Compare with Henri Pousseur about his texture concepts of "overlapping" and "multiplication" in [9, 24ff].

the relation of connectivity and the lack of connectivity convincing? According to experience, problems and questions occur in this process, which were not anticipated. It then poses the question, how to react: Does the new remain a single local incident on its own or does it show consequences for the whole, i.e. for what came before and follows after, and if yes, which ones? Is the elaborated section not too similar to a later one and could it thus be listen to as a "reprise"? Given one accepts the concept of "repetition", would it then not need to be much shorter than originally intended?

Some of these experiences occur repeatedly from piece to piece, others pose themselves only in each individual case. To these repeated experiences belong what could be called the "hardship of continuation" in that there is a developed beginning and the certainty that it has to be the beginning and a vague idea or even already worked-on sketches of a later part, yet one does not, or still not, find a point of contact which evolves into the continuation: the beginning section is "too closed".

Or: two parts in the course of the piece are already developed but it is obvious that "something has to go in between" and it can be a long time before finding out was this something could be and at the end it might be only one sound or a gesture, which was missing. Or one realises that nothing was missing since by now the relation between the two parts has become sufficiently clear. A related case is the experience to realise that a wholly or partly elaborated formal part is in the "wrong place"; in the sextet *schnitte* I couldn't process after two parts until I understood that this was no surprise since with the second part I had already written the end. And after relocating this section the inner blockage was also gone and I could proceed with the composition. In another case it occurs that when working on a section, whose position in the whole is already fixed, either puts its position or the section itself into question since something inconceivable happened, which suggests another direction or even inverts what was originally intended. It can also happen that one suddenly composes into another piece, which is either really new or one, for which sketches have already been made and forgotten and make them now apparent again in retrospective.

One of the most important virtues of composing in this context is patience and awareness: to take time to find out if the struggle with the composing has to do with one's problems or if it points to the structure of the emerging music, which closes itself up against the intention which is put upon it. The material of new music as being atonal does in fact not refer a-priori to a whole and thus is continuable and developable into many directions yet not into arbitrarily many, and not all directions make musical sense. To develop the sensorium, which is "right", which "fits" the matter is what finally distinguishes composition from technically adept harmony and counterpoint exercises.

Project Review by Clemens Nachtmann

I want to be completely honest! As we restricted ourselves to chords after the first tentative attempts as to what could be the right approach for me within this project, I was at first somewhat unhappy. This decision seemed me to be a problematic constriction since when composing I find the vertical dimension far too closely connected with a linear dimension and both linked together with the temporal proportions, as that I would like to rip them apart as we have done it; with all the separation, which also I normally undertake.

However, the concentration onto one aspect was in retrospect very useful and promising, because if one considers a phenomenon only ever as part of larger context then it doesn't get the attention it would deserve, what I experienced/learned in this project.

A very similar experience, which I already had in another context was the one that a computer-based generation of chords at first produces all the combinations matching the defined criteria without regard of preferences, taste, aesthetic criteria etc. The result one gets is first of all complete, i.e. no possibility remains unconsidered and is a-priori or too quickly sorted out. I find that important for several reasons: first a computer-based generation doesn't allow for cheating or overlooking something. One obtains a clear impression of the partly enormous amount of possibilities, which one can reach according to one's own specifications.

When considering and listening to the results individually, one starts to realise that also crystal clear specifications can produce non-intended results. I have explained that already above when considering the 14 chords, which finally remained. Although everything was considered in the end, what according to experience should exclude a tonal touch, some chords still felt like tonal.

Even more surprising was however that also the opposite occurred. In the first working rounds, during which the criteria was not yet as strict final, chords were relatively often generated, which contained individual intervals, e.g. fifths, of tonal interval constellations, which did not feel tonal because they were contrasted by other tones in a way that they look tonal on paper yet de facto appeared very differently. That there is a difference between the scripted and the audible results, that it is thus imperative to listen continuously to what is written down and from which one believes to be able to characterise it clearly in a logical way, i.e. to try and test it by listening, or to phrase it differently to exhaust it with the ear is indeed no new insight yet one which became so clear during the work on the project.

In this context there was an impressive and even humorous incident, which stands quite exemplary for the beneficial function of misunderstandings, a misunderstanding, which had a positive effect, because it yielded unexpected results. In one stage of the project I wanted, against the original definition, to admit minor seconds within the chords, if these were covered by major seconds above or below. The project team however understood to admit minor seconds if they were covered by major and minor seconds. That was indeed not what I had intended, yet in exchange and in particular at higher voice-density very nice ambiguous chords arose from it, which features on the one hand a characteristic interval structure and in which on the other hand at one or several place a quasi cluster-like thickening occurred, which made its way stepwise through the chord in the chosen design. And as components of clusters of seconds the minor seconds got a quality, which they would not have when only occurring at one place in a complex chord.

That a contradiction exists between intention and results, that exclusion-criteria do not necessarily imply the intended elimination but sometimes the opposite; that hence a well-defined and straight approach to the matter leads not inevitably to the goal, is in fact an insight, which is "in principle" not new to me, which nevertheless became never as clear to me as in our project. This experience will certainly continue to occupy me in my composing and the contemplation thereon: it could on the one hand imply a more relaxed choice of the means of achieving a strict goal, i.e. that until now avoided intervals will be permitted as long as certain constellations do not appear tonal. On the other hand such a relaxation would require a larger rigidity, i.e. that one's own sound will need to be listened to even more precisely and strictly instead of trusting in seemingly hard and factual criteria. And here in return I do find the experience confirmed that at the end of the day the mathematical logic, which was applied to the musical material be it technically substantiated or not, must be controlled and absorbed by the critical ear of the composer.

References

1. Adorno TW (2013) Against epistemology: a metacritique. Wiley, Hoboken
2. Adorno TW (1993) Beethoven: Philosophie der Musik: Fragmente und Texte. In: Tiedemann R (ed) Suhrkamp. Frankfurt am Main
3. Adorno TW (1978) Philosophie der neuen Musik. Ullstein, Berlin
4. Boulez P (1979) Anhaltspunkte. Essays, Bärenreiter, London
5. Boulez P (1963) Musikdenken heute, vol 5. Darmstädter Beiträge zur Neuen Musik, Schott, Mainz
6. Boulez P (1972) Werkstatt-Texte. Propyläen Studienausgabe, Ullstein, Berlin
7. Boulez P et al (1976) Wille und Zufall: Gespräche mit Célestin Deliège und Hans Mayer. Belser
8. Broch H (1973) Ophelia. In: Lützeler PM (ed) Barbara und andere Novellen. Eine Auswahl aus dem erzählerischen Werk. Suhrkamp, Frankfurt am Main
9. Decoupret P et al (1990) Henri Pousseur. In: Metzger H-K, Riehn R (eds) Musik-Konzepte, vol 69. edn text+kritik
10. Horkheimer M, Schmidt A, Noerr GS (1985) Protokoll einer Diskussion zwischen Horkheimer und Adorno vom 18. Oktober 1939. In: Gesammelte Schriften: Nachgelassene Schriften, 1931–1949. S. Fischer, pp 493–525
11. Huber NA (2000) Über konzeptionelle Rhythmuskomposition. In: Durchleuchtungen. Breitkopf & Härtel, pp 214–222
12. Kocher P (2004) Das Klarinettenkonzert von Elliott Carter. MA thesis. Musik-Akademie der Stadt Basel, Hochschule für Musik

Eva Reiter/Wire Tapping the Machine

Eva Reiter, Hanns Holger Rutz and Gerhard Nierhaus

Ever since her early childhood, Eva Reiter has nourished an affinity to art and especially to music, in particular to the presentation of musical ideas and art works.[1] For a long time she thought that her professional future was going to be in the realm of fine arts, yet she eventually found her place in music. Reiter has always felt the urge to express and play out her artistic ideas. For a long time the question of genre played no primary role to her. As a child she produced a large amount of paintings, sculptures and music. The value of experience and inner satisfaction that presented itself to her in the following through of firstly an idea, secondly, an arrangement and implementation of this idea, and thirdly, a presentation of the result, was of great importance to her. Whether or not these pieces were able to claim a high artistic significance was secondary to the energy created personally for her in the artistic endeavour. During her childhood and teens she was first taught recorder and piano, and later also learnt viola da gamba.

After finishing school she was confronted with two options. Reiter had prepared both for the music entrance exam at the Viennese University of Music and Dramatic Arts and at the Viennese University of Applied Arts in painting. As the date of the music entrance exam was earlier and she received a place, Reiter took a path in music from this point on.

---

[1] Biographical introduction and texts from the composer translated from the German by Tamara Friebel.

E. Reiter
Vienna, Austria
e-mail: evareiter@gmail.com

H.H. Rutz · G. Nierhaus (✉)
Institute of Electronic Music and Acoustics, University of Music and Performing Arts Graz, Graz, Austria
e-mail: nierhaus@iem.at

H.H.Rutz
e-mail: rutz@iem.at

G. Nierhaus (ed.), *Patterns of Intuition*,
DOI 10.1007/978-94-017-9561-6_4

Although her musical socialisation took part in the framework of "early music", she developed concurrently a firm interest in the language of contemporary music. During the time of her studies in Vienna and Amsterdam she came across a large number of new compositions for recorder—regrettably far too few for viola da gamba—but she found, unfortunately that few were able to connect to the aesthetic direction she was interested in pursuing.

The step towards her compositional activity took place in 1996. In Vienna she had already worked in improvisation collectives with both instruments and it soon became clear that she would devote more and more concentration to this field. During her years of study in Amsterdam she then started to score her sketches. In the course of improvisation Reiter spent a lot of time researching and categorising sound material of various instruments. She increasingly felt the need to define her musical concepts structurally and formally. In the improvisation collective in which she worked in the Netherlands, it was predominately the communication processes, thus the promptness, speed and precision in the processing of impulse and reaction, which made her experience this work as extremely attractive. Nevertheless due to the shortage of time she then left this field progressively more and more behind in order to notate and concretise her concepts, thus to decide in a way for composition. At first sight, the fascinating processes of speed and virtuosity, the formal phenomena of contraction, compression and augmentation, which she could explore many times in the context of improvisation, would now be further pursued and elaborated in a multifariously differentiated way.

The close connection with the instrument, which she considered natural from her work as a performer, also became relevant for her work in composition. As a recorder and viola da gamba player she was from the start very familiar with the basic techniques of wind and string instruments, and this knowledge proved very useful when dealing with other instruments. At the beginning of each piece Reiter investigates the given instruments and equipment, amongst the other fundamental thoughts, positions and ideas. She often takes additional tuition in order to understand the basic requirements of an instrument that may yet be unfamiliar to her. Subsequently she gathers the bulk of the sound material by exploring with improvisation. Of course she is more or less quickly faced with the limitations of her technical abilities and certainly not everything can thus be fathomed or explored. Yet nevertheless this experimental situation became very important as a first step in an encounter and the production of material. Seminal for her composition work was—amongst others—the examination and engagement with the music of Fausto Romitelli, Bernhard Lang, Salvatore Sciarrino, Helmut Lachenmann and Georges Aperghis.

Artistic Approach

Statement

An attempt to capture, in a brief statement, the variety of mental and sensory processes that are active in the work of composition seems to me from its outset an impossible challenge. Every formulation remains insufficient due to a restricting essence, where a metaphorical hint might try to capture all, but loses, in the first utterance of its attempt the empty ideal, embedded in the composer's white piece of paper, which is the state desired at the starting point of composing. Composing is not just a phenomenon or an attempt of creation; it is, in itself, a "state".

Here are some essential thoughts that are currently active forces in my compositions, regardless of all diverse actual phenomena and musical developments within the various pieces. To an extent these thoughts will inevitably remain incomplete as they attempt to describe a process, which is considerably more complex in its effect and operation.

1. *Determining and capturing the starting point*: composing means working with sound material and recognising the quality of the perception that is immanent to the particular material. It comprises an "understanding of listening" as a multi-layered phenomenon and the recognition of one's own preconditions and attitudes. This refers to the insight and the questioning of one's own perspective, which entails the expectation and the desire of the potential "sound-to-emerge".
2. *Development of a possible change of direction*: based on an initial idea, I develop and establish concepts, which allow self-supporting structures to arise, giving new meanings to grasp still yet, unrecognised potentials. In the composition process I follow the tendencies that seem to me intrinsic to the material itself, and I search for musical spaces that invite the listener to take on for some moments an entirely new perspective within an already familiar, yet nevertheless current musical language.
3. *Sorting out the mess*: with the recognition of my own bias and desires and the corresponding personal and collective limitations, the work within an innate resistance starts—the whole construction of a piece appears completely meaningless and futile. At this point the composition process touches the fundamentally operative layers of my being and my personal constitution. Herein lies the dormant and fundamental potential, which opens up renewal towards the work of change and transformation.

The outcome of a composition I consider as relevant and positive if I have gained experience of this profound renewal and if within me, it is able to evoke a continuing fascination which makes me understand the new—but yet again already manifested—material as basis for renewal and continuation.

Personal Aesthetics

In the course of my musical development, several aesthetic perspectives have emerged, from which I observe the world and try to reflect it artistically. As a performer of early and new music, as an improviser and composer, I repeatedly come across new challenges and questions, yet the development of a new piece and its current musical language remains that which centres me the most and which allows the biggest insight into artistically valuable questions. When I began to write music I was looking for a way to challenge anew the listening-expectations of the contemporary-music-audience, which had already been influenced from the many decades of electronic music. I was interested in the small edge between purely acoustic and electronic music. Attention was given to the material, which generates an illusion of electronic music. Thereby it was about creating sound with a complex internal structure by means of simple preparations.

Still today I apprehend sound as a sum of many parameters and try to use playing techniques, which allow the selected layering and mixing of individual parameters. In this, I continuously search for musical material that is capable of development and transformation. In a series of pieces with tape I worked mainly with synthetic sounds of my daily urban surrounding—like the buzz of a power pack, the noise of a ventilation funnel, the sound of engines, printers and copy machines, lifts and elevators. The hum of the ventilation funnel, the noise of a vacuum cleaner, the asymmetric loop-properties of printers and copy-machines, which form the basic structure of the piece *Alle Verbindungen gelten nur jetzt*,[2] or the sound-aura of modern medical devices, that are used in *Biofuge*—all these are the sound of a certain aesthetic. Printing machines from Heidelberg from the 1950s were one of my favourite sound materials for a long time. The previously cool and dry, quasi confronting, raw sound aesthetic of such machine loops were set into the background within the composition process and in the respective performance situation due to the addition of live instruments. I was searching to find an alloy of electronics and instrumental sound—like between synthetic noise and the sound of a flute—thus to create a symbiotic structure, a rigid relationship of dependencies, yet to also separate these layers at a different level again and to put the instruments back in their characteristic positions.

Frequently I have come across remarks that my music was "rough" and "hard", "unbeautiful", and in its construction almost "brutal". In response to these comments I recall again and again an interview with Fausto Romitelli, who said the following remark about his necessarily "violent" music: "Nowadays music must be violent and enigmatic" it was said, "since only in that way it can express the violence of alienation and standardization processes of our environment".[3]

[2] Titles, as they are not wished by Reiter to be given as dual-translations, are available here as footnotes to assist those without knowledge of the German language. *Alle Verbindungen gelten nur jetzt* roughly expresses that "all relations apply just now".

[3] Translated from the French, see [2, p. 76].

Björn Gottstein—editorial journalist for new music of the SWR Stuttgart—has formulated quite accurately in an article about my work: "[...] and Reiter's auscultation of the engine room is both an acoustic metaphor and musical meaning. It is crucial that both semantic levels are not separated, but presuppose and entail each another. One could also formulate it accordingly: the recording of the machine's noise doesn't admit per se a conclusive quality, the attempt to demand from the instrumentalist a machine-like characteristic can also be seen as it's own étude. It is not before the performer has to compete with the dictation of the machine, that the chance to rebel against it or even to surpass it in precision and speed, that the consequences of the conflict become evident."[4]

I was concerned for a long time with the question to what degree can one keep up with these rhythms. Due to high velocity it comes to inevitable moments where reacting is difficult and escalates. After this a point of rest arises and with it a different mode of motion. Today I am still interested in questions of a consequent continuation of the resulting order systems and the development of self-sustained structures to overcome these systems of condensed order. It is these approaches and questions that drive my work although many still remain unanswered. Very often it appears as a highly difficult process to take the next step, yet this room of viable development, as well as the reflection and critical vigilance towards my own creation, remain the biggest and most meaningful challenges.

To My Actual Work

Starting from the inside out, concerning the sound material, I currently concentrate on diverse variations of instrumental articulation. This entails the multiple different constituencies of the beginning of tones—the attack, artificial transient effects, consonants etc.—and of the end of tones—cuts, sudden or slow fading of sounds, as well as specific phrasing, which is often based on mimicking the sound and the melodic formation of human speech. Embedded in between these variants I search for atypical and complex sound forms that are frequently foreign to the instrument and which for instance could be achieved by simple preparations, particular playing techniques or specific placing of microphones.

An essential part of my preferred sound aesthetic is many-varied grades of noise, thus a realm of sound, which aims to dissolve the instrumental characteristics. It is predominately transformation processes, which are currently decisive for the sonic and structural constitution of these pieces. In the pre-compositional process the identification and the investigation of individual parameters of a given sound form the first step. I try to deconstruct the sound in its various elements and to understand and specifically change their mixture.

In doing so I find—as I have already mentioned—more and more from this intense analysis of instrumental possibilities of articulation to a new "sound-speech" in the

[4] Translated from the German, see [1, p. 85f].

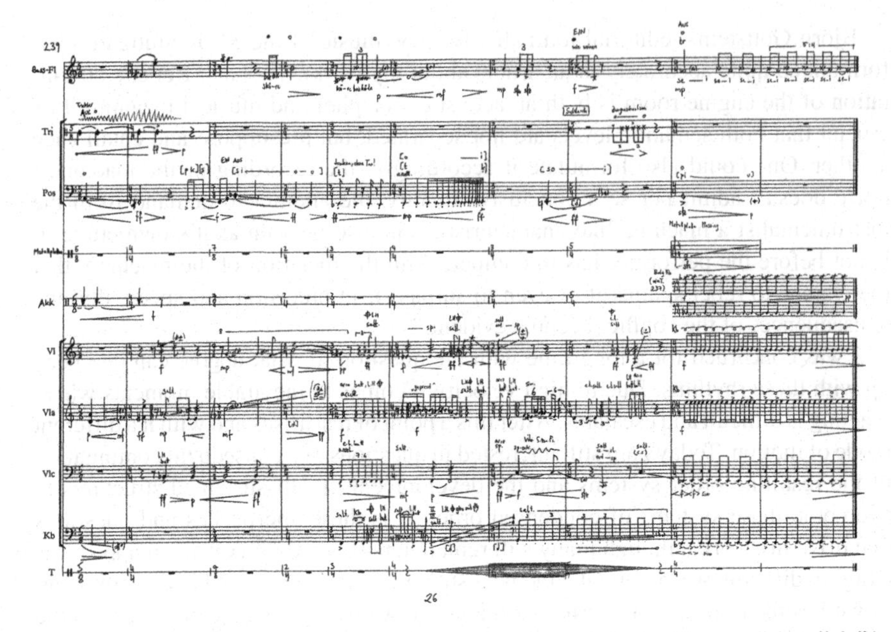

Fig. 1 *Irrlicht* (*Irrlicht*, from the German, refers to a flickering light in the distance, a "ghost light".) for ensemble (2012), bars 239–249

sense of a phonemic structure. From time to time there are actually also—in the case that the instrumentation allows for it—increasingly "spoken" passages (see Fig. 1, flute part). It is a speech of the phonemes, where syllables—as combinations of vowels and consonants—replace notes and the details of their articulation. This rhetorical material organises itself at times into short passages of speech. The articulation of the syllables, the phonemes however is based on the pure musical context. In this transformation process the instrumental sound to human speech and vice versa becomes visible.

A different, currently active form of transformation lies in the transfer of concrete machine sounds—or other similar outdoor recordings—onto an instrumental ensemble. Therefore a pre-recorded audio sample serves as a transfer picture or as its "negative". This undertaking is only relevant in the pre-composition generation of structural relations and serves as a starting point for the actual composition process. Herein I am mainly interested—apart from aspects like pitch and rhythmical analysis—in the search for all those sonic elements of the recorded material, that are difficult to capture. I aim to determine the individual parameters like density, dynamic, timbre and composition of the noise portion and their contrapuntal counterparts.

These sounds—in their original appearance of the audio recording they represent rather rigid relations—are then transferred onto the instrumental ensemble, thus set into motion and rendered alive. A complex sound course, which is gained from the transformation, hereby forms the initial situation. Yet this basic structure is rarely made audible in its original form, it rather serves as a starting point for the emergent process of developing the material. Figure 2 (bars 70–72) shows an extract of the basic contrapuntal loop that immediately marks a process of transformation.

An essential area of work lies moreover in finding an adequate notation, which depicts or captures these sound incidents on the basis of the traditional music notation as clearly as possible.

Formalisation and Intuition

Two contrasting poles essentially determine the working process that defines the creative space of composing. I move between a basic reflective concept—the fundamental systematisation of the composition—on the one hand and the arising, lush creativity on the other, hence between the projectable and the unpredictable turn, giving essential meaning to my work. These two divergent forces are ever present and touch the essential questions of consequence and dedication, as they unfold a framework within which the development of music becomes possible in the first place. The term "compositional system" I understand as the creation and testing of specially designed formal relationships and connections of contents or tonal material. This form of "contrapuntal" thinking plays a very large role in my work. The actual activity of composing is often preceded by an extensive sonic and structural research, in which I try to fathom musical forces of attraction on a still mostly abstract level in order to establish new and surprising contexts. Many multiple complex systems are then generated, which are tested and modified in the actual process of writing until they develop a certain life on their own, which can direct my thoughts into new pathways.

In my composition *Alle Verbindungen gelten nur jetzt*[5] sound particles connect like complementary base pairs with concrete material and form the basic framework of the piece. When two musical strands separate, new complementary "sound partners" find each other and form new bonds. The piece is thus developed from a complex strand similar to a loop chain that stores all the information about the whole piece within it. This basic structure, however, is never made audible in its original form but runs through the music as the code of the musical material, as the "genetic information" of the piece. We hear but details of the strand, partial developments and facets of its encoded intelligence.

This method, underlying the framework of my composition, is derived from the scientific context of medical DNA research, in particular the "molecular matrix", which serves as a template to copy and transmit genetic information. It is a celebrated fact that two DNA strands are complementary to one another. Thus, one strand can serve as the matrix, and a new strand can be synthesised through the mechanisms of "base pairing". *Alle Verbindungen gelten nur jetzt* is structured in a similar way, particularly also with regard to the relationship between the instrumentalists and the recorded part of the composition. The focus is on musical details, the acoustic

[5] *Alle Verbindungen gelten nur jetzt* roughly expresses that "all relations apply just now".

Fig. 2 *In groben Zügen* (In *groben Zügen* refers to something which is "roughly sketched"), bars 57–75

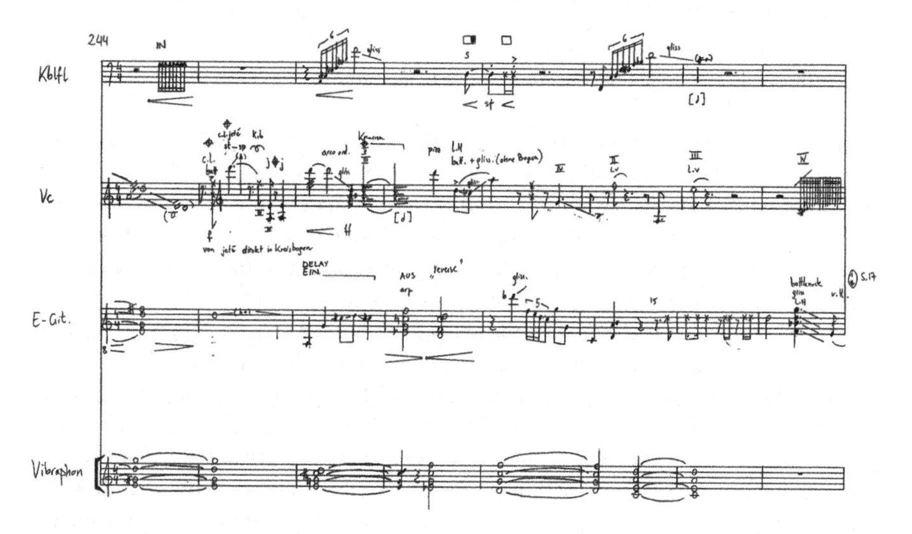

Fig. 3 *Alle Verbindungen gelten nur jetzt*, bars 244–251

molecules themselves, which consist of interdependent atoms bound to each other; similar to an experimental set-up in a laboratory, they are examined microscopically and subject to different external influences (Fig. 3).

The composer describes control, massive compression, successive release—the starting point of the piece *Bénard Experiment #1* could be summed up by two thoughts: Order out of chaos. From chaos to order. How in these grey areas do ordered structures form anew? My reverence for scientific and mathematical fields that study complexity theory obviously shines through the composition. The Bénard experiment examines systems whose dynamics under certain conditions are highly sensitive to initial conditions, rendering a long-term accurate prediction of their behaviour impossible. In the Bénard experiment a thin homogenous layer of fluid is heated from below, while the upper surface is kept at a low temperature. Above a critical temperature difference so-called convection cells will appear, at the edges of which direct exchange between the warmer liquid from the bottom and cooled liquid from the top takes place. Closely linked with these phenomena is chaos theory, which systematically explores such complex, non-linear, dynamical systems. Transferring these phenomena to compositional processes, I created a piece that in no way simply plays out chaos theory but attempts to free musical material from the clutches of order and musically explore this transition into deterministic—creative—"chaos". Almost inevitably, the pressing question arises out of the two steps necessary: from overcoming maximum control and breaking up order to generating new forms of working with musical-temporal structures. The focus of attention is on how to extend the resulting ordered systems in a consistent way, how to develop self-supporting structures (to overcome the very systems of condensed order) and how to write down such phenomena in musical notation. The composer deliberately puts herself and her material in a condition where she can no longer influence musical processes she has put in motion. In the post-chaotic state, the released material is capable of organising itself, so to speak.

The musical idea reflecting forms and orders can thus be formulated as an essential drive of my work in the last years. Starting from a compositional idea I design concepts and modules, which serve as starting point—and in most cases also remain a reference point—for my considerations. I'm interested in how the resulting systems of order can be extended in a consistent way, and self-supporting structures can be developed to overcome the very systems of such condensed order.

Currently I am also fascinated by the analysis of human speech, that is on the one hand the editing and cutting up of certain texts until they become entirely unrecognisable and on the other hand, the development of a new sound speech based on articulation analysis. A further point is the play with illusions, the adversarial behaviour of certain associative relationships or expectations that arise in the listener.

Musical intuition operates within me as a corrective measure of pure thought concepts. Based on musical instinct, it leads me through certain decisions, which I can only comprehend retrospectively. Thus, my ideas develop a certain life on their own, which cannot be calculated. The concepts become an instrument on which I can carry out trials and experiments. In the process of composing, I am always faced with a moment of resistance, leading to a large blackout, where I feel overwhelmed by the inertia of the abyss. At this point, further development of the piece is terribly difficult for me. I seem to reach the limits of my cognitive capacities and trust then more and more blindly my intuition. Looking back, it is during these critical moments that the fundamental creative potential seems to be activated and allows new tendencies to arise.

Evaluation and Self-reflection

In contemplating music and in the critical awareness concerning one's own creation I see the biggest and most meaningful challenge also the most difficult personal task. It is the work against the internal resistance and against one's own inertness and the recognisable "blind spots" that arise in one's world view, that costs an enormous amount of energy and effort and which lets the composition process often seem sluggish and stupefying. The critical and frequently judging voice is therein omnipresent. Yet it mostly happens right in this moment—when the whole concept seems completely meaningless and futile, when I am at the verge of discarding an experiment as a dead end, because the question of the context doesn't reveal itself within my hitherto conceivability—that the composition process touches the fundamentally active layers of my thinking and that the potential of renewal and transformation becomes visible.

I have to have reached this point at least once in order to be able to experience a new piece positively. I want to be enriched in observations, to have changed and gained more clarity about my position in an artistic and social context. In retrospect it is always those spots in the music during the development process in which I get lost and can be "found again" in a modified way, which grab me, carry me along and delight me during the later listening of the pieces.

After the completion of a new piece it often remains impossible for me for months to have a look at it or to listen to an audio-sketch. Everything within me resists against it. After some time however, when I have reached the necessary distance and also when I seem to have already forgotten some thought and decision processes, that I listen to the piece for the first time "from outside". This is the biggest test. If it then happens that I surprise myself then I have "gotten" a piece.

Project Expectations

In participating in this project I see the chance for a possibility to reflect on my own work in an unforeseen way and thus also to enrich it. Like many other composers, during the process of composing, I miss an outsider's perspective, a discourse, a feedback. Apart from the many-faceted concepts and intellectually stimulating models it is mainly the relationship of the intuitive process in making music on the one hand, as well as the "blind spots" in viewing my own work on the other, whose analysis I am intrigued and stimulated by. What are the mechanisms that define a "personal style"? To what extent does this artificial laboratory situation also help to see through one's own barriers and habits, to develop them further and potentially also to overcome them?

Exploring a Compositional Process

POINT: Reiter's pre-compositional process began with an analysis of a number of concrete recordings made from different machines, which served as inspiration for underlying patterns and structural details in the piece.

Reiter: Whenever I develop basic sound elements to be used in a certain composition, I follow an initial pre-compositional thought and concept that would give me direction in search of specific instrumental sound elements in order to determine my final choice of material. I tend to interlink several interrelated compositional ideas that guide and constrain my search. In the case of *In groben Zügen* the translation of pre-recorded audio material to an instrumental body was like a starting point that lead directly to various other aspects of a more complex structural image.

Nine years ago, the Vienna based printing company Walla kindly allowed me to record some of their 1950s Heidelberg printing machines. This has lead to an enormously interesting collection of audio material that after all these years still serves, at times, as a source of inspiration.

During the process of transcribing these machine sounds I follow not only pitch and rhythm analysis but also try to focus on proportionate noise parameters that would determine the sonic appearance of the piece. In general I search mainly for sounds that exhibit a particular mixture of individual, clearly distinguishable parameters. I find them by examining several ideas on the instruments I am writing for in my home studio.

POINT: Reiter was intrigued by the idea to subject the pre-recorded audio files to an "instrumental analysis", which means to recreate the timbral properties by means of a string quartet. She would naturally do this by ear, however it seemed interesting and most possibly inspiring to compare her "natural" analysis to a computer based dissection. So it seemed to be a reasonable first approach to look at possible ways of signal analysis procedures. We started with 45 sound files provided by Reiter, all of which played an important role in the preparatory process of the composition. Some of these sounds were especially recorded for the string quartet, while others were taken from Reiter's sound library, which she has been working with already for several years. The files have an average duration of 8 s, the shortest lasting around 2.4 s, and the longest one lasting 27 s. The material is of mixed origin. Sometimes it is easy to recognise the sample as natural sound, but more often the recording stays unidentifiable due to an unusual, often very close position of the microphone. Nevertheless all recordings are left raw and unpolished. The only transformation tool that is used is an old recording device whose malfunction results in a particular digital distortion added to the sound.

The file names often indicate the sound source, occasionally using metaphors, but more often clearly alluding to the machines from which a particular sound was taken: "Kopierer" (photocopier), "Strickmaschine" (knitting machine), "Rasierer" (shaver), as well as the series "Walla Heidelberger" consisting of eight variations of Heidelberger printing machines.

As sound and timbre are complex phenomena, we decided to begin by examining pitch and rhythmic events.

For our analyses we decided with Reiter to choose a subset of twelve files: four miscellaneous files that had been transformed by the recording device, four examples of the Heidelberger printing machine, two sounds of "cicadas", a recorded knitting machine, and a sounding minor chord played from a gramophone disk which has been turned manually and therefore irregularly, producing unsteady glissandi as a result.

We began to explore the possibility of a music information retrieval toolbox, but eventually stuck to rather simple pitch and onset tracking algorithms.

The pitch tracker is monophonic and based on autocorrelation, searching for an energy peak based on threshold. It allows us to specify the allowed frequency range, a signal to noise ratio, and a temporal smoothing parameter. In order to capture multiple concurrent pitches, we recursively fed the signal into a notch filter driven with the last reported frequency and then repeated the measurement. We allowed for up to four concurrent pitches.

Similar to the pitch detector, the onset detection [4] has threshold, noise floor and temporal smoothing control. Additionally, one can control the density of triggers by specifying a minimum offset between two onsets, and most importantly, there is a selection of seven different algorithms. All of the algorithms operate on the FFT spectrum, but use different techniques to look at the magnitudes and phases.

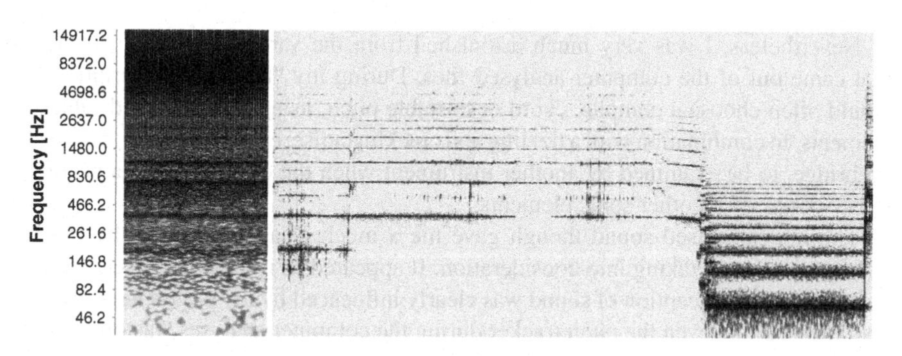

Fig. 4 Sonogram of an excerpt from the original cicada sound on the *left*, and resynthesised "frozen" segments on the *right*

In order to assess the accuracy and granularity of these processes, we devised synthetic sounds so that could be triggered according to the detected onsets of pitch. We then wanted to render these sounds that Reiter could examine them, as they might provide an interesting alternative view on the source material, being stripped down to these two parameters. In order to make the sounds more pleasant to listen to, instead of clicks and sine tones, we used a resonant impulsive sound to re-synthesise the onsets, and a richer timbre for the pitches that can still be clearly heard as the fundamental frequencies.

Figure 4 shows the sonogram of one of the source sounds, the cicadas, on the left side, along with a re-synthesised sound on the right side. Because of the nature of the source material, the pitches could change very quickly, and also the pitch tracker was often not capable of correctly finding pitches or decided to jump between alternative solutions. We turned this disadvantage into an advantage, by allowing the tracker to be "frozen" at specific detected frequencies, so that one can listen to them for longer periods of time in the re-synthesised version. Freezing the frequencies at different detection points allowed us to render different views of the frequency content, for example alternating between the high frequencies and the low frequencies, as can be seen in the sonogram.

A problem with rich harmonic or noisy sounds is that a monophonic algorithm is not really suited to give reliable pitch information as it assumes the presence of a fundamental frequency. Even when iterating the sound, each time notching the detected frequencies, problems remain with the detected pitches having discontinuities.

Reiter: When I received back the examined files it became clear that I had usually already put the aurally extracted pitches in a more complex sonic context of other parameters. Therefore I would naturally leave the initial figure and give space for the material to develop in unforeseen ways. The tendency was to create a tonal framework from the pre-recorded material and while the piece progresses to fill it up with connecting sounds that would also determine the on-going process of the composition.

Nevertheless, I was very much astonished from the variety of "clear" pitches that came out of the computer-analysed files. During my "natural" transcription I would often choose a complex chord or unstable pitch, using glissando and saltato elements, in combination with a fizzling and cracking noise, circular bowing, bowing on bridge, to be examined by another instrument when the original pitch appeared to be "covered" by other sonic elements.

The re-synthesised sound though gave me a much clearer image of the tonal richness I was not taking into consideration. It appeared as if my aural research, or even my initial perception of sound was clearly influenced by my preferences. But I also realised that even the pitch tracker during the computer analyses was often not capable of correctly finding pitches or decided to jump between alternative solutions.

POINT: In order to make the different rhythmic layers from the sound sources clear, we used a number of varied threshold levels, to bring audibility to the starting points of dominant impulses and allow even finer gradations of the rhythmic structures.

A rendering detecting the onsets can be seen in Fig. 5. Here the original sound is shown on the bottom, whereas the upper images show the re-synthesised sound using varied detection thresholds.

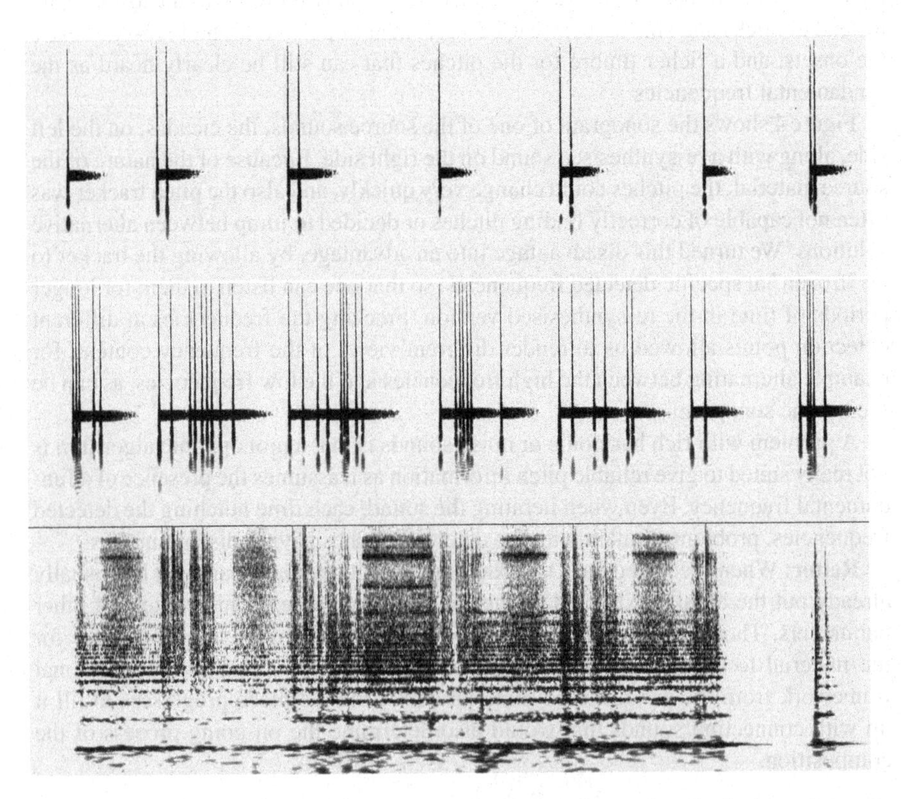

Fig. 5 Sonogram of the original knitting machine (*bottom*), and several synthesised sounds based on onset detection, using low to high thresholds (*top*)

The processed files that were handed over to Reiter are divided into the original sound on the left and the synthetic sound on the right channel, so that by using headphones one could compare them easily.

Reiter: I discovered many analogies when examining the rhythm transcriptions, apart from the pitch analysis. Similar to the computer-operated process I would also first extract the main beats of the machine loop before exploring the smaller, faster and more complex rhythmical structure within. The fascination of the configuration of these machine recordings lies in their irregular and asymmetrical texture. I expanded the rhythmical elements into a contrapuntal set of four string players and successively increased the intermediate rhythmical complexity. I was aiming for a parallel appearance of the augmented and diminished form of an identical element and therefore for creating the possibility to switch between different layers of processing the sonic material.

I started to work against a feeling of "groove" that would certainly be evoked if not restricted on purpose. So even before the listener could possibly ensconce himself in the convenient surrounding of a loop structure I would immediately start to cut it up and deconstruct the established mode, never leaving the contrapuntal connection of the quartet though, in order to give space to new and more interesting procedures. I finally ended up introducing "Breakcore" beats that have strong references to hard-core techno, drum and bass and digital hard-core music.

POINT: For us it was now interesting to make a direct comparison between the analysed rhythmic structures of a sound file and the actual transcriptions from Reiter. We decided to work concurrently on the same source material, one of the Heidelberger printing machines. Reiter made her own translation of the basic material, and then we subjected the file to our signal analysis tools.

Reiter, busy in the compositional work on the piece, now and then reported that she had "gone much further exploring unexpected and musically interesting spinoffs within the piece". We had been aware of the fact that she would not only introduce one, but several interlinked structural and musical ideas to the piece that would be followed and developed in parallel. We had to concentrate on one idea, conscious that it would not be possible to give a clear image of the complexity of the compositional background.

Based on the initial analysis runs which used different onset thresholds or ran the pitch tracker multiple times in recursion, we developed the idea of stratification, whereby the different information thus obtained would be superimposed in one representation. I.e., the signal analysis process would be run multiple times but with different parameters and thresholds, for example leading to a more coarse and a finer grid of onsets. An algorithm would then try to re-align these different onsets and produce a hierarchical diagram, as shown in Figs. 6 and 7.

Reiter: At this particular moment we reached a highly significant point. In my pre-compositional work I relied on my aural transliteration of the "Heidelberger Druckmaschine" to create one instrumental figure that actually, through the analysis, turned out to be part of a more complex strand of various extensive entities similar to a chain of possible instrumental units, the basic structural information behind the piece, so to speak. The transcription of the printing machine was therefore only one

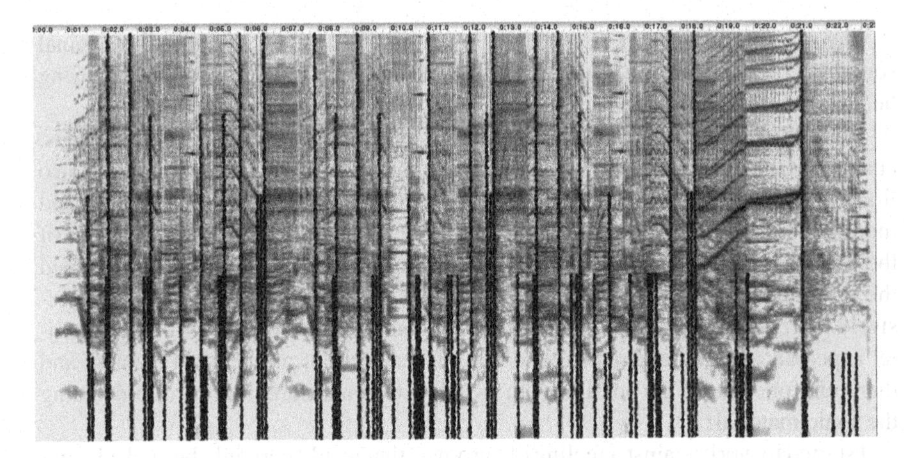

Fig. 6 Multi-resolution plot of the onset detection for the printing machine excerpt from Reiter's rendering

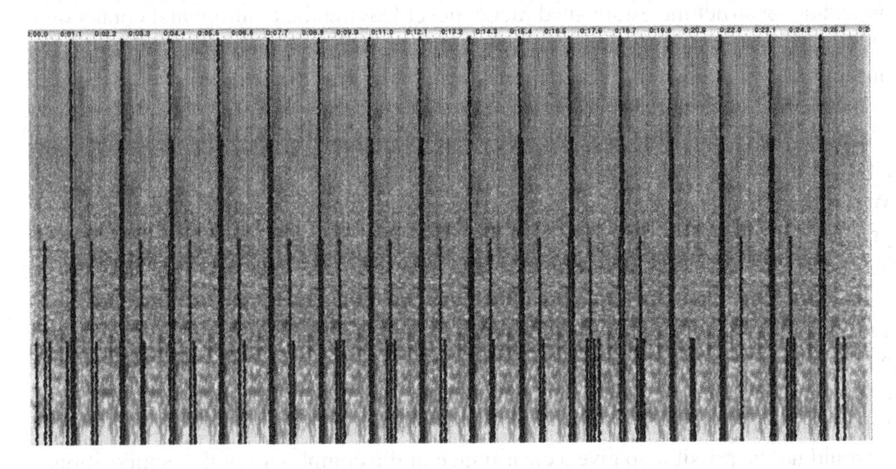

Fig. 7 Multi-resolution plot of the onset detection for the original printing machine file

among various pre-compositional experiments with other samples that I had drawn the material from.

When putting the "substances" together I stick to the initial concept while simultaneously searching for gaps and spots to deliver the material. Each musical gesture has its inherent distinct tendency to spread out differently. Connecting two different pitches of extreme positions through a glissando for instance takes certain time when connected to a particular tonal quality.

So after a while, or sometimes even within the initial "loop", I started to give space to the tonal tendencies of a certain sound and changed the initial frame for it to spread out more and more. Therefore it is easy to understand that the basic rhythmical structures of the analysis move apart. When taking a look at the initial loop on paper, which appears only partly at bar 70, one can see the different materials in time and motion.

Also, during the process of composing, I started to examine the sonic events of various machines that would have just been switched on and off. The starting and stopping of big and heavy engines lead to a remarkable and musically powerful process of a static acceleration and deceleration. I increasingly focused on moments of ever-changing oscillation and decay and was searching for means of translating these technical thoughts onto the musical frame of a string quartet. All four instruments would soon follow an individual scheme of de- and acceleration, producing a highly complex contrapuntal situation in total.

POINT: Indeed, we thought that it would be interesting to model these accelerations and decelerations using a physical model to measure these changes in the detected pitches. In the end this remained a theoretical reflection, as this approach would soon become too involved from the signal processing point of view. For example, in terms of the rhythmic structure we tried to implement a tempo estimation from the detected onsets, so that a meaningful score representation could be rendered (Figs. 8 and 9). Fitting the time instances on a line proved difficult, so the attempt was not made to extend this to curvilinear fitting. We were more successful with the pitches, for which some curvilinear fitting functions were tried, resulting in glissandi of varying steepness. Nevertheless, the fitting algorithm was very sensitive to multiple parameters of the pitch detection, such as spectral range, noise floor, allowed thresholds for discontinuities, etc.

Reiter: What appeared even more interesting to me than pitch and rhythm analysis was the idea to split up the audio material into more refined compositional "ingredients". Pitch and rhythm are the most easy to reproduce and therefore the least relevant to be technically examined. Naturally the rhythmical structure as a basic element of the composition was necessary to examine, but concerning the computerised analyses I would have rather focused on the "hidden" parameters such as proportionate noise, white or coloured noise, including or not including particular pitches. My question was how to reproduce and arrange a fizzling and cracking noise. Ever since

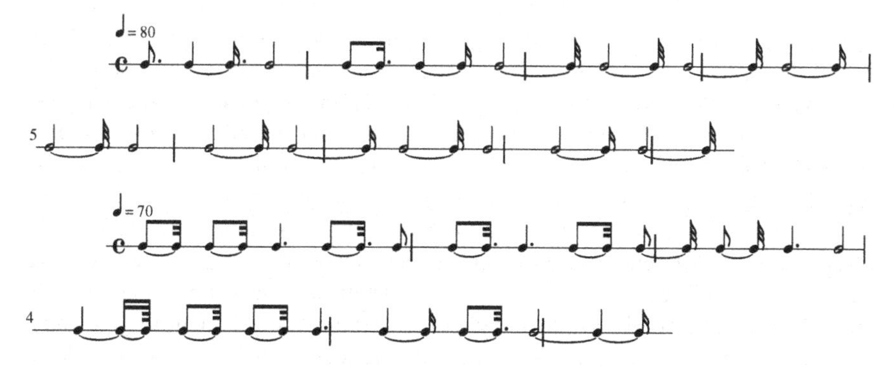

Fig. 8 Detected onsets for the printing machine at high threshold. *Top* original sound file, *bottom* Reiter's instrumentation

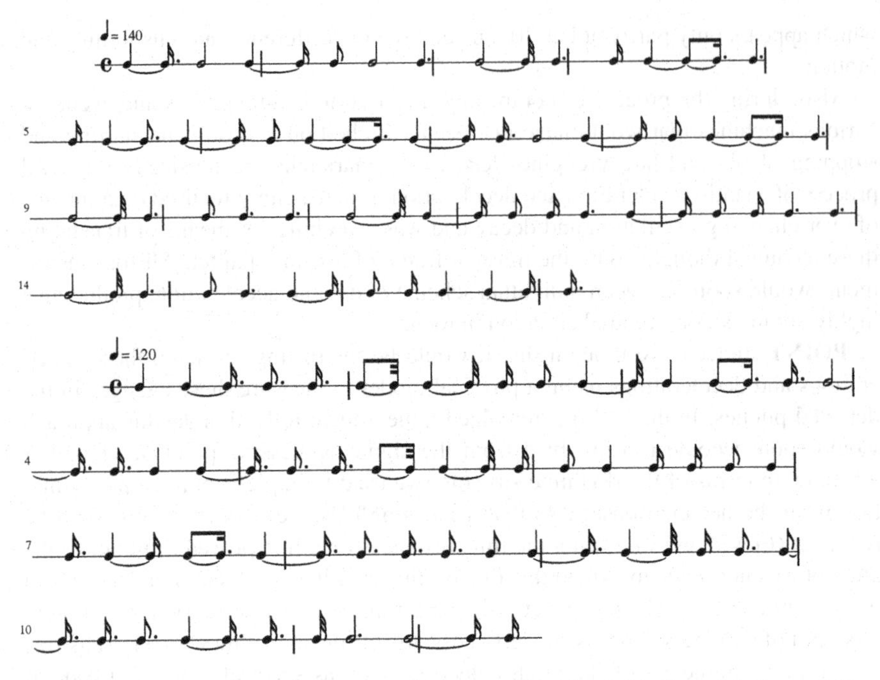

Fig. 9 Detected onsets for the printing machine at medium threshold. *Top* original sound file, *bottom* Reiter's instrumentation

I have been composing I think about transforming and translating various sound phenomena—even more abstract parameters than mentioned above—into a clearly defined sonic frame given by the instrumentation of a certain piece. If I write for a certain instrument I usually spend a lot of time examining its sonic possibilities. Within that process it has become an appreciated tool to expand the possibilities of sound production by means of "simple" preparations on the instrument—in case of the string quartet I worked with thin wooden sticks that would be fixed between the strings, an aluminium foil wrapped around the low string of the cello, polystyrene pieces on the body or between the strings, knitting needles or other metal items fixed beyond the bridge, etc. Adding preparations to the instrument is a simple means of generating highly interesting sounds of complex inner structure, similar to the machine sounds I was trying to reproduce in the quartet. I tend to focus on musical details, for example small rhythmically complex units, the acoustic molecules themselves as it were, which consist of interdependent "atoms" bound to each other; similar to the experimental set-up in a laboratory, they are examined microscopically and subjected to different external influences which would allow me to compose a piece in an unexpected way. I usually start out from the smallest elements, which later on evolve and combine into sound cells. It is always important to give such cells enough space to develop. Controlling and reflecting these processes, indeed cultivating a critical attitude towards my own work, is a fascinating and welcome challenge.

POINT: For us it was clear that most of the complex compositional processes listed by Reiter were mostly not open to a clear formalisation, so as a possible approach, we decided to compare the sound spectra of the sound files and her transcriptions in terms of possible similarities in the sound spectra. Having a sketch audio recording of Reiter's piece along with the alleged source sounds, we re-examined their tonal similarity. A trained human ear can easily identify the sample of the printing machine within the sketch recording. What would a computer make of these similarities?

We had previously used sliding window cross-correlation between matrices made of Mel-frequency cepstral coefficients (MFCC) and loudness contour [3] to detect within a sound file similarities with a given target file. The MFCC describes the spectral content of a signal using a compact representation, typically only a dozen or so bands using the psychoacoustic Mel scale, hence the name. Likewise, the loudness is a feature that tries to estimate the perceived volume of a sound. The cross-correlation between a compound vector that balances these two features yields a single "similarity value". This is a normalised scalar without an absolute scale, however relatively low values indicate low similarity and relatively high values indicate high congruency between two signals.

Reiter produced her own audio rendering of what would eventually become the notated score of the piece, using the prescribed instruments in a recording session at her studio and editing and arranging the material. We then took the printing machine recording as a target file and scanned through the four-part audio rendering of Reiter. The results are shown in Fig. 10a–d, where the similarity curves overlay the notated score.

The results are strongly influenced by the choice of matrix size or selected span in the target sound. If the matrix size is too small, the resulting curve exhibits a lot of fast motions, if it is too long, there are not a lot of pronounced peaks, because there will always be parts of the matrix which do not quite match.

Larger peaks first occur around bars 30–35. They are in fact stronger than those around bar 53 and bar 70 and 71 where Reiter has actually produced an "instrumentation" of the printing machine sound, which can also be verified by ear. Similarly, in bar 104 there is a mild peak, although the human ear can clearly identify the timbre and rhythm of the printing machine, especially in the violins.

Overall, aspects of the printing machine timbre and rhythm are interspersed across various parts of the first part of the score.

Reiter: These observations seem interesting to me. It is impressive to see one aspect of the basic material vanish or reappear within the finalised score. However the analysis appears too simple and one dimensional when considering all the different aspects that had influenced the structure and format of the composition. I had used many different and interlinked audio samples to develop the basic sound material of the composition. The initial instrumental figures I had drawn from were a complex strand of audio files similar to a loop chain that stores all the information about the whole piece.

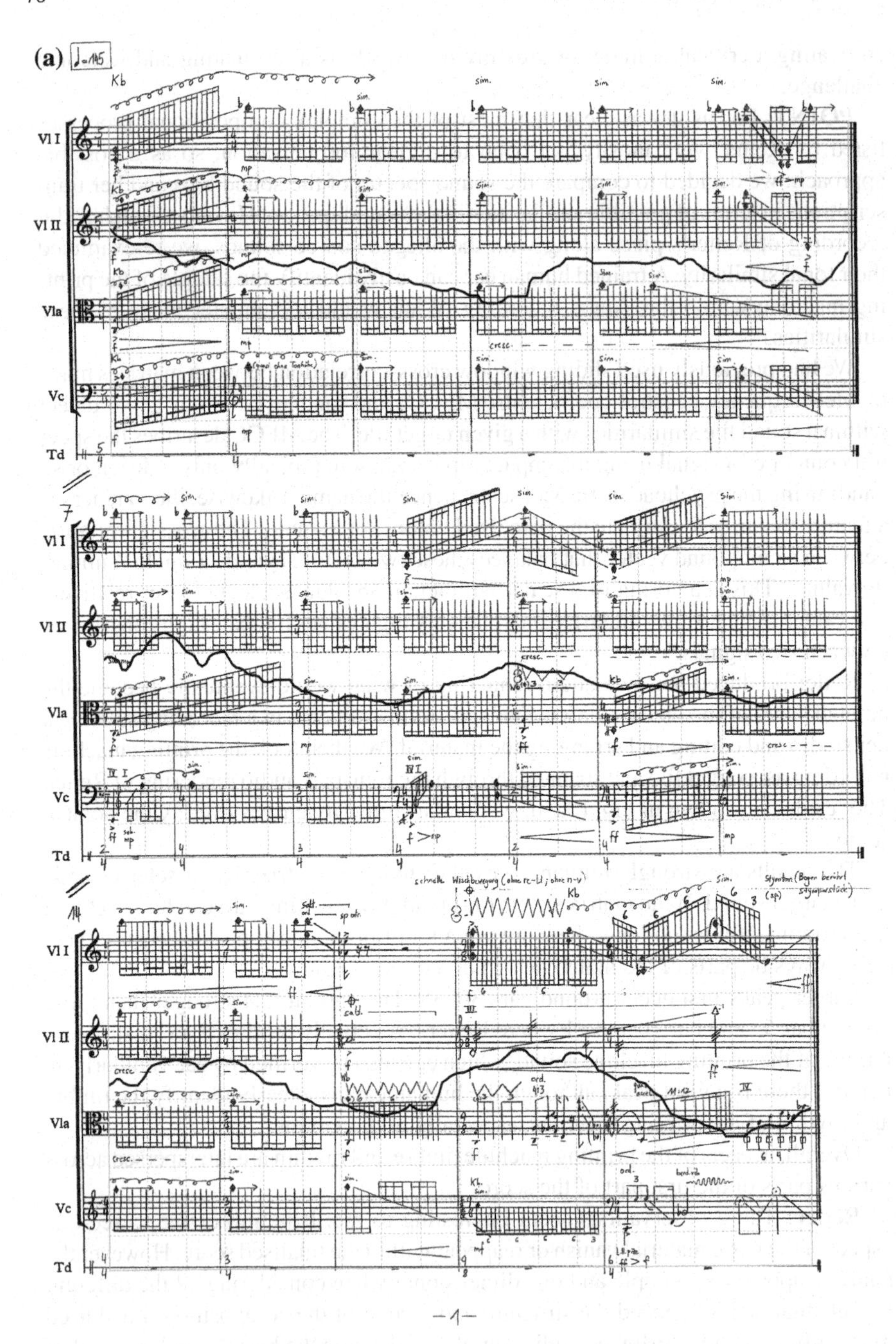

Fig. 10 Figure 10a-d show the first four pages of the score with similarity *curves* of the printing machine file superimposed. The duration of each line is approx. 10 s, each page thus 30 s. Scaling is corrected according to the score and is not completely linear. The higher the *curve*, the stronger the timbral similarity. The *curve* is normalised so that its minimum corresponds with the lowest staff line of the cello, the maximum with the highest staff line of the first violin

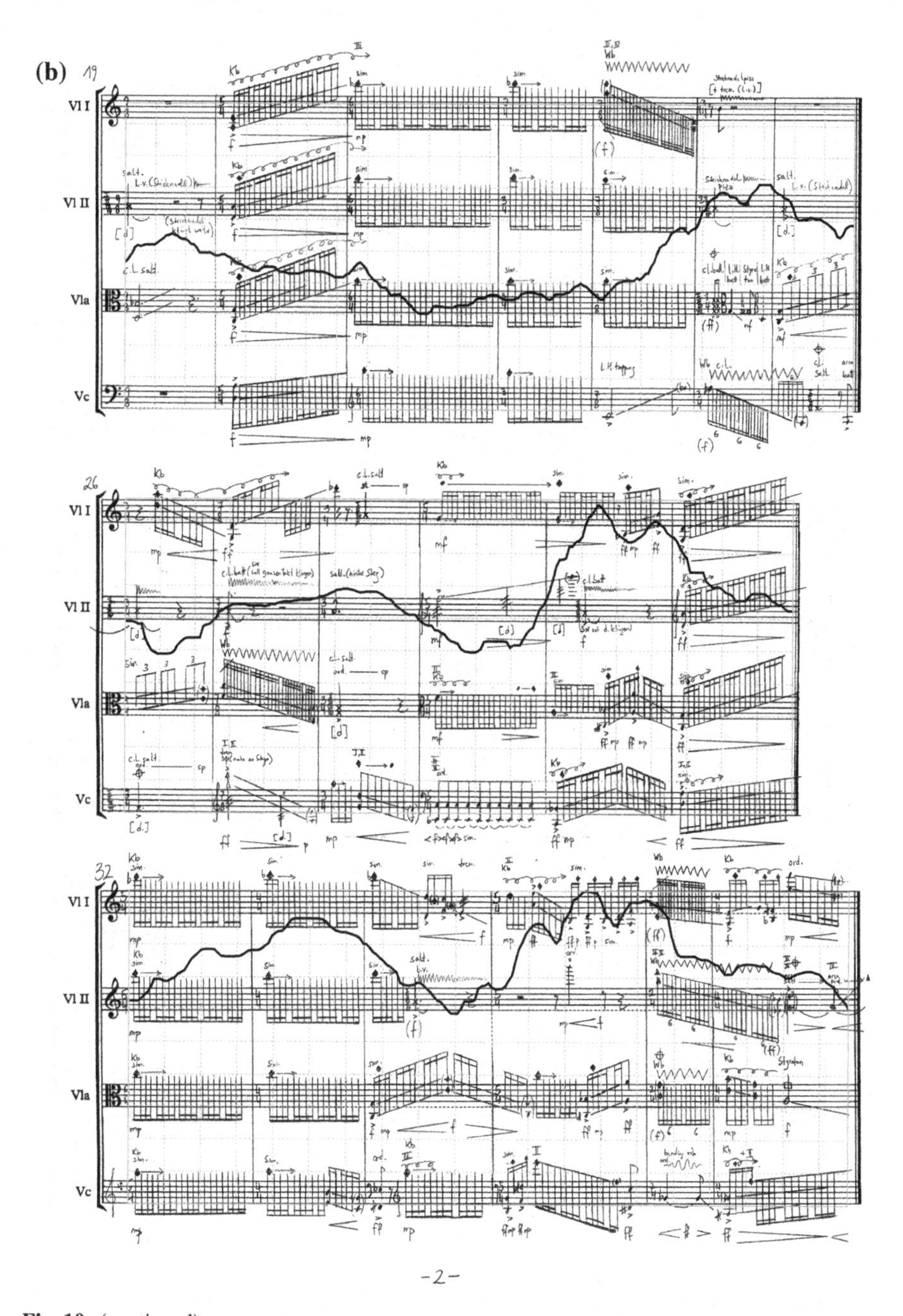

Fig. 10 (continued)

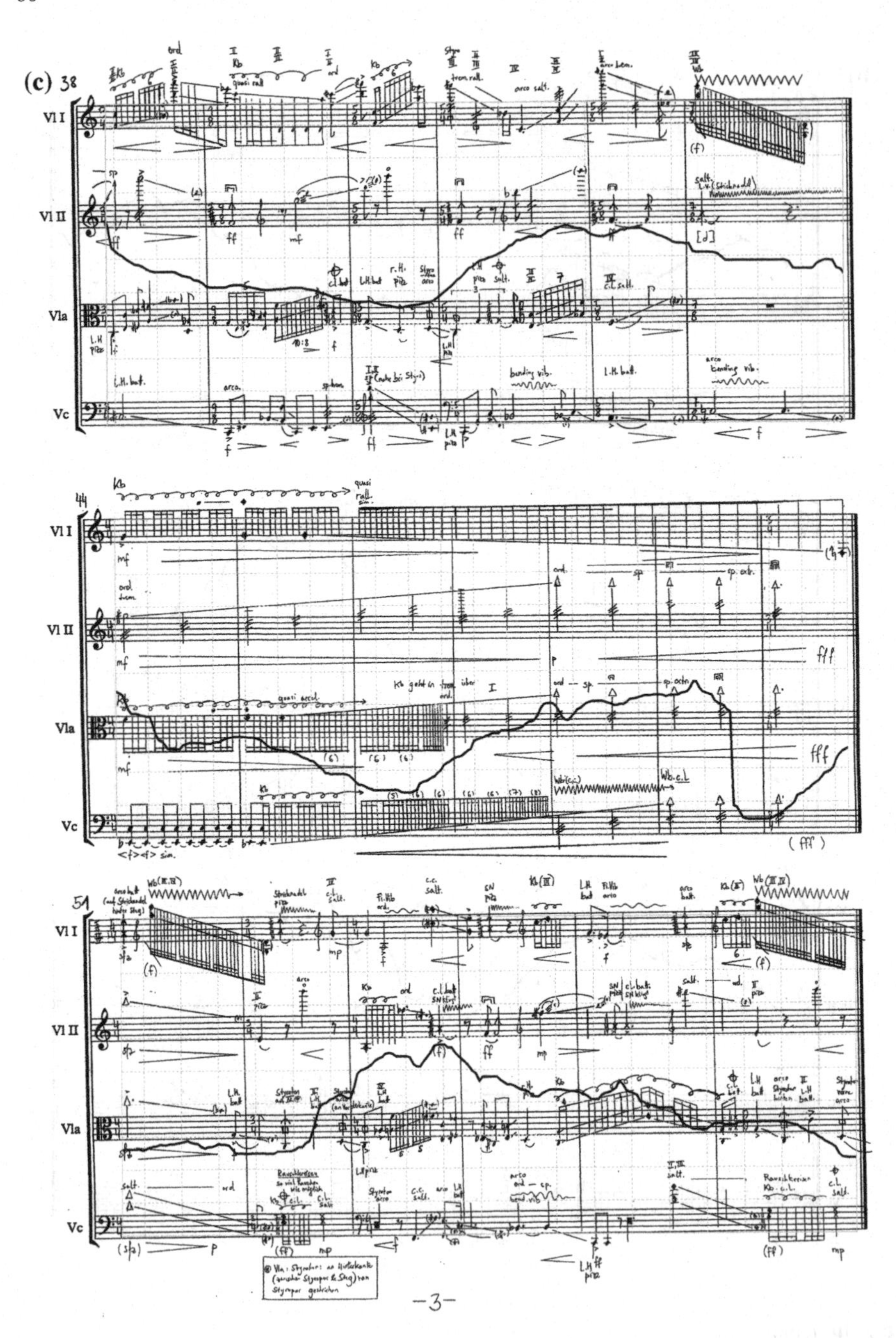

Fig. 10 (continued)

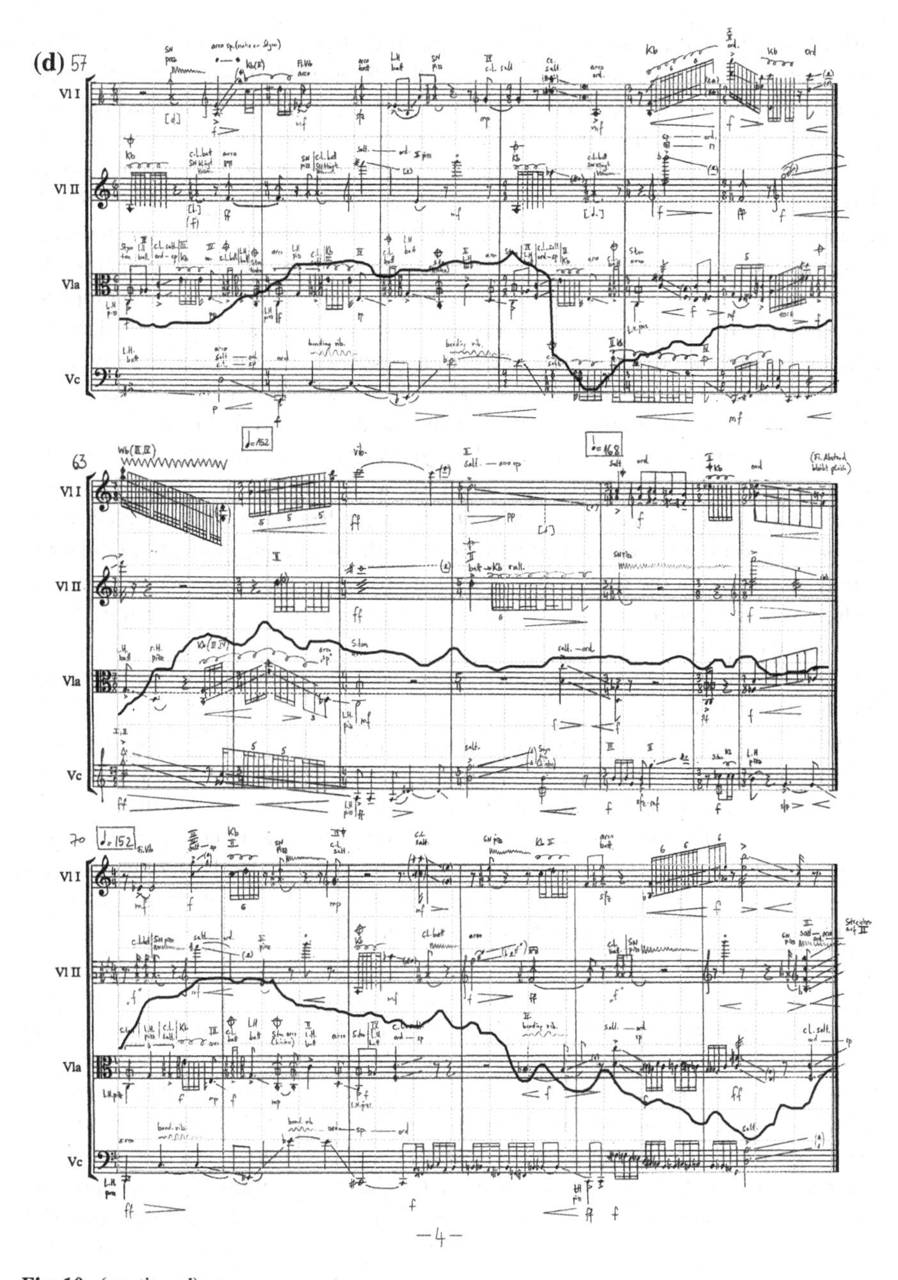

Fig. 10 (continued)

Moreover, the basic sound material of the four strings would on the one hand exist parallel in a contrapuntal frame and on the other hand also evolve in parallel and different directions simultaneously. Furthermore, as I have mentioned before, transforming concrete audio material into instrumental components would also be only one among a few structural ideas that had inspired me. At many positions I have generated new musical materials from the given basis, also invented "rhetorical" elements, etc. So as a summary it seems difficult for me to derive out of these analyses a larger understanding of the cognitive or intuitive processes that are participating in the compositional thinking, other than noting that these processes are indeed complex and difficult to trace or represent. As a matter of fact many basic structural ideas of the string quartet had to remain unobserved. It is now difficult to compare my final score with the electroacoustic analyses since this one examined theme is in fact very much overlaid with other coincidental structural and sonic developments.

I am not sure if it is possible to extend the analysis at this point. Theoretically, all used audio files would have to be accounted for, we would have to rebuild the basic strand of files that underlays the instrumental material. But is this a reasonable investigation? There are many influences—for example the development of the "breakcore rhythm" at the end of the piece—that indeed are rooted in the Heidelberger printing machine sound found at the beginning of the piece, but cannot be analysed without an enormous effort.

POINT: How would you characterise your piece *In groben Zügen* that you have written during this project?

Reiter: The piece moves along the limits of playability without going beyond them. Throughout the whole piece it remains realisable if the player follows a certain choreographical scheme and balances his/her energy. For me as an instrumentalist I like to be under pressure and challenged, this leads to a certain playing passion, an extra "kick", so to speak. The composition intends to evoke this kind of hidden energy. Actually it is also conceived very choreographically. It is about motion and gestures that lead to a specific sonic result. These gestures are cyclical, so they reappear slightly changed and in different contexts. From another point of view, the rigidity of the composition, the precision required for playing, varying and repeating the acoustic modules, and the exact timing of all four parts almost put the musicians into a machine-like state of trance. As to the relationship between man and machine, my music may well be described as a dialectical process. The question is to what extent is it possible to keep up with these rhythms. The fast tempo inevitably leads to moments where it is difficult to react, or where everything escalates, followed by a point of rest or another pattern of movement. I'm interested in how the resulting systems of order can be extended in a consistent way. One example of a "visual" or choreographic aspect of the piece is the instruction of circular bowing; the whole piece starts with this playing technique, which soon becomes extended and enlarged, also leading to an enriched sound experience. At certain moments motion also disengages from sound as to add a purely optical (illusionary) effect. Another example is the throwing of the bow in the air, which is amplified in case of the first violin at the end of the piece.

Finally I had introduced both a sound aesthetic and rhythmical structure that would show particularly strong references to the "Breakcore" genre as I mentioned before. Breakbeat is a style of electronic dance music largely influenced by hardcore techno, drum and bass, digital hardcore and industrial music and is characterised by its use of heavy kick drums played at high tempo. The end of *In groben Zügen* is written in an extremely high tempo, using "breakbeats", unusual "rough" sounds and an almost exclusive usage of "unusual" and unequal metres (especially 7/8 and 11/8 signatures). The musical texture successively compresses, the contrapuntal complexity increases as well as the dynamic level rises, eventually reaching a point of resolution or cut. *In groben Zügen* is a highly energetic piece of music, composed for the fastest possible tempo, at times reaching the maximum limit of playability. It is a game of control and determination.

Project Review by Eva Reiter

When I confirmed my participation in this project my overall interest was in the analysis of the basic models that would unconsciously determine my decision making when composing. This experiment seemed to sharpen my awareness of what I would describe as one of the main conflicts in the fields of art. We find ourselves caught up within this clash between tradition and innovation, between convention as the basis of an alleged successful system, also concerning the creation of music, and the personal need to overcome itself. During my compositional work I am clearly guided by a strong desire for "the new" as something that would surprise my mind set and fundamentally affect my perspective, next to the rising awareness of my personal convention and the convention of this day and age. To which extent is it possible to impulsively follow our primary "creative" instincts and to work on an increasing awareness of our heteronomous state? It is not about the simple polarisation of a "bad" system-oriented convention versus the inconceivable power of unrestricted mobility, but it is the awareness of our position within the power of tradition, the delight in functioning, the forces and attraction of fixation and immobility and our urge to grow beyond this basal strength.

Ever since my fascination with the static immobile sound aesthetics that derive from certain machines of our environment, it has been my ambition to create an instrument out of the transcription of these samples that would possibly be designed like a sound machine itself and then while "playing on it" lead to an unforeseen release and development of the musical and structural material.

I was therefore curious to take part in an experiment that would possibly supervise and systematize what I would regard as the central source of creative work, as providing the only possible power to overcome the limitations and systematisation that work as restrictive forces. I regarded it as a system containing immanent possibility thus it appeared highly fascinating for me from the start.

Concerning my choices of material it was interesting to compare the suggestions of the computer and the way I would use particular sounds in my manual composition, to see if I could incorporate some of the computer analyses or understand why I would make different decisions in a particular case. In the end I could not identify my intuitive choices as a recurring phenomenon, nor could the computer. I still believe that intuition resists systematic analysis by nature. Nevertheless, having said that, I do appreciate the productive discourse we established throughout the compositional work in this project. Not only the computerised feedback opened my mind to certain questions, but also in the first place, there were many substantial conversations I had with the project team.

References

1. Gottstein B (2009) Musikwerdung des Klangs—Über Eva Reiter. In: Polzer BO, Schäfer T (eds) Catalogue Wien Modern
2. Romitelli F (2001) In: Danielle Cohen-Lévinas (ed) Musiques actuelles, musiques savante. Quelles interactions? Entretiens réalisés et présentés par Eric Denut. Paris, L'Harmattan, pp 73–77
3. Rutz HH (2012) Sound similarity as interface between human and machine in electroacoustic composition. In: Proceedings of the 38th international computer music conference. Ljubljana, pp 212–219
4. Stowell D, Plumbley M (2007) Adaptive whitening for improved real-time audio onset detection. In: Proceedings of the 33rd international computer music conference (ICMC). Copenhagen, pp 312–319

Clemens Gadenstätter/Hidden Grammars

**Clemens Gadenstätter, Daniel Mayer,
Thomas Eder and Gerhard Nierhaus**

Clemens Gadenstätter grew up in an environment sensitive to the arts.[1] His parents organised readings and ran a gallery which, among others, exhibited the works of Arnulf Rainer, Günter Brus and Hermann Nitsch. Nitsch later on became a paternal friend of Gadenstätter. The musical education of Gadenstätter began with the flute, and encouraged by his teacher, he started to explore the repertoire of the 20th century. Subsequently, Gadenstätter occupied himself, mainly auto-didactically, with composition, musical analysis, score reading and jazz guitar. At the department of music pedagogy at the Orff institute in Salzburg, he received training in music theory and aural skills. At the age of 18, he began studies at the University of Music and Performing Arts in Vienna. He studied flute with Wolfgang Schulz and composition with Erich Urbanner. The flute lessons established the relationship between body, instrument and interpretation—this complex field of embodiment which assumes a central role in his later works. Whilst it is Urbanner who provided him with a solid foundation of composition technique, it is with Helmut Lachenmann during a three year postgraduate course where he faced his main challenge.

[1] Biographical introduction and texts from the composer translated from the German by Tamara Friebel.

C. Gadenstätter
Institute for Composition, Music Theory, Music History and Conducting,
University of Music and Performing Arts Graz, Graz, Austria
e-mail: clemens.gadenstaetter@kug.ac.at

D. Mayer · G. Nierhaus (✉)
Institute of Electronic Music and Acoustics, University of Music
and Performing Arts Graz, Graz, Austria
e-mail: nierhaus@iem.at

D. Mayer
e-mail: mayer@iem.at

T. Eder
Department of German Studies, University of Vienna, Vienna, Austria
e-mail: thomas.eder@univie.ac.at

G. Nierhaus (ed.), *Patterns of Intuition*,
DOI 10.1007/978-94-017-9561-6_5

In this course, everything he thought and composed was questioned—a necessary but often arduous procedure, which on reflection greatly helped him to become critical and more aware of his own concepts and work. When asked about his role models at that time, he responds with: "Beethoven, Bruckner, Mahler, Schönberg, Webern, Berg, Nono, Stockhausen, Musil, Kraus, Proust, Tizian, El Greco, Nitsch, Rainer, Eisenstein, Hitchcock, Hendrix, Led Zeppelin, Monk, Coltrane…".

Gadenstätter often works with artists from other media such as video, installation, dance or literature. He writes essays and publishes in renowned music magazines. He also works as editor and teaches music theory and composition at the University of Music and Performing Arts in Graz.

A focal point of his artistic work is to write music in which the reflection and the transformation of its own prerequisites are deeply embedded. The prerequisites are manifold and can be identified in many different contexts: historical, social and of course, musical concerning material, form and instrumentation…, where the aspects of embodiment, perception and cognition become increasingly important.

Clemens Gadenstätter's compositional work is rooted in the everyday sounds that surround us. Instead of an indiscriminate use of these sounds, a thorough analysis of their connotations and a musical reinterpretation is applied. The embedding of apparently habitual sounds in unusual contexts paves the way to a radically new listening experience. This approach requires the intensive studying of the disposition of hearing which guides our everyday perception of sound objects. This complex disposition may be broken up into several questions: How does the historical or contemporary context determine the hearing of the objects? How is the network of connotations established? How do we classify sounds depending on the situation in which we hear them? And to what extent is this classification dependant on cognition and processes of embodiment?

Central here is the exploration from the manipulative potency to the commercial function of sound events. Clemens Gadenstätter puts himself in the position of being a test subject to study this aspect. The conscious examination of his own hearing eventually allows him to transform the material and shift its context, freeing the sound objects from their habitual understanding. The bare sound objects are thus susceptible to reflection and aid in the unmasking and change of traditional patterns of perception.

The process of re-contextualisation is a process of many layers that does not follow a fixed schema or set of rules, which are generally applicable. The starting point is elaborated for each new piece and depends on the signal-like energy and connotation of the individual sound events, as will be seen in the further course of this text. The contextual network, the semantic layers, and the loci of remembrance must be identified and transformed to empower the sound events with the pristine energy which is dissipated alongside the historic trajectory.

Artistic Approach

Statement

I believe composing means to work on an understanding of listening, an examination of the manner in how we take in and perceive acoustic information and make it a part of us. This understanding goes far beyond a conceptual realm; it is only achieved through the act of perception, an act that takes place on many simultaneous levels. Embodiment, tactility, space, memory and mimesis are just a few aspects that are relevant in this process and as a starting point, the act of composing is an engagement with the conditions of listening as a process which *includes* the listening as a necessary precondition.

The meaning of composing lies in the attempt to fathom levels of senses and meaning, to find meaning besides the established norms, to transform sound processes in so far as that composing is perceived as containing, rather than conditioning possibilities. This process is for me an act of creative research. The boundaries between music and self become transparent, permeable and delocalised. The energy of this process manifests itself in every new composition under entirely different conditions—the expression of the particular work arises via the confrontation of the known with its new-contextualisation, a dynamic process to discharge its tension finally in the sound.

Personal Aesthetics

My engagement with the multi-layered aspects of listening starts with the fundamental properties of the sound material and probes in the following the conditions of listening, which are imprinted by subjective and collective memories and are associated with very specific fields of meaning/connotation. This becomes especially clear with so called banal sound events, which we, following a process of getting used to, no longer hear, which to not "tell" anything in particular anymore. To break open these pre-described fields of meaning and the associated ways of hearing is still a central aspect of my composing. I do not simply want to accept this pre-described pattern but I want to probe and evoke other connotations and perceptions.

I would describe this as a "cheerful sceptical process" because it requires a self-reflection, which has quite a strong affinity to irony and humour. This process entails also a change of my perception and composition preferences, intuitive aspects become aware and permit a new and altered access to compositional decisions.

The necessary next step is to advance further into the conditions of listening via corporal mimicry of the sound event, embodied perception, and by fathoming synaesthesia and weak synaesthesia, which leads to the development of new composition tools. What drives me is to research the "comprehension by listening" and its development: to write music, which permits a new understanding via listening.

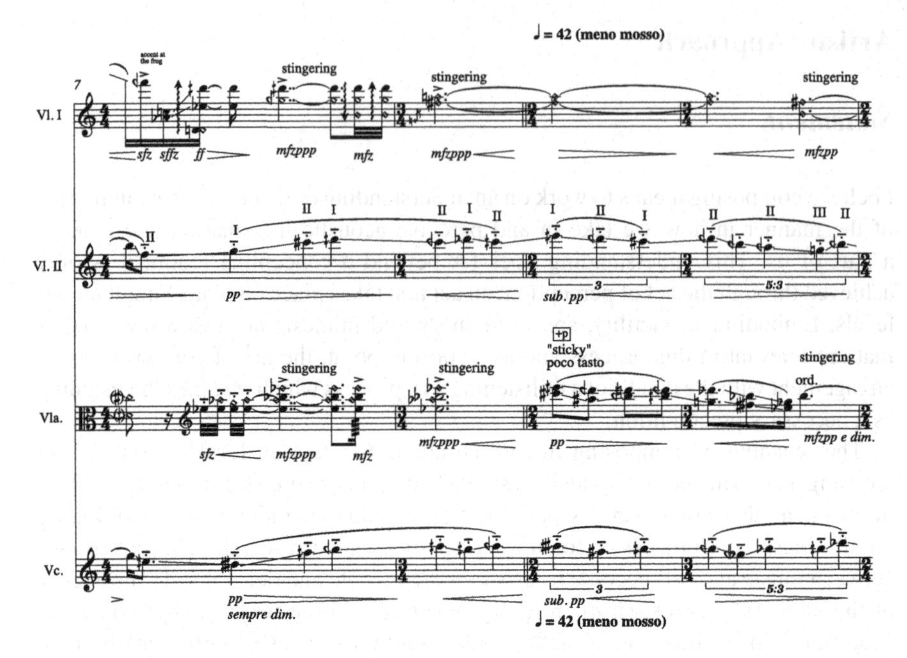

Fig. 1 *häuten/paramyth 1*, bars 7–11

Figure 1 shows an excerpt from my string quartet *häuten/paramyth 1.*[2] In these bars, the quality of sounds which have been produced are indirect translations from sensations of the skin and the corresponding touch reflex, through the movement of the bow and at the same time the translation of the sensations into sound events. It is clarified in the associative performance instructions next to the technical requirements.

Formalisation and Intuition

I work with and on structure-forming methods predominately to achieve results which lie outside of my imagination and possibilities of thoughts. Given concrete situations these (structure-forming methods) are edited via relation-creating, sometimes also abstract systems. This work prompts directly the experience that the sounding unlocks itself only in a specific context of collective memory, personal recollection and a series of other aspects. The composition work is then a probing via the "comprehension by listening" concrete material is simultaneously investigated, edited and reformulated on various levels. This material-based, archaeological process sparks itself off from the familiar and takes me away to the "other", the unpredictable.

[2] "*häuten*" roughly translates "to skin something".

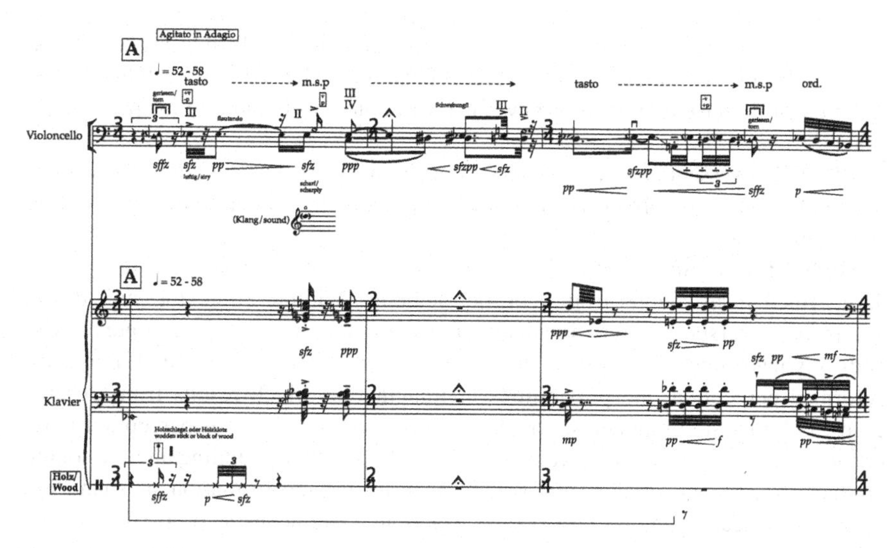

Fig. 2 *bersten/platzen—paramyth 4* for cello and piano, bars 1–3

Figure 2 shows an excerpt from *bersten/platzen—paramyth 4*[3] for cello und piano. The gesture here serves as a means to "tear away" from the initial spark (bar 1) and the bouncing of the bow illustrates the nucleus of the piece: a form of sound obtained through embodiment, bursting, projected into the piano (sostenuto pedal) through the transfer of the dry cracking wood, the wooden blocks on the frame as beats, projected as chords, events, which is then transformed into a bursting gesture of the cello, among other things connects the sound to the domain of frequencies, the resonance body of the piano is extended, sending you the link to a melody suggestive of the events that the cello produces.

Evaluation and Self-reflection

If I should have the feeling at the end of the process of composition that I didn't become somebody else then I can start afresh—that is the measure of quality which I apply to the work on a piece. If after the work I hear, think, feel differently, then I am happy on the one side yet already again unsatisfied since what I have done already became again obsolete, insufficient for me. And so on and on. It keeps my busy. Each piece is also linked to my observation and research of the perception of music as a sociological phenomenon and as an emotional object of trade as well as to many other aspects. It is essential to have here an interrelation to something, which I would call the reality of perception, of sensation, of thinking and determination.

[3] "*bersten/platzen*" roughly translates to "bursting/exploding".

The inspection of my music is naturally also influence by further aspects: how, for instance, my sociologic circumstances is changing, what a different status art has acquired in the last years and in how far I was able/will be able to adapt to these changes.

POINT: An essential aspect in your compositional work is a new-contextualisation of sound events, including every day occurrences, to allow them to be perceived in an unfamiliar way.

Gadenstätter: Structuring, also in the sense of contextualisation is the procedure, which form sounds from acoustic elements. Without context I can't relate to me or understand sound incidents; I myself am thus the first context: My body, my neurological conditions, my memories, my learned, by cultural processes deformed perception, the possibilities of imagination which results from experiences, polymodal components, which projects every incidents of perception onto images of other sensations. Moreover, the conditions of the sounding and the sound production are contextual levels, which are used for the formation of sound by building particular, unique networks of relations. The acoustic properties of the sounding are removed from their "naturalness" by the process of putting-into-relation and transferred into the "artificiality" of a formed structured sound. At first it is important to perceive the levels on which a sound event is effective. The example of an emergency car siren entails amongst others the following aspects:

A siren, which is contextualised in a very specific way in every day life.

A siren, which evokes associations to the past also by its particular historic connotation.

A siren, which triggers certain body-emotions also via its acoustic quality.

A siren as a single interval, which can appear as an instrumental gesture, an upbeat, as part of a tonal space.

As concrete sound with beating waves and the Doppler-effect as the emergency car passes by.

As mimic aspect: The siren as a mimicry of the instrument horn.

As part of a located event in a scene on a street.

All these aspects are now used in the sense of a new-contextualisation for the process of composition.

Hence, contextualisation appears in my works at different levels. The most important levels are, the acoustic level, acoustic similarities, the mimetic level where an incident is emulated in a graded way. Relating to the emergency warning of the sirens that would mean that the original sound is gradually imitated by a sample or several instruments etc. or also just by intervals, until only the sound figure itself is left, the semantic level, the level of meaning. The sirens as emergency sound can thus be related to a warning sound and sounds with similar meaning like diaphones, warning bells, warning screams, the situational level, the contextualisation of the siren in relation to sounding in a concrete environment, the level of embodied perception, the level of the corporal feeling and re-feeling of incidents, which I would also call mini-mimesis, the level of weak synaesthesia.

Weak synesthetic assignments of certain sensory stimuli to other modalities of perception, one could call this also polymodal mapping. A high pitch or a rough sound, for example, corresponds to a spatial or tactile experience, respectively, or everything which is "spicy/sharp" forms a field of contexts, the level of space: everything which sound in a certain space, either real or artificial, is put into relation with each other, the level of the source of sound due to the idiomatic of the instrument, of the human voice, the level of collective memory: sound incidents of historic or cultural memory fields are used for re-contextualisation in order to generate from it specific sound structures, the level of temporality: the quality, how sound incidents are projected in time is used to compile/develop structure and context. The level of emotions: nuances, of "how sense I an incident?", "which emphatic qualities have incidents?", or "what is triggered by incidents?" are used to structure sound events. Relating to the siren this can mean that further incidents are assigned contextually, which carry also within the emotion of something shrill, warning, signal-like, far-reaching, cutting, etc.—sounds therefore, which provoke a similar "emotion" also if they originate from a very different context.

These and further fields of contextualisation are then processed to "succession-structures". The term succession-structure means here that a succession of incidents forms shapes, which specify the respective way they appear, sharpening their quality, meaning etc. and making them unique.

The goal of the polyphonic arrangements of such succession-structures, quasi "melodies" as result of my search for a genuine non-tonal "melodic" structure, is then the development of sound via the contemporaneity of various levels of contextualisation mostly around a single material particle.

Sound is thus formed from the analysis of the material on the various levels of their levels of appearance, effect, and contextualisation. Sound formed in such a manner is then "polyphonic" analysis, which is applied simultaneously to the above mentions levels in order to then fathom the "depth" of the possible forms of appearances of sound incidents.

Each incident is then newly contextualised by temporal structuring (quasi melodic successive and polyphone-simultaneously): in the example of the siren this is then, for example, embedded in a harmonic context, the quart (mostly detuned and with beat frequencies) is part of a pitch process, also possibly part of a historically imprinted harmonic progression etc. Or: The sound figure of the rhythmically oscillating quart interval becomes an upwards-directed sound figure, thus part of a figurative upswing, the "original" or the sampled siren can the be contextualised in the sound space of the brass—the siren as an exaggerated signal trumpet (or vice versa). In this manner, various new-contextualisations are thus created—preferably every sound element is thus newly determined in appearance, meaning, function, etc. in my pieces.

POINT: Can you give us a concrete example from your work, with this principle of new-contextualisation?

Gadenstätter: Figure 3 shows an excerpt from *Fluchten/Agorasonie 1*[4] for soloists, orchestra and space. Here solely different configurations of sirens and horns are produced, transformed into pitch movements and harmonic fields from which form the sound event of the excerpt.

[4] "*Fluchten*" roughly translates to "alignments".

Fig. 3 *Fluchten/Agorasonie 1* for orchestra und soloists, bars 273–275

Fig. 4 *Figure/Iconosonics 1*, bars 38–40

A more abstract example of a contextualisation is shown here in Fig. 4 from *Figure/Iconosonics 1* for clarinet, string trio and piano. Typical sound gestures were brought together from the pool of "theory of characterisation", those which were connected to an experience of a "storm": fast, rumbling movements (viola, cello), lashing gestures (violin, clarinet), penetrating elements (piano, cello). All sounds were obtained via various translations and mechanisms of contextualising the instruments themselves, in order to trigger the experience of a storm.

Project Expectations

I expect of the project a new way of mirroring of myself, of my work, of my thinking and eventually also of my intuition, thus of my surly often also very unconscious "hand", which forms the details of the sounds. This mirroring shall also allow me to transform this hand, if this can work is for me not so crucial: important is the effort to try it. The attempt makes the music appear differently, lets me experience myself differently—this energy leave then an imprint behind, this is at least what I am hoping for, in me and in my music, which will then be created.

Exploring a Compositional Process

POINT: In the work of Clemens Gadenstätter semantics of sound and gesture play an important role. Gadenstätter emphasizes that perceived sounds are not purely sounds but are embedded in a cloud of associations. This semantic aspect is one he is working on in his pieces and which he thematically explores in his compositional process. In the early stages of a composition Gadenstätter explores a catalogue of gestures and words in a brainstorming manner as seen in Fig. 5.

At some point, groups of musical gestures, motifs and procedures become associated with certain terms. Combinations and clashes of words have the potential to then cross-breed and trigger themselves, generating further related gestures and terms. By doing so, musical gestures are by far not subordinated to language and least of all are they thought of as their illustration.

When studying his sketches we decided to concentrate on Gadenstatter's grouping and structuring of terms, here in a more polished version. Table 1 shows an excerpt from Gadenstätter's analysis of a voice from *Semantical Investigation 2*, bars 1–6.

We thought that a general *formal grammar* would be an appropriate way to frame the language-centered aspect of his compositional approach, in addition we saw that this calculus is general enough to decide for a specific usage and computational implementation. A formal grammar is a set of *production rules* for words or *strings* based on a *alphabet*, a set of atomic items. A *formal language* is the set of *strings* that can be built from the vocabulary using the productions rules. Formal grammars have attracted interest in connection with Noam Chomsky's theories on language since the 1950s [1]. Chomsky especially emphasised the importance of syntax and mathematical tools for their investigation. The following is a formal definition:

A vocabulary V is a finite non-empty set of symbols.

A tuple $(x_1, \ldots, x_k)$ with $x_i \in V$ for $i = 1, \ldots, k$ is called *word over* V *of length* k.

The *empty word* ε has length 0.

V^* denotes the *Kleene closure* of V, which is defined by all words over an alphabet V.

A subset of V^* is called *language* or *formal language*.

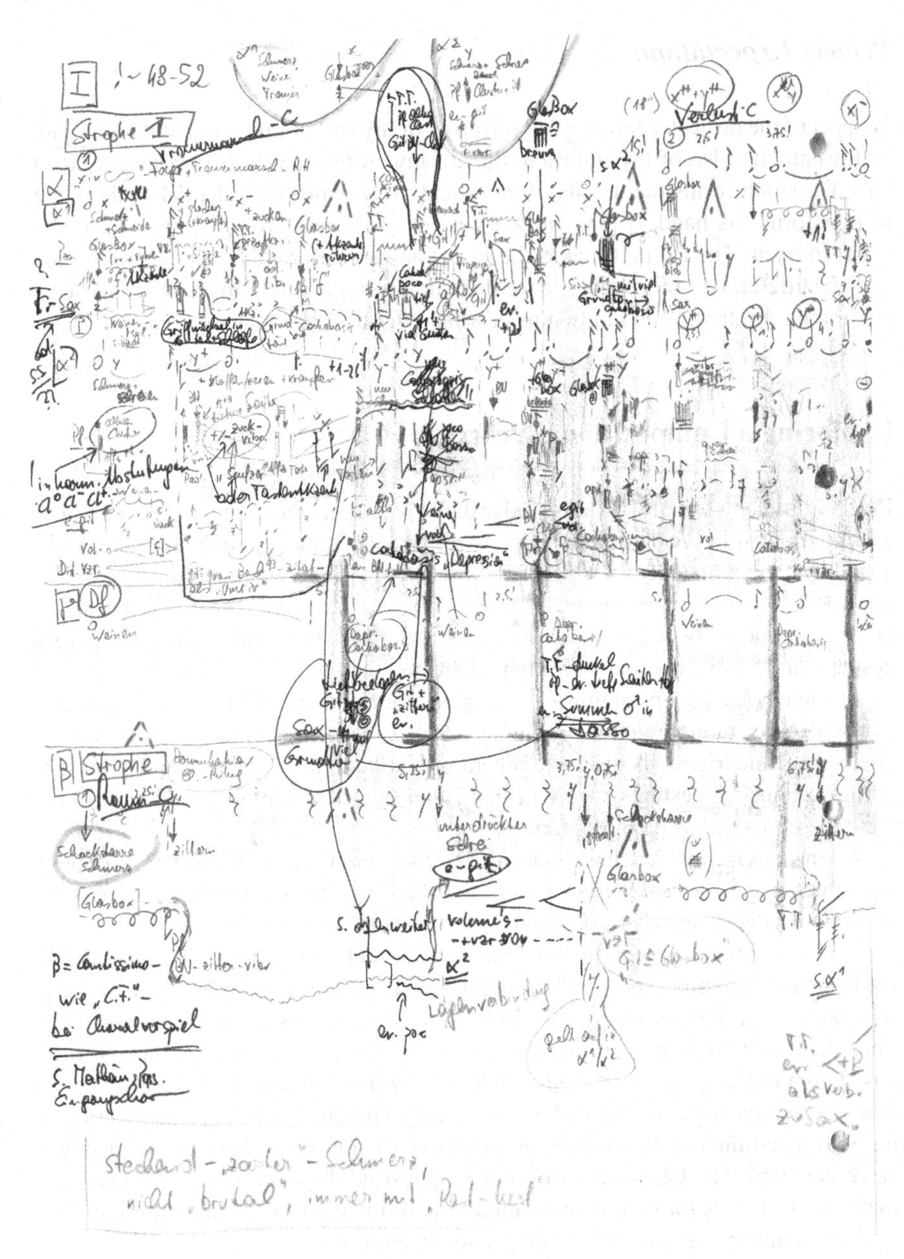

Fig. 5 Compositional sketch from Gadenstätter

Table 1 Excerpt from Gadenstätter's analysis of *Semantical Investigation 2*

Synthesefolge/synthetic sequence		
Takt 1/bar 1		**Takte 2 + 3/bars 2 + 3**
"a"	Non a → "Gestalt 1"/shape 1	Synthese (synthesis) "a/non a"
Fahrradgl./bike bell	Ministrantenglocke/handbell	Pf. + Git./piano + guitar
Einzelklang/single sounds	3/4-Klang (3 or 4 part chord)	Impuls—tremolo, Interval/interval
Rh.	Trem.-Ten.	Rh. Modell in Wiederholung/repeating model
Analysefolgestruktur/structure of resulting analysis		
Takt 4a/bar 41	**Takt 4b/bar 4b**	
synth a → "synth a1" → Gestalt 2	a → a1	Non a1 → non a2 → Gestalt 3
Pf. + Git/piano + guitar	Perc. Git. Pf/perc. git. piano	Streicher/strings
Rep. in 16tel—Vielklang/multiphonic	Einzelereignis/single event	ricochet à saltando
Negationsfolge/negation result		
Takt 5/bar 5	**Takt 6 + 7/bars 6 + 7**	
a2	Non a2 → Gestalt 4/shape 4	
Handgl./handbell	Türklingel/doorbell (sample)	
Pendelbew./pendulum movement	Tenuto	

$V^+ := V^* \backslash \{\varepsilon\}$ denotes the *positive closure of* V.

A formal grammar G, also phrase structure grammar, is a quadruple (V_N, V_T, P, S) where

1. V_N and V_T are finite non-empty and disjoint sets.
   V_N is called the set of *variables* (non-terminals), often denoted with capitals.
   V_T is called the set of *terminals*, often denoted with small letters.
2. $S \in V_N$ is the *start symbol.*
3. P is a set of *production rules* of the form $\alpha \to \beta$, where $\alpha \in (V_N \cup V_T)^+$ and $\beta \in (V_N \cup V_T)^*$.

Table 2 gives an example for a formal grammar that shows why also the term phrase structure grammar (PS-grammar) is used. For VN let's take the symbols S (sentence), NP (noun phrase), VP (verbal phrase), ADJP (adjective phrase), N (noun), V (verb), ADJ (adjective), P (preposition), ADV (adverb), DET (article). Let us define the following production rules (rewriting rules), from the definition of a formal grammar it becomes clear that the order of rules doesn't play a role here, they can be chosen for rewriting in arbitrary order to produce words of the formal language:

A valid word in the formal language—in analogy to a sentence in the natural language—can be derived like this by rewriting:

Table 2 Example for a formal grammar

S → NP VP
VP → V
VP → V NP
ADJP → ADJ
ADJP → ADV ADJ
NP → N
NP → ADJP N
NP → DET ADJP N
NP → DET N
N → composers
N → musicians
N → ideas
N → chords
DET → the
V → generate
V → analyse
ADJ → witty
ADJ → beautiful
ADV → quite
ADV → less

S → NP VP → N VP → N V NP → N V ADJP N → N V ADV ADJ N → composers V ADV ADJ N → composers generate ADV ADJ N → composers generate quite ADJ N → composers generate quite witty N → *composers generate quite witty ideas.*

The formal language's word *composers generate quite witty ideas* is a semantically meaningful sentence in the natural language, though other syntactically correct derivations might, depending on the context, be semantically ambivalent (or nonsense): S NP VP → ADJ N VP → ADJ N V NP → ADJ N V ADJP N → ADJ N V ADV ADJ N → witty N V ADV ADJ N → witty ideas V ADV ADJ N → witty ideas analyse ADV ADJ N → witty ideas analyse less ADJ N → witty ideas analyse less beautiful N → *witty ideas analyze less beautiful chords.*

With his hierarchy Chomsky distinguishes four types of formal grammars (type 0 to type 3). The higher the type the more restrictions are defined for the production rules, vice versa the *generative capacity* of the grammar, the amount of valid words that can be derived by rewriting rules is decreasing.

Let $G = (V_N, V_T, P, S)$ be a formal grammar, $V = V_N \cup V_T$. G is of

- type 0 (unrestricted grammars/languages) if and only if all kinds of production rules, as defined above, are admitted.
  Generative capacity: very high.

- type 1 (context-sensitive grammars/languages)
  if and only if each rule is of the form
  $\alpha_1 A \alpha_2 \rightarrow \alpha_1 \beta \alpha_2$
  with $A \in V_N, \alpha_1, \alpha_2 \in V, \beta \in V^+$,
  as only exception $S \rightarrow \varepsilon$ is allowed, but in that case S is not allowed in any right-hand side of a production rule.
  Generative capacity: high.
- type 2 (context-free grammars/languages)
  if and only if each rule is of the form
  $A \rightarrow \beta$
  with $A \in V_N$, $\beta \in V^+$ and the same exception as with type 1.
  Generative capacity: middle.
- type 3 (regular grammars/languages)
  if and only if each rule is of the form
  $A \rightarrow aB$ or
  $A \rightarrow a$
  with A, $B \in V_N$, $a \in V_T$ and the same exception as with type 1.
  Generative capacity: low.
  Regular grammars of this definition are called *right-linear*, which is equivalent to *left-linear* grammars (each rule of the form $A \rightarrow Ba$ or $A \rightarrow a$), however a mixture of right-linear and right-linear rules doesn't guarantee a regular grammar.
  Generative capacity: low.

For $i < j$ grammars of type j are also of type i. The inclusion is strict, for $i = 0$, 1, 2 there exist grammars/languages which are not of type $i + 1$.

The interpretation of formal grammars has changed over time, before building a grammar modelling Gadenstätter's approach we take a short survey. In *Syntactic Structures* Chomsky distinguishes between the *deep structure* and the surface structure of a natural language sentence, which can be regarded as a word over an alphabet in terms of a formal grammar. According to Chomsky's theory (at that point of time) the deep structure consists of the main semantic relations and is transformed into the surface structure by a set of transformation rules. Introducing this view Chomsky connected the fields of semantics and formal grammars, whereby the latter can be treated purely mathematically. However the connection between both disciplines and its history is complicated. Since *Syntactic Structures* [1] Chomsky's theories on language have changed a lot, the antagonism between deep structure and surface structure has been continuously overruled by the concept of *logical form* and *phonetic form* (LF/PF) [3, 10], with *Minimalism* [2] deep structure and surface structure have been completely dropped. Moreover the interpretation of central terms of Chomsky's early theory has changed over the years—meaning was bound to deep structure first, but extended to include surface structure later on. Chomsky has also influenced the development of new models for musical understanding, amongst the most famous one is *Generative Theory of Tonal Music* (GTTM) by Fred Lerdahl and Ray Jackendoff [6]. However, as we don't focus on the syntactical structure of musical gestures itself and Gadenstätter's music is situated outside the tonal idiom, we don't refer to this theory here. In order to build a modelling grammar we looked at Gadenstätter's sketches. He made lists of certain adjectives and verbs, which could also be in the form of past participles, and gave us some examples how he would link them. In his compositional approach these terms serve as placeholders for musical gestures.

POINT: Can you tell us something about your motivation and use of these terms?

Gadenstätter: I work with these terms because they are a shorthand for sonic phenomena, of instrumental cells of sound and gestures, that can describe short musical processes and structures.

Furthermore, a reference to the world is only possible through concrete sounds from our environment, but also in using such terms it can trigger unfamiliar weak synaesthetic sound phenomena.

Another reason for the use of these terms is my involvement with the forms of characterising within the history of music, whose meaning is taken from the rhetoric and facilitates listening and comprehending such characterisations of sound.

By using the shorthand of sound events I have a practical means to outline the material and its overall form at an early stage of composing. In this case, however, this pre-structuring is of course still very flexible and open—both for changes as well as for interpretating my terms. This interpretation may in the process of working entirely change for a later excerpt of a composition, and then send you back recursively to the development of the form. By naming the concepts and by simultaneously holding open their interpretation I invite the chance to access the levels of the sonic effectiveness of my structures.

A musically new and re-interpretation of the terms used is to some degree a paradoxical undertaking, as this can naturally also trigger known tonal associations. However, I start my work again and again with the goal to advance from the known to the unknown, to make sonic events tangible, without the usual associations.

POINT: Within each category, adjectives and verbs, Gadenstätter defined an order from one to eight, that way clauses with shape or tendency could be demanded. We designed a phrase structure grammar for building clauses with a main term and an arbitrary number of adverbs (the rule ORDADV_LIST → ORDADV_LIST ORDADV, where ORDADV_LIST denotes a list of ordered adverbs, serves as prolongator). The main term can be an adjective, a verb or a past participle. When using adjectives of categories C and D as adverbs we simply added suffix *-ly* in the result. To be very exact: adverbial forms could be invented with their own categories and rules, but, as the issue was obvious, we omitted additional rules for the sake of a better overview. Moreover the problem doesn't arise in the original examples in German language where the adverbial use of adjectives doesn't change them.

To depict the rewriting system (Table 3) in a more compressed form we oblige to the notation of logical OR (v), so V → A v B is equivalent to writing the two rules V → A and V → B. In case of sequential application of production rules we write → ... for one single application. With the grammar defined in Table 3 we can e.g. derive the following phrases (Table 4).

POINT: How do you judge the resulting phrases?

Table 3 Rewriting system in a compressed form

A 1 → rip	A′ 1 → ripped
A 2 → rupture	A′ 2 → ruptured
A 3 → frazzle	A′ 3 → frazzled
A 4 → explode	A′ 4 → exploded
A 5 → diverge_scales	A′ 5 → diverged_scales
A 6 → atomise	A′ 6 → atomised
A 7 → expose	A′ 7 → exposed
A 8 → cruelly_analyse	A′ 8 → cruelly_analysed
B 1 → grout	B′ 1 → grouted
B 2 → pressure	B′ 2 → pressured
B 3 → crunch	B′ 3 → crunched
B 4 → mash	B′ 4 → mashed
B 5 → abrade	B′ 5 → abraded
B 6 → scrape	B′ 6 → scraped
B 7 → graze	B′ 7 → grazed
B 8 → destroy	B′ 8 → destroyed
C 1 → angry	D 1 → burning
C 2 → panic	D 2 → vitriolic
C 3 → orgiastic	D 3 → toxic
C 4 → sadistic	D 4 → flashy
C 5 → tremoring	D 5 → screaming
C 6 → gazing	D 6 → glistering_cold
C 7 → crashing	D 7 → longing_cruel
C 8 → unconscious	D 8 → tender_cruel
S → ORDADV_LIST ORDMAIN	
ORDADV_LIST → ORDADV_LIST ORDADV	
ORDADV_LIST → ORDADV	
ORDADV → ADV INT	
ORDMAIN → MAIN INT	
INT → 1 v 2 v 3 v 4 v 5 v 6 v 7 v 8	
MAIN → V v ADJ v PASTPART	
V → A v B	
ADJ → C v D	
PASTPART → A′ v B′	
ADV → C v D	

Gadenstätter: The results of the system were all basically useful. A detailed analysis then showed me that some effects were more suitable than others—especially because of the need for the idiom of the sound event to deal with certain instruments or sources.

Table 4 Derived phrases from the rewriting system

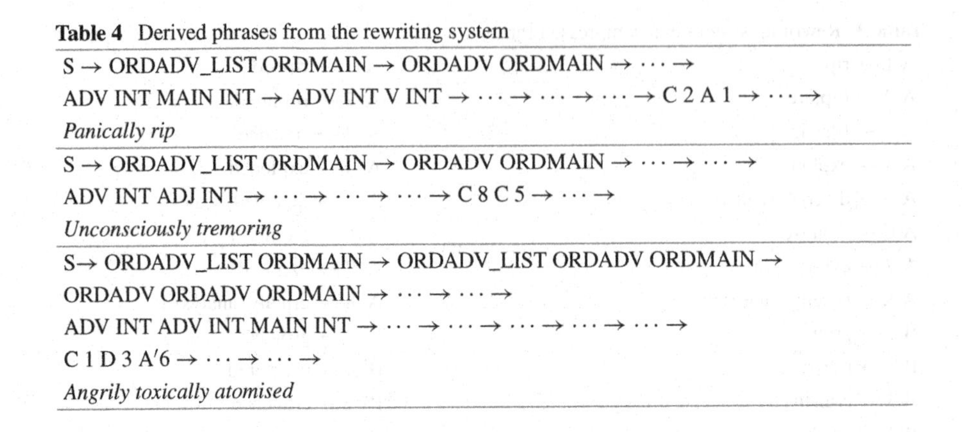

S → ORDADV_LIST ORDMAIN → ORDADV ORDMAIN → ··· →
ADV INT MAIN INT → ADV INT V INT → ··· → ··· → ··· → C 2 A 1 → ··· →
Panically rip

S → ORDADV_LIST ORDMAIN → ORDADV ORDMAIN → ··· → ··· →
ADV INT ADJ INT → ··· → ··· → ··· → C 8 C 5 → ··· →
Unconsciously tremoring

S→ ORDADV_LIST ORDMAIN → ORDADV_LIST ORDADV ORDMAIN →
ORDADV ORDADV ORDMAIN → ··· → ··· →
ADV INT ADV INT MAIN INT → ··· → ··· → ··· → ··· → ··· →
C 1 D 3 A′6 → ··· → ··· →
Angrily toxically atomised

POINT: In the grammar, see Table 3, size and form were flexible, we revised it in order to model Gadenstätter's compositional principles more closely. Matrices of adjectives and allocated verbs of a specific form did come nearer. Rather than building arbitrarily structured but singular clauses we agreed to use groups of "clauses" of same kind as in Table 5. For each row Gadenstätter chose a characteristic adjective to be used adverbally with all consecutive verbs. The principles after which they are selected, are described later on.

POINT: Why would you rather use three rows of three groups of three terms, i.e. a matrix structure of phrases?

Gadenstätter: I limited myself to the simplicity of three groups because I also often work with three groups as the smallest units. Groups of two were also possible, but it seemed to me too unspecific for this experimental setup. Such resulting structures would then be "synthesised consequences", attempted to be recreated into an algorithmic arrangement. One reason for the multiple use of three terms resulted from my polyphonic compositional technique: all group terms are not contextualised only successively but also simultaneously. These group terms are also yet another set of a "polyphonic axis" whose structure is generated not only by the material but primarily by their qualitative contextualisation. I find this polyphony similar to the hand and eye perception—an object is sensed, the tactility is felt, the temperature is known—the eye follows the shape of the object exactly as the movement of the hand palpitates.

POINT: In the grammar (Table 3) we took arbitrary selections of integer numbers in the context of the grammar: arbitrary choices of production rules for the variable INT on the left-hand side. By using the defined ordering of terms within categories, Gadenstätter chose integer tuples of certain shape to build phrase constellations with specific tendencies. E.g. he calls the tuples (1 2 4 7) and (8 7 5 2) "parabolic".

Table 5 Groups of clauses

ADV | V V V | V V V | V V V
ADV | V V V | V V V | V V V
ADV | V V V | V V V | V V V

POINT: Why do you work with such tendencies?

Gadenstätter: I work with such tendencies, because they give me the opportunity to find a connection to a material context in a certain way. The ones we have used here are the ones that could be formalised the easiest, but of course I also experimented with tendencies of other appearances.

The sound events are then brought into relationship with each other by proximity, distance and movement of form through the material field. When such structured sequences are actually used in a piece, even as a concrete theme of a piece, the material is always "owed": the consequences are always created from the already concrete aspects of the topic or the materials inscribed, developing into the resulting form.

POINT: To distinguish such shapes we can take over the terms concave and convex from functions within the domain of real numbers.

A tuple $(x_1, \ldots, x_k)$ with $x_i \in \mathbb{N}$ for $i = 1, \ldots, k$ is called *concave-ascending* (Fig. 6) iff $x_i \leq x_{i+1}$ for $i = 1, \ldots, k-1$ and $x_{i+1} - x_i \leq x_i - x_{i-1}$ for $i = 2, \ldots, k-1$.

A tuple $(x_1, \ldots, x_k)$ with $x_i \in \mathbb{N}$ for $i = 1, \ldots, k$ is called *concave-descending* (Fig. 7) iff $x_i \geq x_{i+1}$ for $i = 1, \ldots, k-1$ and $x_{i+1} - x_i \leq x_i - x_{i-1}$ for $i = 2, \ldots, k-1$.

A tuple $(x_1, \ldots, x_k)$ with $x_i \in \mathbb{N}$ for $i = 1, \ldots, k$ is called *convex-ascending* (Fig. 8) iff $x_i \leq x_{i+1}$ for $i = 1, \ldots, k-1$ and $x_{i+1} - x_i \geq x_i - x_{i-1}$ for $i = 2, \ldots, k-1$.

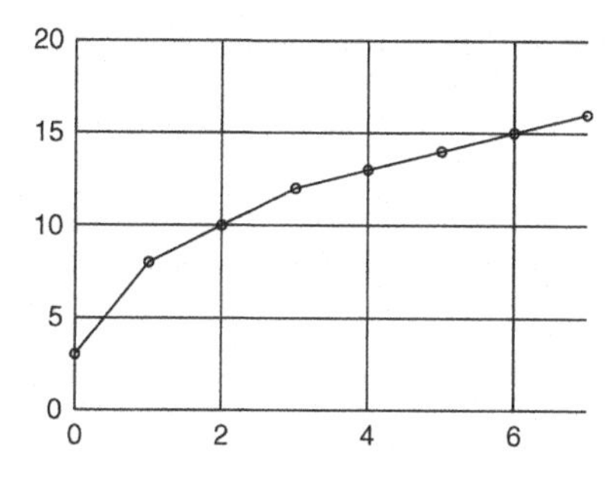

Fig. 6 Concave-ascending tuple (3 8 10 12 13 14 15 16)

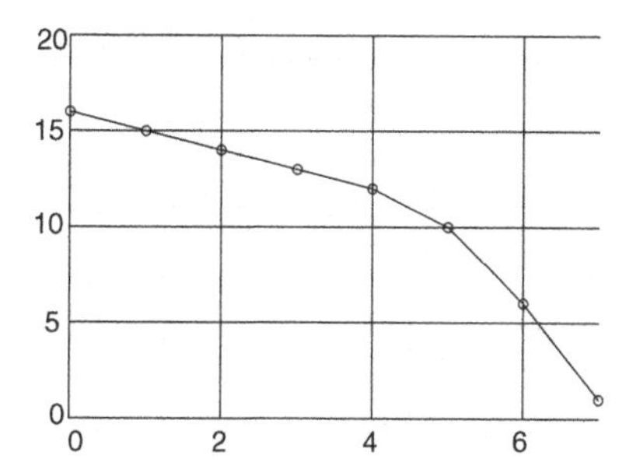

Fig. 7 Concave-descending tuple (16 15 14 13 12 10 6 1)

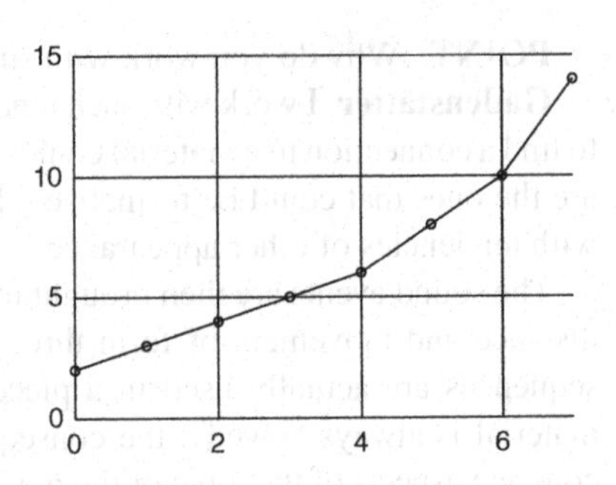

Fig. 8 Convex-ascending tuple (2 3 4 5 6 8 10 14)

A tuple $(x_1, \ldots, x_k)$ with $x_i \in \mathbb{N}$ for $i = 1, \ldots, k$ is called *convex-descending* (Fig. 9) iff $x_i \geq x_{i+1}$ for $i = 1, \ldots, k-1$ and $x_{i+1} - x_i \geq x_i - x_{i-1}$ for $i = 2, \ldots, k-1$.

Above definitions could be extended to strictly convex and strictly concave shapes, in this form a straight linear shape is also included.

A tuple $(x_1, \ldots, x_k)$ with $x_i \in \mathbb{N}$ for $i = 1, \ldots, k$ is called a *zigzag* iff $\text{sign}(x_{i+1} - x_i) * \text{sign}(x_{i-1} - x_i) = -1$ for $i = 2, \ldots, k-1$.

A zigzag tuple $(x_1, \ldots, x_k)$ with $x_i \in \mathbb{N}$ for $i = 1, \ldots, k$ is called *strictly converging* iff either: $x_2 < x_1$, the odd partial tuple is strictly descending and the even partial tuple is strictly ascending (Fig. 10) or: $x_2 > x_1$, the odd partial tuple is strictly ascending and the even partial tuple is strictly descending (Fig. 11).

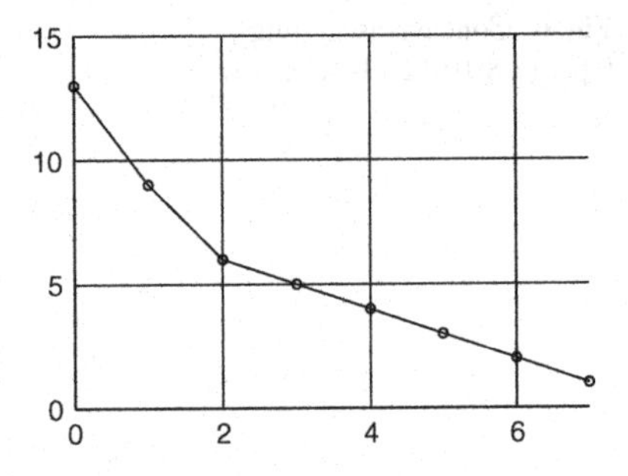

Fig. 9 Convex-descending tuple (13 9 6 5 4 3 2 1)

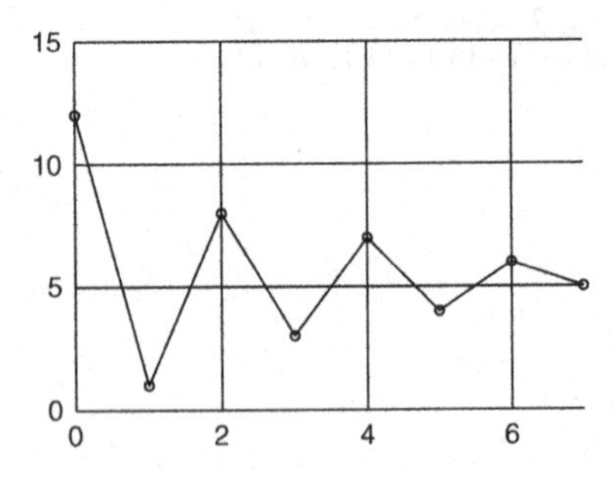

Fig. 10 Strictly converging zigzag tuple (12 1 8 3 7 4 6 5)

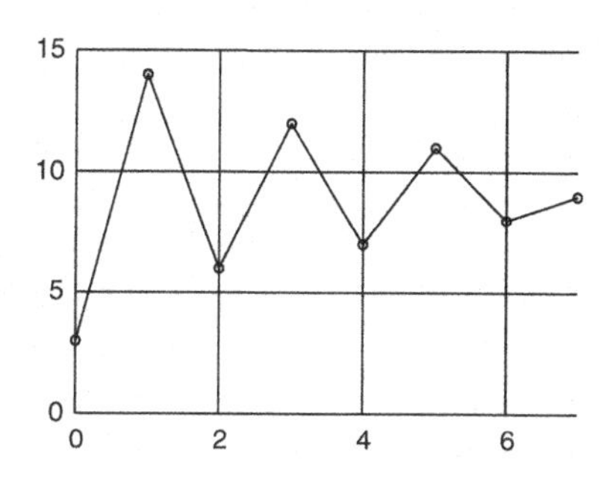

Fig. 11 Strictly converging zigzag tuple (3 14 6 12 7 11 8 9)

A zigzag tuple $(x_1, \ldots, x_k)$ with $x_i \in \mathbb{N}$ for $i = 1, \ldots, k$ is called *strictly diverging* iff either: $x_2 > x_1$, the odd partial tuple is strictly descending and the even partial tuple is strictly ascending (Fig. 12) or: $x_2 < x_1$, the odd partial tuple is strictly ascending and the even partial tuple is strictly descending (Fig. 13).

Specific shapes of convex, concave and zigzag types were chosen, respectively arbitrarily selected, terms and their ordering were changed, see Table 6.

Characteristics of groups of three verbs are abbreviated in the first part of Table 7. The general shape is determined by operations that choose a group of described type according to prefixes UP, DOWN and ZIGZAG_CONV (converging zigzag). Suffixes SAME and PERMUTE describe if the categories given as argument will be taken for all verbs or will be chosen. An optional second argument determines the order of the group's first item. Concave and convex shapes are mixed here.

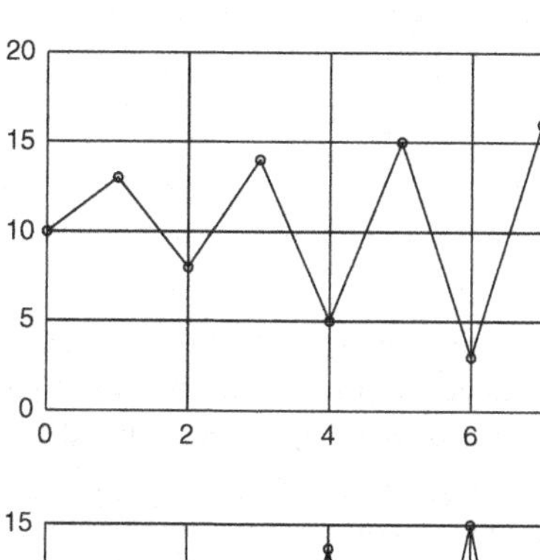

Fig. 12 Strictly diverging zigzag tuple (10 13 8 14 5 15 3 16)

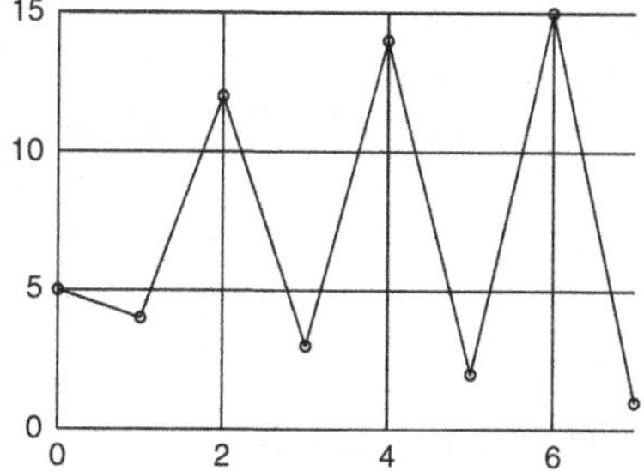

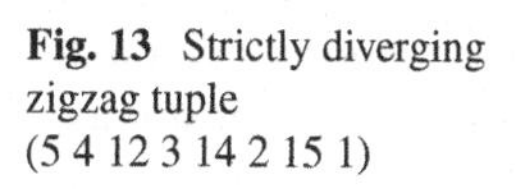

Fig. 13 Strictly diverging zigzag tuple (5 4 12 3 14 2 15 1)

Table 6 Rewriting system in a compressed form

A 1 → rip	B 1 → hieratically_hold	C 1 → graze
A 2 → frazzle	B 2 → pressure	C 2 → scratch off
A 3 → explode	B 3 → amalgamate	C 3 → abrade
A 4 → break off	B 4 → cruelly_synthesize	C 4 → separate
A 5 → pierce	B 5 → implode	C 5 → expose
A 6 → slit	B 6 → mash	
A 7 → prick	B 7 → sew	
A 8 → cruelly_analyse		
A 9 → clasp		
A 10 → destroy		
A 11 → strike		
D 1 → tender_cruel		
D 2 → longing_cruel		
D 3 → cold_cruel		
D 4 → screaming_flashy		
D 5 → vitriolic_toxic		
D 6 → burning		
D 7 → angry		
D 8 → panic_desperate		
D 9 → orgiastic		
D 10 → tremoring		
D 11 → convulsing		
D 12 → gazing		
D 13 → fizzy_quenching		
D 14 → crashing		
D 15 → unconscious		

In the lower part of Table 7 an example is shown for a larger number of groups we produced and from which Gadenstätter chose appropriate ones. With the final kind of phrasing we achieved to build a semantic retriggering structure that approached the composer's rather free search attempts, as he confirmed. To shed light on further compositional aspects which could no longer be derived from an algorithmic modeling, we invited the linguist Thomas Eder to discuss Gadenstätter's compositional approach[5] from the perspective of metaphor theory.

[5] Gadenstätter and Eder have worked together on several occasions to explore the creative use of metaphor theory in the context of contemporary composition.

Table 7 Rewriting system in a compressed form

D 1 | UP_PERMUTE(A, 1) | ZIGZAG_CONV_SAME(B) | DOWN_SAME(C, 5)	
D 5 | UP_SAME(B, 1) | ZIGZAG_CONV_SAME(C) | DOWN_PERMUTE(B, 7)	
D 3 | UP_PERMUTE(C, 1) | ZIGZAG_CONV_SAME(A) | DOWN_SAME(A, 11)	
→	
D 1 | A 1 C 4 B 7 | B 5 B 1 B 3 | C 5 C 3 C 1	
D 5 | B 1 B 4 B 6 | C 3 C 1 C 2 | B 7 A 4 C 1	
D 3 | C 1 B 2 A 4 | A 2 A 4 A 3 | A 11 A 5 A 2	
→	
tender_cruelly	rip, separate, sew
	implode, hieratically_hold, amalgamate
	expose, abrade, graze
vitriolic_toxically	hieratically_hold, cruelly_synthesise, mash
	abrade, graze, scratch off
	sew, break off, graze
cold_cruelly	graze, pressure, break off
	frazzle, break off, explode
	strike, pierce, frazzle

Thomas Eder: A View from Cognitive Metaphor Theory and Weak Synesthesia

The approach to a formal description of how different operations with concepts in the context of a *rewriting system* can be represented, has been found to be highly plausible when, syntactically, only a view of the formal criteria is taken. The system was designed in such a way to produce "correct" solutions, but of course it is not able to feature all the criteria of "semantic" consistency, which in turn has led to a process of reflection in Gadenstätter's work, with respect to his own semantic criteria. A first attempt to approach this semantic dimension could be seen in view of Chomsky's paradigm with its approach of transformational grammar, in terms of deep and surface structures. As Gadenstätter is just interested in the semantic aspects of sound that only a formal grammar—with production rules for words (strings) based on a *vocabulary* (*alphabet*), a set of atomic items—cannot be handled.

Also Chomsky's approach of a transformational grammar has been fiercely criticised by George Lakoff, one of his former students, as being Cartesian and rationalist. The following debate has entered the literature under the label "Linguistic Wars". Lakoff attacked Chomsky's transformational grammar for not coming to grips with the relation between syntax and semantics and for confusing the issues of structural, behavioral and rationalist approaches to linguistics: "those working in the area have found that many of the most basic assumptions of transformational grammar were inadequate and have rejected them, including the following of Chomsky's fundamental assumptions: that syntax is independent of human thought and reasoning, that there exists a syntactic deep structure, that transformational rules are fundamentally adequate for the study of grammar, that syntactic categories are independent of the categories of human thought, that language use plays no role in grammar, that syntax is independent of the social and cultural assumptions of speakers, and many other central positions of Chomsky's that many of us find inadequate, especially in the light of recent research" [4].

Chomsky and his colleagues, among them above mentioned Ray Jackendoff, of course, have replied and rejected Lakoff's critique as being misinformed and as misunderstanding Chomsky's approach. The relation between syntax and semantics is at stake in Chomsky's theory, as he formulates in one of his books: "[T]here are striking correspondences between the structures and elements that are discovered in formal, grammatical analysis and specific semantic functions. [...] These correspondences should be studied in some more general theory of language that will include a theory of linguistic form and a theory of the use of language as subparts. [...] An investigation of the semantic function of level structure, as suggested briefly in Sect. 8, might be a reasonable step toward a theory of the interconnections between syntax and semantics." [1, 101f].

It would go too far to judge the combatants of the Linguistic War—but some of the main issues are important with regard to the research goal of tracing Gadenstätter's production process by formal methods. What is important here is the fact that Lakoff went way beyond Chomsky's analyses by concentrating on the role of metaphors as tools of thought. Lakoff and followers focused on what they have called the "embodied mind".

The essential approach of theory of metaphor after the so-called "cognitive turn", which moved the question of the interdependence of language and thought more in the direction of their independence, unlike language philosophy with its "linguistic turn", is George Lakoff's and Mark Johnson's theory, which states that metaphor is not a matter of mere language but one of the "conceptual system" that is a basic and widely universal feature of the human species. To learn something about metaphor is therefore to learn something about our conceptual system. Metaphor is not only a figure of speech (trope) but a specific form of "mental mapping" that decisively influences the way people think and conclude (infer) in everyday life, that is, not only in poetic or non-literal contexts. According to George Lakoff, mapping is a relation (not, as in mathematics, a function, which would mean that for every element in one set there is exactly one single element in another set) of assigned correspondences between two different "conceptual domains". E.g. LIFE IS A JOURNEY [5, p. 26], one is alive and one is travelling. One's purposes are destinations; i. e. destinations are mapped from the source domain "journey" onto purposes in the target domain "life". Linguistic metaphors are not mere ornamental, communicative devices to describe topics inherently difficult to describe in literal terms. Verbal metaphors reflect underlying conceptual mappings in which people metaphorically conceptualise vague, abstract domains of knowledge e.g. time, causation, spatial orientation, ideas, emotions, concepts of understanding, in terms of more specific, familiar, concrete and concrete knowledge (e.g. embodied experiences). Physically accessible domains are mapped to abstract domains of knowledge (e.g. the concrete, familiar "destination" to the abstract "purpose").

So, basically at stake is the relation of the metaphorical expressions to the respective sounds: how does it come that most people ascribe some sort of sounds to "rip" as opposed to different sounds connected to "frazzle"? I regard the approach of the project very convincing with regard to categorizing complex combinations of single elements and the reception and production of their complex combinations. Additionally the relation of the basic single units to the respective sounds (their semantics in the broadest sense) is in my view best explained by a theory of metaphors and embodiment.

Moreover, the notion of weak synesthesia may be fruitful. The suggestion of (weak) synaesthesia as an appropriate basis for an explanatory theory (production and reception-oriented) goes back to Martino/Marks [8, 9]: The psychological mechanisms termed "weak synaesthesia" are responsible for experiencing analogies of perceptions in different sensory modalities (e.g. auditory and visual).

A currently discussed hypothesis which explains the phenomenon of weak synesthesia assumes that "all of these interactions take place at a post-perceptual locus, after information is encoded or recoded into an abstract format common to both perceptual and linguistic systems" [9]. Other researchers [11], in turn, assume that weak synesthesia has the same neural basis that strong synesthesia has, but just in a milder form.

This weak form of synaesthesia may be the basis for more abstract processes (concept formation). It is the psychological basis for aesthetic processes explaining how formal elements, such as pure sounds, are semantizised. In contrast to the demographically very rare strong "synesthesia" sensu strictu, "weak synesthesia" describes different gradations of a general phenomenon in human cognition. One can create, identify, and appreciate cross-modal connections or associations even if one is not strongly synesthetic, it is no involuntary, automatic process, but it occurs on request. One famous example: Given a set of notes varying in pitch and a set of colors varying in lightness, the higher the pitch, the lighter the color paired with it [7]. This pitch-lightness relation resembles the one observed in strong synesthesia, with one notable difference. In weak synesthesia, the correspondences are defined by context, so that the highest pitch is always associated with the lightest color. Here lies a distinction between strong and weak synesthesia: Although crossmodal correspondences in weak synesthesia are systematic and contextual, those in strong synesthesia are systematic and absolute (display a one-to-one mapping). "Weak synesthesia is characterized by cross-sensory correspondences expressed through language, perceptual similarity, and perceptual interactions during information processing. Despite important phenomenological dissimilarities between strong and weak synesthesia, we maintain that the two forms draw on similar underlying mechanisms" [9]. The issue of sound symbolism, which is closely related to the the characterization of sounds in this project, leads to the following line of thought: sounds of speech sometimes serve in and by themselves to evoke meanings—as if the sounds that constitute a word form part of the semantic content [7, 75f]. Sound symbolism transcends the relatively simple process of onomatopoesis. In onomatopoesis, consonants and vowels of speech actually mimic some naturally occurring, nonspeech sound; well-known examples are onomatopoeic words like "buzz", "crackle", "swish", and "meow". Sound symbolism proper enters the scene when sounds and referents differ, when sounds express some nonacoustical property of nature. Speech sounds can convey sensory meanings, whether visual, tactile, gustatory, or olfactory. This is a question of cross-modal translation, according to which vowels and consonants suggest referents in other sensory modalities by dint of certain psychoacoustic characteristics of speech through the operation of suprasensory attributes (size–intensity–brightness) [7, p. 76]. This notion of suprasensory attributes is at the core center with regard to weak synesthesia. The relation between sound and meaning, in this project between the atomic items and their "meanings", may be an intrinsic one, with meanings coming either from sensations aroused in moving the mouth and tongue when the sounds are spoken, by a kinesthetic process; or from the sounds themselves, by an acoustical process. Sounds can suggest meaning through those of its perceptual attributes that are suprasensory.

From this it could be concluded about the functioning of Gadenstätter's work: a sound event can be intense, so we accept his description (as in Gadenstätter's abbreviations for sound events) in a metaphor, which transfers the terms of intensity from another modality into the sphere of acoustics where we can perceive them as "convincing". It follows that the manipulation of linguistic metaphors and the so-called sound events, as proposed in the present project on the basis of combinations in the phrase structure grammar of Chomsky, relevant aesthetical results are produced. Moreover, the concept of weak synesthesia allows a good approximation of the semantic dimension of Gadenstätter's work with respect to his sound events and their specific descriptions.

Project Review by Clemens Gadenstätter

To accompany my compositional work with analytically-reflexive processes or to sharpen the compositional process with such means is a long-standing practice for me. A new aspect, through the work in this project was my work with inter-related terms—as the trigger for different tonal events—in a sense to observe a formalised approach on a metalevel.

Both the possibilities and different solutions that resulted from the formalisation of my approach within a generative grammar also led to an interesting outsider's perspective for dealing with linguistic metaphors, as well as to subsequently transform them into a number of different tonal events. Where the possible "rewrites" of the system on the whole offer quite useful solutions, i.e. have shown a combination of terms for offering another musical "interpretation", for me it was that only certain combinations of terms were clearly preferred, in others words, were found to be "consistent".

Where the possible "rewrites" of the system on the whole offer quite useful solutions, i.e. have shown a combination of terms for offering another musical "interpretation", for me it was that only certain combinations of terms were clearly preferred, in others words, were found to be "consistent".

Thinking about the combinations of terms, which although they functioned within the framework of the system, I would not have intuitively generated, has sharpened my view on my selection criteria and on the limits of formalisation in mind where the intuitive decision no longer wants to be "rationalised". Through these processes, works have also emerged in parallel with the project, where a new approach arose for me: the concept of combinations representing musical material have undergone an expansion, more complex structures are now manageable for my compositional thinking.

The focus of this project on the various possibilities of combinations of terms, which is an essential starting point in compositional work for me, also resulted in an intense debate on the further steps, such as global form building, structuring arrangements in time, whose polyphonic processes and contents can remain in context. The work on weak synesthesia and the incorporated perception (embodiment) has been in the last years for me of eminent importance in my compositional practice. First, it was the awareness of the multifacted levels of meaning of sound events, from which then I developed a technique of composition which acts directly at the level that triggers the sounds in us. The goal was then to enable the performance of this sound experience, far from the normal labels which are formed from the usual connotations and levels of meaning. In this process it became clear to me that sounds could transform from their traditional layers of meaning through recontextualisation and editing, with the added value that hearing both the original level of meaning as well as the newly developed were then present. This yields, even in the case of a single sound event, the paradox of a polyphony of different levels of meaning and perceptive modalities.

The work in the project and the discussions with Thomas Eder, which began in 2008, opened in me a greater sensitivity with respect to the weighing of different levels of meaning in a sound event. Discussions with Eder were inspirational, as they brought me closer to the idea of transformative performance, which is an adaptation of semantics at the level of the weak synesthesia of the incorporated perception through the perspective of metaphor theory. Basically it is always my goal as much as possible to know about my actions, but also knowing that the greatest part of me will nevertheless remain hidden.

Knowing more about what I'm doing, allows more reflection in the work, about what I'm really doing. This is essential to me: I need to know what my set up really "causes", I have to get to know this potential, that I can determine and in the work lead it to a further "existence". Only in this way can I be responsible for my own actions that it may be aesthetically effective: every tool that helps overcome a blindness towards one's self is of utmost artistic importance.

References

1. Chomsky N (1957) Syntactic structures. Mouton and Co, The Hague
2. Chomsky N (1995) The minimalist program, vol 28. Cambridge University Press, Cambridge
3. Jackendoff R (1972) Semantic interpretation in generative grammar. MIT Press, Cambridge
4. Lakoff G (1973) Deep language. In: The New York review of books 20(1)
5. Lakoff G, Johnson M (1980) Metaphors we live by. University of Chicago Press, Chicago
6. Lerdahl F, Jackendoff RS (1983) A generative theory of tonal music. MIT Press, Cambridge
7. Marks LE (1978) The unity of the senses: interrelations among the modalities. Academic Press, New York
8. Martino G, Marks LE (1999) Perceptual and linguistic interactions in speeded classification: tests of the semantic coding hypothesis. Perception 28:903–924
9. Martino G, Marks LE (2001) Synesthesia: strong and weak. Curr Dir Psychol Sci 10(2):61–65
10. May RC (1978) The grammar of quantification. PhD thesis. Massachusetts Institute of Technology
11. Ramachandran VS, Hubbard EM (2001) Synaesthesia–a window into perception, thought and language. J Conscious Stud 8(12):3–34

Dimitri Papageorgiou/Interlocking and Scaling

Dimitri Papageorgiou, Daniel Mayer and Gerhard Nierhaus

Dimitri Papageorgiou spent the first years of his life under a dictatorship in Greece[1] and developed an aversion towards oppressive authority. He grew up in a house decorated with paintings of his uncle, a talented young artist who, due to financial hardship, had to abandon his dream of becoming a painter and migrated to Australia. Highly impressed by the work of his uncle, Papageorgiou dedicated himself as a young boy to painting, including 3 years of tuition, but a shift to music happened at the age of nine, when he started to play the guitar at the local music conservatory. He was initially drawn to improvisation because of his attitude where he wanted to reject following orders. So, instead of focusing on learning a piece, he would use it as raw material for improvisation.

As his ability to play the guitar developed, he began his first attempts at composing. Some years later he found himself playing blues and rock in various bands, heavily influenced by the music of Pink Floyd, King Crimson, Camel and especially Frank Zappa. An interview with Zappa, in which he expresses his admiration for Stravinsky, Boulez and Stockhausen, introduced him to the world of contemporary music, before he had even had harmony and counterpoint lessons at the conservatory. Listening to the music of these composers was a jaw-dropping experience and his interest in rock declined in favor of contemporary music.

[1] The Greek military junta of 1967–74.

D. Papageorgiou
Department of Music Studies, Aristotle University of Thessaloniki,
Thessaloniki, Greece
e-mail: pdimitri@mus.auth.gr

D. Mayer · G. Nierhaus (✉)
Institute of Electronic Music and Acoustics,
University of Music and Performing Arts Graz, Graz, Austria
e-mail: nierhaus@iem.at

D. Mayer
e-mail: mayer@iem.at

G. Nierhaus (ed.), *Patterns of Intuition*,
DOI 10.1007/978-94-017-9561-6_6

After graduation from high school Papageorgiou moved to Austria to pursue studies in composition and first had lessons with Zbigniew Bargielski.[2] One year later he applied for composition with Hermann Markus Pressl,[3] who introduced him to the concepts of controlled aleatorics and nihilistic mysticism. Another inspiring teacher was Andreij Dobrowolski.[4] Papageorgiou raised some eyebrows when he composed a piece for solo contrabass with a duration of 99 years. Through the influence of Dobrowolski, Papageorgiou explored various concepts of serialism and new and inspiring approaches of musical analysis. By this time he was a founding member of "die andere saite", a composers' society for avant-garde music that had been initiated by the Austrian composers Bernhard Lang and Georg Friedrich Haas. Papageorgiou's compositions were performed for the first time in the concert series of "die andere saite" and he composed music for a theater play, which was commissioned by the Austrian Broadcasting Corporation (ORF) and the "Grazer Gruppe: Forum Stadtpark Literatur". After he graduated in composition at the University for Music and Performing Arts Graz, Papageorgiou returned to Greece. Although he made a living teaching music theory at a conservatory level, this period was not fruitful in regard to his compositional work—cultural life in Thessaloniki was almost non-existent, his works were considered "too difficult" and did not receive performances. To make things worse, shortly afterwards Papageorgiou had to complete a 23 month long mandatory army service.

Things changed when Papageorgiou became acquainted with David K. Gompper, president of the Society of Composers, Inc. (SCI) at that time, professor of composition and director of the Center for New Music at the University of Iowa. Papageorgiou accepted Gompper's offer for a presidential fellowship of the Graduate College of the University of Iowa and moved to the USA to pursue a doctoral degree in composition. In the next years he worked at the University of Iowa as a teaching assistant, taught courses in composition, instrumentation and orchestration, was the assistant of the director of the Center of New Music and was instrumental in organising various festivals, especially for Austrian and Russian contemporary music. His works were performed at various festivals and concerts in Iowa, Illinois, Michigan, Ohio, Florida, California and New York.

After his Ph.D. Papageorgiou returned to Greece—having his first residency in mind—he was now determined to change things. He co-founded the dissonArt ensemble, the first professional non-state sponsored ensemble in Greece, which is dedicated solely to avant-garde music. The ensemble "dissonArt" meanwhile performs all over Europe and through its connection to the Department of Music Studies at the Aristotle University of Thessaloniki has helped to create a new music scene in Thessaloniki and provided student composers an opportunity to collaborate with a professional ensemble. Papageorgiou now works as an Assistant professor of

[2] Zbigniew Bargielski (*1937), Polish composer.

[3] Hermann Markus Pressl (1939–1994), Austrian composer and professor at the University of Music and Performing Arts Graz.

[4] Andreij Dobrowolski (1921–1990), Polish composer and professor at the University of Music and Performing Arts Graz.

composition at the Aristotle University of Thessaloniki. Through various artistic projects and performances he still keeps close ties with Austria. In the last years his works were performed at various concerts and European festivals for contemporary music, most prominently at the 46th Summer Course at Darmstadt (2012) and the Klangforum concert series at the Wiener Konzerthaus (2013).

Artistic Approach

Statement

In my music, I have always had a certain predilection towards various forms of similarity, which are expressed as multiple representations, reconstructions and regenerations in a continuous or discontinuous manner which re-use the same form over and over again. In such a context, structural redundancy and parsimony stand at the epicenter of my interest, as cyclic patterned recurrences of patterns that share the same underlying structure unfold in time, elaborating centripetally on the same structural complex or similar structural features again and again, creating, maintaining and exalting one or various states from a single or from a variety of perspectives.

In recent years, I have created patterns using an additive and combinatorial generative process of interlocking. In each recurrent cycle, however, some elements are retained whilst others are discarded, re-created and re-generate multiple approximations of the original form at various levels and scales, ranging from subtle variations to skewed rearrangements beyond recognition: repetition as a form of change. The power of reiteration—the transformation of the fragmentary into an enduring thing—is not meant to create an "illusion of the pleasant". On one hand, the reiteration is often not overt but rather obscured. The opposition between unity and multiplicity, in terms of the "dialectic of the repeated and the non-repeated" [26, p. 91] on the other hand, often results into a loss of causality and distortion of linearity, thus disrupting musical developments and constructing forking auditory paths for the listener.

Personal Aesthetics

The core of my work revolves around specific themes: memory, time, identity, repetition and order—the fragility of order in the creative process. I have been using repetition, in a variety of ways, since the beginning of my career: repetition as a way of preserving abstract identity; repetition as a means of emphasis; repetition as a launch pad for yet another variation of a musical idea, etc. It is no surprise, as my education stems from the compositional circle of Hermann Markus Pressl and I am, also, a deep admirer of Morton Feldman's music, of Samuel Beckett and Jorge Luis Borges' thematic use of repetition.

In my recent works, however, the use of repetition shifted its focus when I became intensely interested in the subjective act of remembering the past, the fuzziness of recollective experiences, the imperfections and vulnerability of mnemonic phenomena.

"Memories are often ephemeral and distorted, . . . yet subjectively compelling and influential", writes Schacter [28, pp. 20–21]. During the last century, the accuracy of the information conveyed by memory has been severely questioned and often discredited by a number of scholars. In the 20s and 30s, sociologist Maurice Halbwachs and psychologist Frederick Bartlett laid the foundation of modern research in memory. In his Cadres Sociaux de la Mémoire [14] Halbwachs analysed the influence of socio-cultural contexts on the individual memories and proclaimed the idea of memory as a collective phenomenon. In recollection, we do not retrieve images of the past as they were originally perceived, he claims, but rather as they fit into our present conceptions, which are shaped by the social forces that act upon us. In 1932 Bartlett [2], one of the first who examined reconstructive memory, conversely pointed out that remembering is not a simple retrieval of a past event. On the contrary, he claimed that remembering relies on interest-determined "schemas" of the past and memories are often altered or distorted reconstructions derived from these schemes.[5]

Cognitive and developmental psychologists are largely in agreement about the constructive and reconstructive nature of the mnemonic processes. The scientific interest in memory distortion has a long history and a detailed review here would be out of context.[6] For Candau [8, p. 5], memory is more a constantly updated reconstruction of the past than its faithful reconstitution of it. Instead, it is an active, subjective, malleable, and creative process. Schacter asserts that memory stores only fragments of the past that later serve for the mental reconstruction of past experiences [29, pp. 112–116, 345–350], which are susceptible to various kinds of biases and distortions. Bernecker [3, 4] has shown recently that the material in memory may be subject to selective and elaborative processes. Loftus and Loftus [16] reviewed the scientific evidence relevant to the two options regarding how memory works given to subjects in their survey. They concluded that "contrary to popular belief, the evidence in no way confirms the view that all memories are permanent and thus potentially recoverable" [16, p. 420].

Accumulating research has revealed that memory is not always reconstructible in its entirety. Today, it is widely accepted that memory does not operate as a full storage model, keeping carbon copies of the original experiences. Roediger [25] recently proposed an alternative model of memory, focusing on what he calls memory illusions, or memories that depart from the original event. Memory is thus a dynamic and evolving phenomenon. It seems to be in a continuous state of flux: we are temporally tied to the present, therefore we are only able to perceive our past in the light of the present and as we change—the way we perceive our world around us changes—elements that we never paid much attention to may come into the foreground, altering the remembrances of the past.

---

[5] For corroborative evidence of Bartlett's theory, see [7, 23, 34].

[6] See, however, [5, 25, 28] for reviews.

My current work focuses foremost in the generation of tangled structural units that consist of interlocking cells. These cells are atomised to a certain degree, so the structures can be regenerated anew as various re-combinations of their constituent cells on the basis of a highly structured musical lexicon. The intrinsic conflict among the factual and its short- or long-term memory vestiges results in a series of present-minded reconstructions (structural repetitions) of the sound imagery of the past in present contexts. The initial structures (fragments of experience that change over time) undergo perpetual transformations, transfigurations but also distortions, corruptions, even bastardisations or counterfactual simulations (how past experience may have turned out differently) and, ultimately, dissolve in a movement of vestiges and shadows, which emerge and perish immediately in a tragic dance of the impermanence. It is, in other words, like drawing "on the elements and gist of the past, and extract, recombine and reassemble them into imaginary events that never occurred in that exact form" [30], "[...] not to preserve the past but to adapt it so as to enrich and manipulate the present" (Lowenthal, cited in [10, pp. 79–80]). It is all about what is remembered and how it is remembered.

The above are just general consistent characteristics that reflect my musical aesthetics. In each particular work, however, or series of works, I try to employ a different set of principles of design, ranging from very strict structural prerequisites, at times to complete absence of any given rules at the pre-compositional stage of the work.

Formalisation and Intuition

Intuition is a core faculty of human consciousness which has indissoluble bonds to creativity, inspiration, and artistic expression. I have often experienced moments of a *heureka*, a state of awareness—outside of the rational, linear thinking process—of immediate, effortless apprehension of an initial musical idea, an insight to the solution of a particular compositional problem, a "revelation" of the shape or formal process of a work, etc. At times of creative blockage, it is not uncommon for me to immerse into already generated material, until some kind of an idea or concept begins to take shape.

Although intuition is a key element in holistic problem solving, an uncritical reliance on gut-level impressions alone is highly problematic. I do not favor the need for rigor as opposed to open-mindedness. On the one hand, even though I do not disregard the innate sense of what feels right, I am not willing to take it up without protest, without first considering other, opposing ideas. Spontaneous creative displays may be disrupted by emotions and instincts, or may be derivatives of our memory, preconceptions and habitual thinking. On the other hand, I don't think that creativity emerges out of the blue, as some tend to believe. "In reality", as Csikszentmihalyi and Sawyer poignantly observe, "most creative ideas, especially of a discovered kind, are the result of multiple cycles of preparation, incubation, insight, and elaboration, with many feedback loops, [...]" [11, p. 344]. In my view, creative compositional decision-making relies on the conflux of all corridors of insight: rational, intellectual, imaginative, intuitive, experiential and emotional.

Can one formalise intuition? To some extent, I think it is possible. A lot of compositional procedures, rules and laws and abstractions of musical ideas can be expressed in a formalised manner and can be implemented as algorithms. "Music", however, "cannot be confused or reduced to a formalised discipline, even if music actually uses knowledge and tools coming from formalised disciplines, formalisation does not play a foundational role in regard to musical processes" [35, p. 54]. On one side are the "flash out of the blue" moments of creativity that can barely be put in words—at least I am not able to describe them. On the other side, there is the urge of the creative mind to defy the rules and work against methodologies (creative license) in pursuit of originality. In that sense, it seems that the mechanical aspects of composition are well-suited to formalisation, whereas the creative decision-making processes would seem, at least to my understanding, far from being comprehended and formalised and, in fact, I would be very surprised if they ever did.

Examples from Recent Works

There are some basic formal principles that permeate the majority of my recent works, the most important one being an avaricious economy of structural means, which is based on a very limited amount of building blocks: usually, not more than two or three brief and ordered pitch patterns that serve as the sole source out of which the material of the entire work is made. The initial building blocks are subject to an interlacing technique: the building blocks (germinal cells)—including a limited range of possible transpositions, inversions, and retrogrades—are constantly braided (interlocked) with one another in various possible ways, thus creating a larger structural entity. Figure 1 shows the construction of the first six measures of my trio, *In the Vestige of the Present* (2008). The initial building blocks are tuples of pitch classes $a = (0, 1, 2, 3, 4)$ and $b = (0, 10, 1, 11, 2, 0)$. The possible transpositions of the blocks are, for the most part, limited by the factors of transposition 0, 3, 6, 9.

One can easily notice that the blocks remain always unaltered (ordered), but they are broken down into chunks of variable sizes (1, 2, or 3 notes). Chunks of a pair of blocks are interlocked and, ultimately, a larger structural entity is composed by concatenating several such interlocked pairs of blocks. The building blocks themselves never appear in their original form. They are like proper encodings of all melodic constructions of the work and, as they are retrieved to be interlocked in pairs, they generate a plethora of inexhaustible variants of similar as well as more complex melodic patterns, constantly revealing different perspectives within the development of the musical discourse.

The corruptive dimension of the reconstructive nature of memory serves as a model for the unraveling of the form, as the original formative idea is constantly recalled in a series of reconstructions of its past (non-faithful recall of the original) in constantly

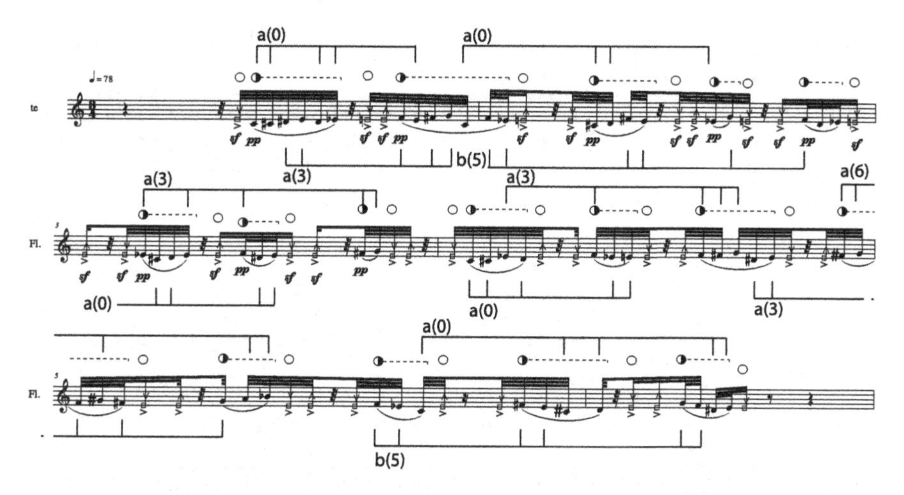

Fig. 1 *In the Vestige of the Present*, bars 1–6

present contexts having, at times, more fidelity to the original, and at times growing "less similar to the original experience" [27, p. 29]. These reconstructions by means of gene recombinations leave only memory vestiges of the original by constantly shifting, diffusing, distorting, transforming, omitting or adding events according to a model of retention/corruption.

Figure 2 shows in a schematic way some examples of reconstructions of an original formative idea made of a series of interlocked pairs of germinal cells.

Such corruptive techniques, however, are not used in a simplistic way. Most of the times, a re-occurrence may be corrupted in many ways. Indicatively, an example of a distorted re-occurrence of the initial structural entity is given below. In Fig. 3, the initial structural entity in bars 1–6 is shrunk to entail only the last three interlocking pairs in bars 17–20 which, in turn, are partly transposed.

Original Formative Idea						**Method of Transformation**	**Reconstructions**					
a(0)	a(0)	a(3)	a(3)	a(6)	a(0)	retention	a(0)	a(0)	a(3)	a(3)	a(6)	a(0)
a(0)	b(5)	a(0)	a(0)	a(3)	b(5)	corruption	a(0)	Rb(5)	a(0)	Ra(0)	a(3)	b(5)
a(0)	a(0)	a(3)	a(3)	a(6)	a(0)	retention and shrinkage				a(3)	a(6)	a(0)
a(0)	b(5)	a(0)	a(0)	a(3)	b(5)					a(0)	a(3)	b(5)
a(0)	a(0)	a(3)	a(3)	a(6)	a(0)	retention	a(0)	a(0)	a(3)	a(3)	a(6)	a(0)
a(0)	b(5)	a(0)	a(0)	a(3)	b(5)	permutation	b(5)	a(0)	a(0)	a(3)	b(5)	a(0)

Fig. 2 Examples of transformations

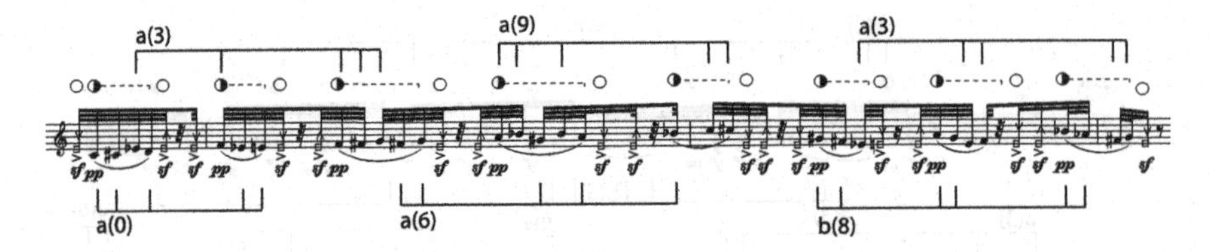

Fig. 3 *In the Vestige of the Present*, bars 17–20

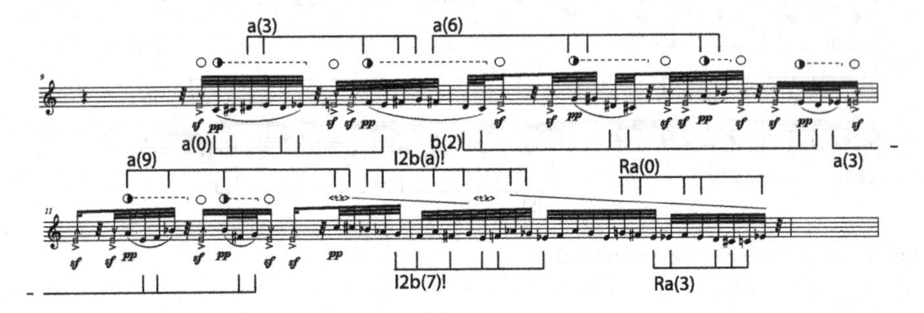

Fig. 4 *In the Vestige of the Present*, bars 9–12

The first pair is retained in its original form, the other two interlocked pairs, however, have been transposed by a minor third higher.

bars 1–6	a(0)—a(0)—a(3)—a(3)—a(6)—a(0) a(0)—b(5)—a(0)—a(0)—a(3)—b(5)
bars 17–20	a(3)—a(9)—a(3) a(0)—a(6)—b(8)

The above example, Fig. 4, exhibits a close connection to the previous one, in that it retains its structure in part, see a(3)/a(0)—a(6)—a(3), twisting and extending it with the addition of a "cadential" downward run, I2b(a)!/I2b(7)![7]—Ra(3)/Ra(0).

Evaluation and Self-reflection

I consciously employ a bottom-up approach to my compositional process, some kind of explorative composing, in which "form is a result—a result of a process" [36]. I start just with approximate ideas about the desired goals. Even the basic material is loosely defined from the outset, and may be subject to minor or even significant

[7] I2b(a)!/I2b(7)! are not strict transformations of cell b. They constitute a deviation from the norm, as they rather present a mixture of inversion and retrograde. This deviation has been used at this point because it seemed to better serve the melodic flow of this "cadential" figure.

modifications, as goals emerge gradually during the construction of a composition in a trial-error process. Decision-making comes in moment-to-moment steps, exploring many possibilities, searching and trying out combinations at various structural micro/macro levels that are deduced from a generative element. This process runs in cycles of generating, planning-drafting, revising-refining, putting together and, in that sense, composing for me is an activity, which is relived over and over again until I am satisfied by the by-product of this process. Post-composing-session reflection and evaluation of the generated material accompany each cycle of the process.

At moments when intuitive compositional decisions start shaping the material towards a certain direction, I need to step back for a moment and try to evaluate my course of action. I start imagining how things can be done another way and explore alternative perspectives of shaping the same material.

Criteria of self-evaluation:

1. The primacy of the ear—the inner ear is the primary force for evaluating a composition, music is ultimately an aural art that confines to an aural logic.
2. Aspects of novelty—non-reliance on musical memories and influences.
3. Economy—do it with the least possible means.
4. Clarity of formal procedures.

Furthermore, self-reflection and evaluation is enhanced by feedback by fellow composers, performers and audience. After the first performance of the piece, I take some time to sum up the entire experience, as I reflect on what I did well and what I need to improve on in future works.

Project Expectations

The present research project gives me an opportunity to reflect on my own work (prior to, during or post-composition) formally, explicitly and consciously. This fact is rewarding per se in the sense that I am hoping to gain valuable insight and better awareness—more than I usually have when I work alone—of the inner workings of my compositional practice and re-examine my personal views as a composer. The feedback that arises from this procedure will hopefully help me to improve my expertise, refine my compositional tools and my own problem finding and solving. Self-reflection is essential for artistic and creative development. The process of rationalisation and externalisation of compositional strategies not only has widened my horizon as a composer, it has also been useful to me as an educator, for it furthered my ability to discuss with my students about their development of their own compositional strategies. Self-concerned motivations aside, I'm expecting that our research project will give us a better understanding on the ways in which intuitions operate in the context of musical composition, a topic which only recently has gained significant scientific attention.

Exploring a Compositional Process

POINT: What is the actual focus of the compositional work, the starting point for formalisation within the project?

Papageorgiou: In my attempt to coin a personal way of musical expression, for quite a while now, I have been intensively concerned with the systematisation of the structural use of the material in my compositions. In this intellectual engagement with the musical material, my attention and inquiry is mainly focused on creating variance out of invariance, accumulations of reformulations of the same or similar material, based on a twofold axis: (a) the use of a limited set of germinal cells as the source of the whole piece, which is enriched, transformed and proliferated by means of an interlocking technique—some sort of a cross-breeding method—and (b) a basic macro/micro-rhythmic structure, which is based on a series of a durational ratios, relationships of attacks that permeate compositional aspects at various temporal levels by means of scaling.

In the following, I will touch on some of the key concepts and issues of my compositional techniques. Specific examples will be drawn primarily from my work *Intrascalings* (2013), a trio for clarinet, double bass, and marimba, which was commissioned by the ensemble Et cetera (USA). In this work I make extensive use of algorithmic routines that have been implemented in the frame of the project *Patterns of Intuition*, before applying them to the composition of my *Quasi* (*ébauche*), for string quartet, the work which is actually commissioned for the current research project.

POINT: In order to model Papageorgiou's compositional approach we developed routines to generate interlockings in one or more layers with constraints as well as routines for scaling and concatenating rhythmic structures. What is the difference between the use of pen-and-paper procedures and computer algorithms in composition, especially concerning your own work?

Papageorgiou: I use algorithmic tools, developed in the course of the project, to generate material at a micro-structural and meso-structural level, that is then used for the preparation of instrumental scores. "It takes two to invent anything", observes the poet Paul Valèry, "The one makes up combinations; the other one chooses, recognizes what he wishes and what is important to him in the mass of the things which the former has imparted to him" [13, p. 30]. The computer makes up the combinations and I make the choices. As humans, we incorporate external tools in our activities all the time, in order to overcome our physical or mental limitations. Tools are not passive or neutral agents through which our creativity flows unimpeded. Quite on the contrary, they have an impact on us and exert a significant influence on our decision process.

Both the interlocking and scaling techniques might have also been implemented via non-digital means. The framework, however, within in which we work enables us to follow certain directions while, at the same time, it prevents us from even thinking to pursue others. Due to the combinatorial nature of the techniques I am using, non-digital compositional methods, although possible, are far from being a suitable not to mention an optimal solution to harness such a labyrinthine terrain of innumerable possibilities. The invention of the "accordion-like" technique for example (see below) was a product of this human-computer interaction. The creation of dilations and contractions by constantly accumulating or discarding material poses a combinatorial problem that cannot be easily solved within a non-digital framework.

Definition of Germinal Cells

As far as the pitch content is concerned, the principal technique of interlocking is based on a set of germinal cells. Given a limited material as input—usually two or three different ordered pitch cells—linear arrays of cells are formed which are subsequently braided with one another, constantly regrouping into new formations in order to engender inexhaustible variants at the output.

The basic material of *Intrascalings* originated from an 8-note series with a zigzagged melodic curve, that occurred to me intuitively. This descending pitch sequence exhibits an interesting intervallic constellation, as it consists of two intervalic cells, x and y in Fig. 5, and contains all chromatic pitches within the range of an augmented fourth (C-F#), with the exception of the pitch E. Furthermore, intervallic cells x and y are related to each other by intervallic diminution, as all intervals of y are by one semitone smaller than those of x: $x = (4, 3, 2)$, $y = (4 - 1, 3 - 1, 2 - 1) = (3, 2, 1)$.

At first, I have developed a chordal progression, which is based on superposition of various forms: Original (O), Retrograde (R), Inversion (I), Retrograde Inversion (RI) of the two initial intervallic cells x and y (3 intervals and 4 pitches) as shown in Fig. 6.

Fig. 5 Initial intuitive idea

Fig. 6 Marimba chord progression

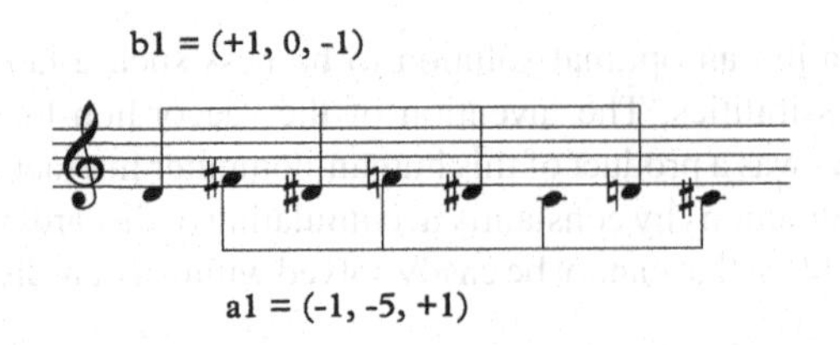

Fig. 7 Derivative interlocking cells

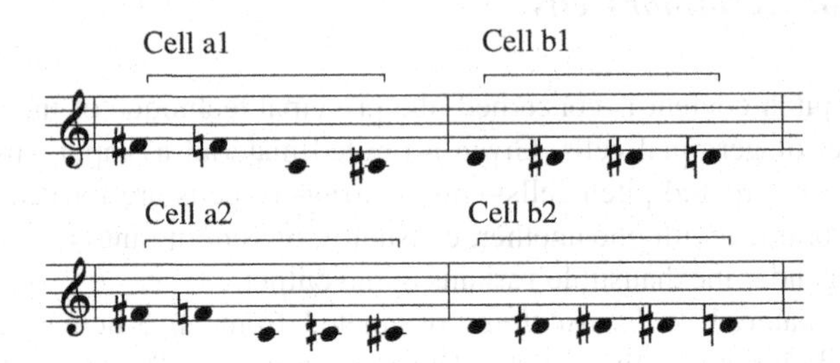

Fig. 8 Final versions of the germinal cells

Upon further thought, however, I deviated from the original idea as I extracted out of this 8-note series two interlocking and symmetrical, non-invertible, intervallic cells (Fig. 7): a1 $= (-1, -5, +1)$ and b1 $= (+1, 0, -1)$, from which I derived the raw material, the germinal cells of the work. Finally, I inflected the germinal cells with microtonal passing notes, extending them by one note, thus creating two variants, a2 and b2 (Fig. 8).

There are some basic principles and structural constraints that permeate the use of the basic melodic cells and their proliferation when interlocked with each other:

1. The order of the elements of each individual cell is important. Cells appear always in ordered form.
2. The germinal cells may appear in prime, retrograde, inversion, and retrograde inversion.
3. Erroneous recollections of the original cells can also be deployed, in the form of shortened versions, without disrupting the order of the pitch material, however.
4. The cells are subject to a rather limited number of admissible transpositions.

The precepts for the employment of two versions of the germinal cells is practical, of course, due to the fact that the marimba can only play chromatic material, whereas the double bass and the clarinet lines can be inflected by quarter-tones. Only four distinct transpositions are allowed of the germinal cells, namely by 0, 3, 6 and 9 semitones, upwards or downwards, as well as their compound versions. The reason behind this choice of a systematic transpositional pattern becomes evident upon a closer look. The concatenation of Cell a (a2) and Cell b (b2) results into an ascending quarter-tones segment that ranges from C to D#. Subsequent transpositions of the

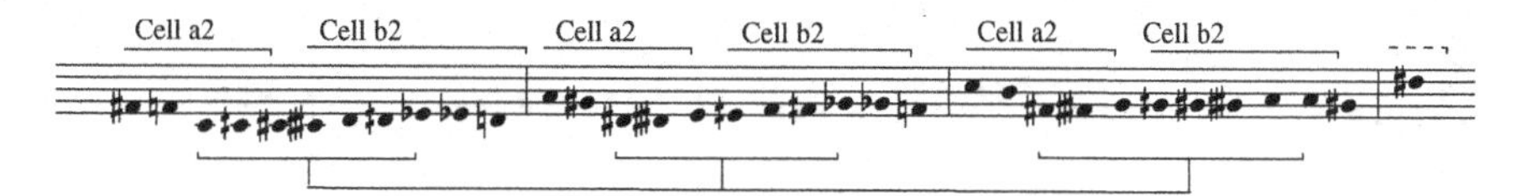

Fig. 9 Transpositional scheme of Cells a and b

germinal cells by multiples of 3 will fill in the octave, after three steps, with an underlying quarter note scalar arrangement, as you can notice in Fig. 9.

POINT: What are the principles you use to develop one monophonic line out of a few germinal cells?

Papageorgiou: Linear events in *Intrascalings* that have been constructed with the use of algorithmic routines, are based on two germinal cells (a1 and b1) and their microtonal versions (a2 and b2), as shown in Fig. 8. There are two types of monophonic line construction in *Intrascalings*: (a) serial (static), and (b) dilating-contracting (dynamic). Serial (static) type is based on the following generative premises:

1. Constructed pitch arrays of various sums by concatenating pairs of germinal cells.
   - Arrays are decomposed by chopping up their content into chunks of various sizes.
   - The chunks are then reassembled by interlocking the groups of chunks seriatim in an orderly fashion (begin with chunk 1 of group a, then chunk 1 of group b, followed by chunk 2 of group a, then chunk 2 of group b, etc.).
2. The interlocking arrays are put in to order to create the full-blown form of the section.

An example of serial construction can be found in the first part of *Section I* (bars 1–41). The construction of the pitch material in that section is based on three series, S, T, and U. All three series are formed by interlocking triplets of concatenated pairs of germinal cells, which entail diverse combinations of transformations (P, R, I, RI) and transpositions, as shown in Fig. 10.

Fig. 10 Series S, T, and U

Fig. 11 Five similar variations of series S

The construction of the three series S, T, U is based on only five distinct concatenated pairs of cells, which are indicated as A, B, C, D, and E in Fig. 10. The attentive reader may have already noticed that each new series begins with the last two concatenated pair of cells of the previous one. The interlocking of the three pairs of cells in each series results into some thousands of possibilities. After reviewing the generated material, I chose a limited number of possible interlockings for each series. To help the reader understand the similarity of the results that the interlocking technique generates, Fig. 11 puts on display the five interlockings of series S that have been selected from a pool of 7,966 possibilities. Notes with stems up belong to the pair of cells A, those with stems down belong to B, and these with no stems to concatenated pair C. Each aggregate is subdivided in chunks of 1, 2, or 3 notes and the chunks are subsequently interlocked. All variations are quite similar. They begin and end with the same notes. Obviously, they contain the exact same notes but in different arrangements. They share same or similar note groups of 3, 4, and sometimes more pitches.

The same procedure has been followed for the construction of three interlocking possibilities of Series T and four possibilities of Series U. Subsequently, the three series were put in order, to construct the full-blown form of the monophonic line in *Section I*, as follows:

S1—S2—T1—U1—S3—S4—T2—U2—U3—S4—T3—U4

Fig. 12 Beginning of the monophonic line

An excerpt from the beginning of the monophonic line, which in *Section I* is given almost exclusively to the double bass, is displayed in Fig. 12.

The second type of monophonic construction appears in *Section II* of the work. Contrary to the procedures in *Section I*, in which only triplets of pairs of cells have been interlocked with one another, the cells here are interlocked in sequences of variable numbers of concatenated pairs, using an additive-subtractive technique that creates dilations and contractions of linear statements—essentially an "accordion-like" structure. In the following abstract example (Table 1), the reader may notice that the number of concatenated pairs A, B, C, . . . , M increases, resulting into 3-ply, 4-ply, 5-ply, 6-ply, and 7-ply interlockings.

The unfolding of such accumulative or dissipating structures does not need to be linear, that is, constantly accumulating or constantly dissipating. Instead, although the unfolding of such structures may have a clear accumulative or dissipating tendency, it may follow a more twisted path by occasionally reversing the tendency.

The pitch structure of *Section II* is based on the five concatenated pairs used in *Section I* (A, B, C, D, and E, see Fig. 10) with the addition of two more pairs (F, G). This series of seven pairs is repeated twice, each time permuted and at the same time transposed a minor third higher, as follows:

Table 1 "Accordion-like" structure

3-ply	4-ply	5-ply	6-ply	7-ply
[A, B, C]	[B, C, D, E]	[C, D, E, F, G]	[E, F, G, H, I, J]	[G, H, I, J, K, L, M]

$$A, B, C, D, E, F, G$$
$$B+3, C+3, D+3, E+3, F+3, G+3, A+3$$
$$C+6, D+6, E+6, F+6, G+6, A+6, B+6$$

Then, segments of this series containing a variable number of concatenated pairs, ranging from a minimum of two to a maximum of eleven pairs, are interlocked with each other:

2-ply	3-ply	3-ply	5-ply
[C + 6, E]	[E, D + 6, D]	[F, C + 6, E]	[A + 3, F, C + 6, E, D + 6]

7-ply	4-ply
[G + 3, G, A + 3, F, C + 6, E, D + 6]	[G, A + 3, F, C + 6]

POINT: What are the specific aspects of performing interlockings with computer algorithms?

Papageorgiou: The interlocking technique is a systematic transformation plan that takes some raw material and creates structurally and aurally similar pitch arrays. These are then combined with each other or contrasted by other groups of interlocked material, thus creating the full-blown form of the work.

In its most basic form the algorithmic routine that creates interlocked linear structures is set up to produce a lot of raw material. In some cases, thousands of possibilities are engendered. Generated interlocks are more or less within a range of acceptability, some of their implementations, however, sound more interesting than others. A great number of possibilities are reviewed and analysed. From them some are selected, creating many, choosing the preferred or the best for further manipulation and some sub-routines have been added to facilitate the review and selection process:

(a) Random: the routine may be instructed to generate one random possibility at a time. This facility is used only in the initial stages of the reviewing process, just to get a feel of the generated possibilities.
(b) By range: since possible interlockings are sorted out in an orderly manner, adjacent entries are quite similar. To avoid the burden of having to review all entries, one by one, it is more efficient to review non-adjacent parts of the whole by selecting specific ranges, i.e. from 10 to 15, from 35 to 41, from 63 to 67, etc., skipping the possibilities in-between the selected ranges. This statistical reviewing method gives adequate information about different variations of the interlocked structures. Of course, ranges can be fine-tuned over and over again, until sufficient material has been reviewed and enough information has been gathered so as to make informed choices of specific implementations.
(c) By precision: any entry from the total pool of possible results can be brought forward and stored aside for further use.

Alternatively, various constraints can be introduced in the algorithmic routines in order to restrict the enormity of possible solutions according to specific criteria. The more constraints we introduce the more regimented the sorting of structures become. In general, the choice of constraints is very much a creative process inherent to composition. Possible constraints include:

1. The number of subarrays in which given arrays can be subdivided.
2. The possible lengths of subarrays in which given arrays can be subdivided:
   - An array of admissible lengths may affect all given arrays, or
   - Lengths can also be defined individually for each array.
3. The order of the interlocked arrays may be defined, e.g. when interlocking two arrays m and n there are two distinct possibilities of arranging the interlocked segments, either starting with m or n.

Some criteria for the selection of arrays:

1. Difference in the length of the selected aggregates (different sums).
2. Avoidance of sequential or quasi-sequential structures.
3. Melodic contour.

Criteria for grouping a number of different variations according to their associations or disassociations:

1. The retention or the avoidance of similar intervallic structures.
2. Avoidance of often recurrences of the same beginning and same ending.

Scaling

POINT: What are your guiding principles on the level of "time organisation"?

Papageorgiou: I am very interested in the conflict between variance and invariance among different—two, three, or even more—superposed rhythmic structures. It is very appealing that the antagonistic relationship of such perpetually interwoven rhythmic structures and their metric ambiguity results from a limited number of sets of invariant data, each one set independently in motion from the others. In this respect, work by composers who have extensively explored different aspects of polyrhythms, such as Messiaen [20, pp. 22–24] with his strict use of rhythmic periods, superposition of a rhythm upon its different forms of augmentation and diminution, superposition of a rhythm upon its retrograde, has been influential. Ligeti's *pattern-meccanico*, [9, 194pp] Steve Reich's phasing and resulting patterns [32, 33], etc. All of them were led to work with polyrhythmic structures under the influence of exotic musical paradigms from diverse heritages such as African, Indian, Balinese. In the theoretical field, on the other hand, Simha Arom's monumental treatise *African Polyphony and Polyrhythms* [1] sheds extensive light into the use of polyrhythms in the indigenous vocal and percussive music of Central Africa.

I, on the other hand, often work with polyrhythms that are based on self-similar temporal schemes, whose structure recurs over differing time scales and different parameter ranges: a structure which is based on a series of durational ratios relationships of attacks that permeate compositional aspects at various temporal levels by means of scale invariance. Although fractal thought can be traced back to Koch's snowflake (1904) [37], the notion of self-similarity was first conceived by Richardson [24], as Mandelbrot points out in his paper "*How Long Is the Coast of Britain? Statistical Self-Similarity and Fractional Dimension*" from 1967 [18] and it was Mandelbrot [19] who later elaborated further this notion into a theory. In several disciplines, as diverse as mathematics (e.g. mathematical models such as Cantor sets and Weierstrass functions), biology (e.g. the formation of stripes in zebras and numerous other patterns, the design of the snail's shell, etc.), physiological ecology [12], etc. or the financial world (e.g. random walks in the stock market [6, 17], objects (e.g. the Romanesque Broccoli), laws and natural processes (e.g. the clustering of galaxies, turbulent flows, shapes of clouds, rain areas, etc.) have often been found to exhibit *self-similarity*, an invariance with respect to *scaling*.[8]

POINT: You speak about polyphonic coordination, which seems to be a central aspect for you. Do you have something like rhythmical germinal cells in a monophonic sense?

Papageorgiou: In my works, scaling as a compositional technique is employed so that temporal structures of actions remain invariant over substantial changes of speed scale and so, the statistical structure of the work remains consequently the same at different measurement scales. Currently, I limit myself to using a single series of durational ratios that organises the rhythmicity of various streams of actions. Having said that, I have to make clear at the same time that, my concern with regard to scaling is not to find a linear problem-solving strategy that depends on a single logical generative construct. This kind of automatism is alien to me. I am not interested in implementations that apply a uniform scaling pattern rigorously and monotonously to all temporal components of a work. On the contrary, different parts may be scaling differently depending on compositional situations, processes and strategies that evolve locally within a piece.

In order to outline some aspects of rhythmic organisation strategies, which have been implemented with the assistance of algorithmic routines that have been developed during the present research project, I will refer to my work *Intrascalings*.

Polyrhythmic Construction of Section I of Intrascalings

POINT: How does polyrhythmic construction work, is it bound to a number of different instruments?

Papageorgiou: Not necessarily. It can also be implemented on solo instruments, so as to create a kind of compound polyphony. As a matter of fact, this type of construction was first employed in my work "*…anD…*" (2012) for solo viola. In

[8] The reader should consult the book *Fractals, Chaos, Power Laws: Minutes from an Infinite Paradise* [31], which is an comprehensive source of information with regard to self-similarity.

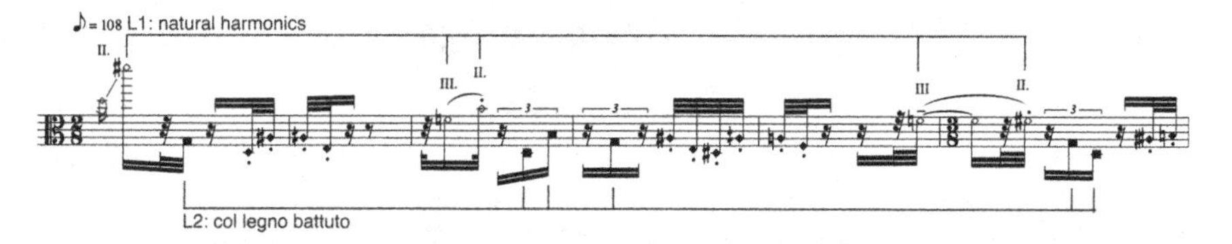

Fig. 13 "*. . .anD. . .*" for solo viola: layered patterns of activity

that piece the pitch content is based on a monophonic linear structure, an array of interlocked arrays of germinal cells, out of which predetermined pitches are extracted and given a specific timbral quality (natural harmonic, thus creating different layers of activity. Figure 13 shows three such layers: (**L_0**) spiccato articulated riffs and runs in capriciously pulsating 32nds , (**L_1**) staccato natural harmonics, and (**L_2**) percussive sounds (col legno battuto).

The contrasting timbres that have been assigned to the design layers, as well as their different registral positioning, create an effect of independent interwoven lines at extremely precise points, which are in constant conflict with one another, thus maintaining an intense and escalating energy clear until the end of the work. Although the work was successful, in my opinion at least, I found this particular structural solution not completely satisfying, for the rhythm of the independent layers resulted as a random by-product of the conjunction of pitch and timbre. The optimal solution to this problem would have been a device that would allow me to control the rhythmical aspect, as a means for the arrangement of layers and the distribution of various elements (timbres, articulations, dynamics, etc.) that can be woven into the texture. Moreover, this structural device should be somehow similar to the way I treat the pitch by manipulating a very limited material at the outset.

POINT: How is the polyphonic rhythmical grid combined with the monophonic sequences of pitches?

Papageorgiou: In pursuing this thought even further, it didn't take me long before starting to consider polyrhythms as a possible solution to my structural concerns. Since the various rhythmic figures within such a polyrhythmic context should rather be derived from a single temporal germinal cell, scaling came up in the line of reasoning as well, so that, based on a single series of durational ratios, two or more superposed rhythmic figures at various time-scales, each of which is so articulated that its onsets do not coincide with those of other rhythmic layers (hocket), create an interwoven effect of dilating or contracting layers of textural activity.

This type of construction has been implemented for the first time in *Intrascalings*. The texture of *Intrascalings* is based on a constant flow of a 1/32 pulse that is confronted by multiple, polyrhythmic temporal streams of different and varying densities, so that continual interruptions of the underlying 1/32 pulse occur as the various lines of activity intersect, see Fig. 14. This kind of construction, with its general rules and constraints:

(i) rhythmic manifestations derived by scaling,
(ii) principle of interweaving and interlocking,

Fig. 14 *Intrascalings*, bars 1–9

(iii) combinatorial problems resulting from the superposition of different figures,
(iv) coordination with the monophonic layer,

that generate complex data out of rather simple input material and simple rules, could be better served by an algorithmic, as opposed to the traditional paper and pencil approach.

For reasons of brevity, I will focus on the construction of the polyrhythmic layers in *Section I* of *Intrascalings*. The basic level of micro-rhythmic organization throughout *Section I* is a continuous pulsating thirty-second raster, which I label as $\mathbf{L_0}$. In addition, the texture of *Section I* is composed of layered patterns $\mathbf{L_i}$ of activity, partly recurrent and intersecting lines of conflict that draw their pitch content from

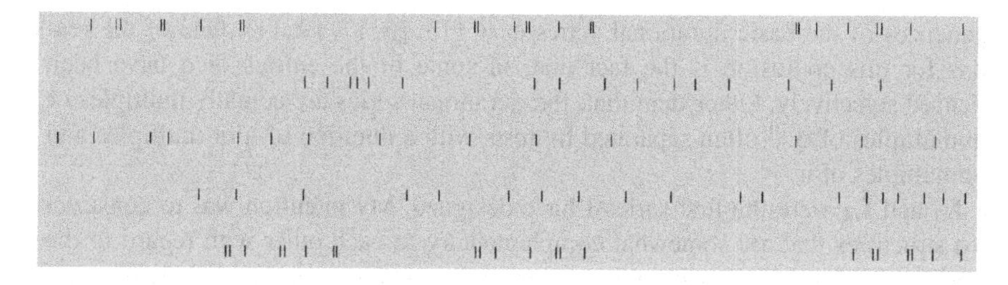

Fig. 15 Distribution of events in layers L_1–L_5

the monophonic pitch structure I described above and interrupt its flow or set accents against the prevailing pulse. Such rhythmic constructions focus on the hierarchical opposition between unity and multiplicity. Whereas multiplicity is produced by the superimposition of several periodical or quasi-periodical structures, unity is ascertained by the continuous underlying pulse and the high degree of similarity of the interlocked melodic/harmonic structures. Ultimately, the whole structure is made up of discreet strips of "polyrhytmia", "eurhythmia", and "arrhythmia"[9] [15] woven individually and then assembled side by side to provide the finished form of the section.

Section I is composed of five discrete layers of events over the underlying $\mathbf{L_0}$.

- $\mathbf{L_1}$: marimba chords derived from the chordal progression of Fig. 6.
- $\mathbf{L_2}$: sforzando articulations.
- $\mathbf{L_3}$: octave dislocations of selected pitches of the underlying pitch structure, which appear as natural harmonics in the double bass line.
- $\mathbf{L_4}$: octave dislocations of selected pitches of the underlying pitch structure, which appear as staccato notes in the altissimo register of the clarinet.
- $\mathbf{L_5}$: marimba staccato notes.

Figure 15 shows how the events in the five layers are distributed in the course of *Section I*. The reader may notice in the graph below that, the overall distribution of events occurs roughly in five stages. *Section I* begins with a gradual exposition of the layers until a significant density of layers is reached with the presence of $\mathbf{L_1}$, $\mathbf{L_3}$, $\mathbf{L_4}$, and $\mathbf{L_5}$ (stage 1). With $\mathbf{L_2}$ entering for the first time, the density decreases rapidly (stage 2) and begins a new build up until the middle of the section (stage 3), in which the maximum density of the section is reached with the presence of all layers simultaneously. In stage 4 $\mathbf{L_1}$, $\mathbf{L_3}$, and $\mathbf{L_5}$ cease, leaving $\mathbf{L_2}$, and $\mathbf{L_4}$ alone, before a new, rapid build up (stage 5) concludes the section.

At first sight, the structure of these series may seem confusing or even arbitrary. Each one of them, however, is based on a simple formal scheme using nested

[9] In his book [15] Lefèbvre refers to three theoretical categorisations that can be applied to aggregate rhythms: "polyrhythmia" (two or more rhythms that are not perceived as deriving from each other and are, therefore, in conflict); "eurhythmia" (rhythms characterized by a harmonious relationship); and "arrhythmia" (rhythmic irregularity).

sequences of the basic durational series $q = (13, 23, 17, 19, 11)$. One of the reasons for this confusion is the fact that, in some of the entries of q have been silenced selectively. Other than that, the durational series are actually multiples or submultiples of q,[10] often separated by rests with a duration of q or multiples and submultiples of it.

L_3 and **L_4** were the first series I have designed. My intention was to construct two structures that are somewhat complementary to each other with regard to the distribution of the events within the section. Whereas $\mathbf{L_3} = (1/2.5, 1/2, 1, 1) \times q$[11] exhibits a gradual decrease of density of the events, as it becomes apparent from the reduction of the divisors from 2.5 to 2 and, finally, to 1, $\mathbf{L_4} = (-1/1.5, 1, 1/1.5, 1/1.5, 1/2, 1/2.5) \times q$[12] has an even distribution of events almost throughout the section and almost up to the end of it, at which point its density increases as the divisor increases to 2.5.

L_1, on the other hand, is a symmetrical construction that employs three concatenated pairs of q, although asymmetry is infused into the structure by using different scaling factors of q (divisors 1.5, 2, and 2.5), so that the concatenated pairs have different densities in comparison to each other. The three rhythmic figures are separated by two silences, each of them having a total duration of q. Left aside the silencing of some of the attacks positioned differently in each q, the structure can be rewritten as $\mathbf{L_1} = (1/1.5, 1/2, -1, 1/2, 1/2.5, -1, 1/2.5, 1/1.5) \times q$, which is almost symmetrical. Moreover, some of the onsets of the rhythmic figures are silenced, thus resulting into an increase of density towards the end of the section.

$\mathbf{L_2} = (-2.5, 1/2, 1/1.5, -1, 1/1.5, 2, -1/1.5) \times q$ begins later in the section (see the long -2.5 rest in the beginning) and ends before the end of the section with a rest. In between, there are again two concatenated pairs of different scalings of q, that are separated by a rest with the length of q. Here, too, some of the elements are silenced, selectively. Finally, $\mathbf{L_5} = (-1.5, 1, -1, 1, -2, 1) \times q$, the last of the series I have designed, in chronological order, is a simple construction that does not need any further elucidation. All in all, the five temporal series are constructed by repetitions of onset sequences or silences of q, which have been dilated or contracted by four multiplication/division factors (1, 1.5, 2, 2.5), organised in different combinations.

Since, on one hand, all layers consist of very short events, and the routine that has been employed results into long or longer sustained notes, the need arose for programming another routine, of a mere practical purpose—to avoid manual manipulation after importing the results into a music notation software—which constrains all long non-rests entries to a desired length—in this case 1/32—thus leaving only the onsets.

---

[10] To avoid too complicated rhythmic subdivisions, I traded off accuracy against flexibility, in as much as scaled durational patterns could be normalised at a 1/32 metric level.

[11] '×' denotes a concatenated scalar product, e.g. $(1, 1/2, 1/3) \times (a, b, c) = (a, b, c, a/2, b/2, c/2, a/3, b/3, c/3)$.

[12] The minus sign indicates a rest, here with a duration of $1/1.5$ of q's duration $= (13 + 23 + 17 + 19 + 11) * 2/3 = 166/3$.

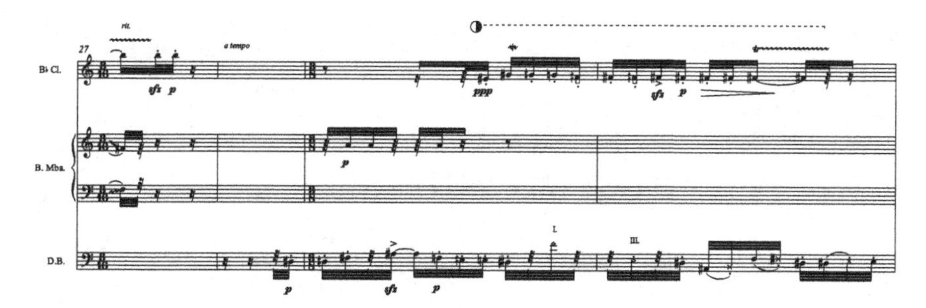

Fig. 16 *Intrascalings,* bars 27–29

POINT: What were your experiences in finding appropriate scalings for your polyrhythmic structures?

Papageorgiou: Although the routine that I have employed for the calculation of the distribution of the event in the five layers of *Section I* is very powerful, the construction of the durational series was a time consuming process. It took me about three weeks of experimentation and countless revisions before I reached the final version of the construction. The main problem was to find a proper syntax for the design of the different dilations and contractions of event distributions at the desirable timing. At the same time, it was equally important to avoid a feeling of a mechanical or even processual construction and achieve some variety in the groupings of the layers. To get a better insight of the polyrhythmic technique, a rather lengthy example is given in Figs. 16, 17, 18, 19, 20, 21, 22, 23 and 24, showing the polyrhythmic distribution of layers 1–5 (marimba chords, sfz articulations, octave dislocations either as double bass harmonics or clarinet notes in altissimo register, and marimba), and the resultant dilations and contractions of the layers.

POINT: The relation between identity and variance has often been discussed in philosophy and music history. We referred to that discourse in the context of Alexander Stankovski's working with mirror techniques. As we know "Einheit in der Mannigfaltigkeit" (unity within variety) was especially praised in the classical era. Do you feel obliged to an ideal of getting everything out of one, would a break

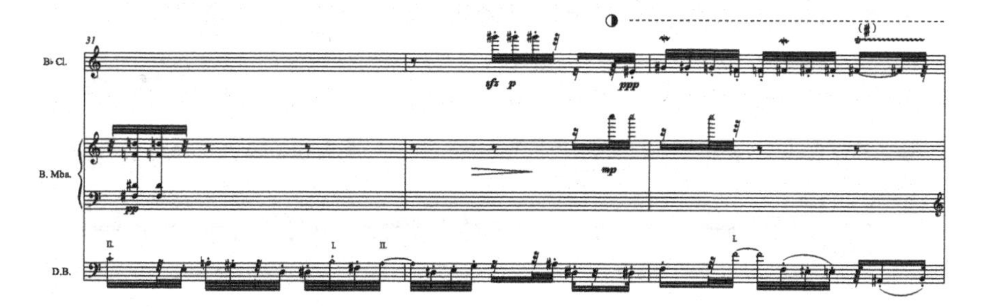

Fig. 17 *Intrascalings,* bars 31–33

Fig. 18 *Intrascalings,* bars 34–37

Fig. 19 *Intrascalings,* bars 38–40

Fig. 20 *Intrascalings,* bars 41–43

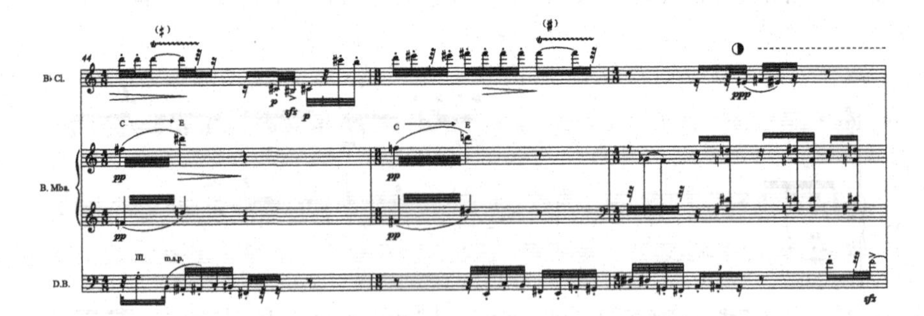

Fig. 21 *Intrascalings,* bars 44–46

Fig. 22 *Intrascalings,* bars 47–49

Fig. 23 *Intrascalings,* bars 50–52

Fig. 24 Figures 16, 17, 18, 19, 20, 21, 22, 23 and 24 show the polyrhythmic distribution of layers 1–5 from *Intrascalings*, bars 27–57

with the system (e.g. cutting interlocking, suddenly inventing an other composition technique) be possible at some point or would you prefer to invent a disruption by maximal variance of the system?

Papageorgiou: In general, I have a rather elastic relationship towards compositional techniques and rule setting. "Works of art make rules; rules do not make works of art." as Debussy remarks. To me, formal and structural procedures are purely utilitarian, in the sense that they serve the composer as agents that may facilitate the compositional procedures. In the case of *In the Vestige of the Present* the material is derived strictly by using the interlocking technique to generate pitch sequences that exhibit a great degree of similarity, as I described above. The whole piece is constructed by using very similar musical gestures. Some few ruptures in the form are achieved by introducing textural changes or changes in density.

In other works, however, such as *Enlacées* and *Effluénces*, the musical gestures are more diverse. In those works, I used strands of different types of structural entities (heterophonic, contrapuntal, etc.) of arrays of interlocked cells, and subsequently I have interlocked these strands of structural types in order to create a more malleable and less repetitive form. In the middle sections of both of these works, I used maximal variance of the interlocking technique to corrupt the material to a point that the listener barely hears a connection among that and the initial section. Finally, in *...anD...* for viola, in addition to interlocking pairs of cells, I have used an additive technique in certain parts of the work, starting with one interlocked pair of cells and each time adding an additional cell, thus creating interlocking units of 2, 3, 4, ..., n cells ("accordion-like" structure).

In recent times I have been experimenting extensively with the idea of allowing different degrees of formal or structural strictness of the components within a single composition. I am currently working on a piece for flute and piano, which is based on the idea of lapses of attention—"*action slips*" in psychological jargon—the familiar phenomenon of walking from one room into another and forgetting what task brought you there if the first place.[13] So, the work will consist of relatively rapid changes of textures and materials. The sudden textural changes and structural shifts attempt to "purge" information related to previously exposed components, so that we can listen to reiterations of components afresh.

My preoccupation with the interlocking technique is relatively recent. The scaling technique is even newer. I feel that I have just started to scratch the surface. At the present stage of my research, it is more important to me that I fully explore the possibilities these techniques have to offer, either each one of them or both of them in combination.

POINT: Did computer feedback influence your decisions?

Papageorgiou: The relationship between the composer and the computer is an interactive one, a cooperative agency of human and machine rather than a master and slave relationship. Although the computer did not impose its choices or certain choices on me during the compositional process, it would be wrong to state that it didn't influence my decisions. Most certainly it has changed my behavioural patterns. By enabling me to define complex rhythmic, harmonic, and polyphonic structures that would have been almost impossible to design otherwise, it gave me an enormous freedom of decision and, at the same time, the interaction with the computer forced me to reflect on the process of formalisation to an extent that I was not used to.

In addition, the computer gave me freedom to modify the design at any formal level:

(i) Modification of the musical material (pitch cells or durational relationships), while the form is preserved.
(ii) Modification of the algorithm (syntax), while form and temporal pace remain unchanged.

---

[13] See [21, 22].

(iii) Re-organization of formal elements, while micro-structural elements are kept the same.
(iv) Any combination of (i), (ii), (iii).

These facts widened my space of experimentation allowing me to create, compare, and evaluate a great number of variations of a structural idea, sometimes leading to slight or more significant alterations or modifications of a formalism, sometimes even to rejection. Working in a widened field of possibilities, allowed me to optimise my choices.

POINT: What do you think about computer feedback during the compositional process in general?

Papageorgiou: Algorithms encapsulate in an abstract form knowledge accumulated from previous experiences. Such an accumulation, as an extension of the human factor, can be used—with all potentials and limitations—as a foundation upon which more and more complex inquiries can be pursued, each time aggregating more layers of experience in the revised or anew constructed algorithms, thus constantly opening new paths for a compositional technique to evolve. I feel that, so far, I have managed to explore only a few of the implications of the techniques of interlocking and scaling.

To offer some reflections on my experiences during the creative process, I should stress that the use of algorithmic tools enabled me to investigate several aspects of my compositional strategies, test the validity of my compositional schemes and their stylistic properties, and develop and evolve a repertory of techniques. Not only that, but they also helped me explore new territories. Innovation not only comes from within, but also from our interactions, influences, external stimuli of our physical, social, and, in our times, also virtual environments. The computer and its algorithmic tools provide modern composers with flexible and powerful tools that can open up new horizons, as the orchestra has done in the romantic and the piano during the classical eras or the organ in the baroque.

Project Review by Dimitri Papageorgiou

The current project, where it solved some important issues to the extent that the algorithms served my current needs perfectly, also created new questions and problems, that I am planning to address in the near future. To outline some central concerns pertaining to future plans, one emergent issue is the formalisation of some syntactic devices derived from patterns of meso-structural organisation that surfaced while working with these techniques. Another important issue would be to investigate how the two techniques, interlocking and scaling, which now operate independently from each other, can be unified as to control pitch and rhythm simultaneously, and if such a unification makes sense at all. My overall experience of exploring music though algorithmic abstractions has been extremely positive. Not only am I going to expand

and refine my system but, despite the steep learning curve, one of my plans in the near future is to learn how to program algorithmic functions on my own.

References

1. Arom S et al (1991) African polyphony and polyrhythm: musical structure and methodology. Cambridge University Press, Cambridge
2. Bartlett FC (1995) Remembering: a study in experimental and social psychology. Cambridge University Press, Cambridge
3. Bernecker S (2009) Memory: a philosophical study. Oxford University Press, Oxford
4. Bernecker S (2008) The metaphysics of memory. Philosophical Studies Series, vol 111. Springer, New York
5. Best DL, Intons-Peterson MJ (2013) Memory distortions and their prevention. Psychology Press, Hove
6. Bouchaud J-P, Potters M (2000) Theory of financial risks: from statistical physics to risk management. Cambridge University Press, Cambridge
7. Bransford JD, Franks JJ (1971) The abstraction of linguistic ideas. Cogn Psychol 2(4):331–350
8. Candau J (1998) Mémoire et Identité. Presses Universitaires de France
9. Clendinning JP (1993) The pattern-meccanico compositions of György Ligeti. Perspectives of New Music 31(1):192–234
10. Conway MA et al (1992) Theoretical perspectives on autobiographical memory. Kluwer Academic in Cooperation with NATO Scientific Affairs Division, London
11. Csikszentmihalyi M, Sawyer K (1995) Creative insight: the social dimension of a solitary moment. In: Sternberg RJ, Davidson JE (eds) The nature of insight. The MIT Press, Cambridge
12. Ehleringer JR, Field CB et al (1993) Scaling physiological processes: leaf to globe. Academic Press, San Diego
13. Hadamard J (1949) The psychology of invention in the field of mathematics. Princeton University Press, Princeton
14. Halbwachs M (1952) Les Cadres Sociaux de la Mémoire. Presses Universitaires de France, Paris
15. Lefebvre H (2004) Rhythmanalysis: space, time and everyday life. Continuum, London
16. Loftus EF, Loftus GR (1980) On the permanence of stored information in the human brain. Am Psychol 35(5):409
17. Mandelbrot BB (1997) Fractals and scaling in finance: discontinuity, concentration, risk. Springer, New York
18. Mandelbrot BB (1967) How long is the coast of Britain. Science 156(3775):636–638
19. Mandelbrot BB (1982) The fractal geometry of nature. Freeman, San Francisco
20. Messiaen O (1956) The technique of my musical language (trans: Satterfield J), vol 2. Alphonse Leduc, Paris
21. Radvansky GA, Copeland DE (2006) Walking through doorways causes forgetting: situation models and experienced space. Mem Cogn 34(5):1150–1156
22. Radvansky GA, Krawietz SA, Tamplin AK (2011) Walking through doorways causes forgetting: further explorations. Q J Exp Psychol 64(8):1632–1645
23. Read SJ, Rosson MB (1982) Rewriting history: the biasing effects of attitudes on memory. Soc Cogn 1(3):240–255
24. Richardson LF (1961) The problem of contiguity: an appendix of statistics of deadly quarrels. Gen Syst Yearb 6(13):139–187
25. Roediger III HL (1996) Memory illusions. J Mem Lang 35(2):76–100
26. Ruwet N (1972) Langue, Musique. Poésie, Editions du Seuil, Paris
27. Santayana G (1905) The life of reason. Constable, London

28. Schacter DL (1995) Memory distortion: history and current status. In: Memory distortion: how minds, brains, and societies reconstruct the past. Harvard University Press, Cambridge, pp 1–43
29. Schacter DL (1996) Searching for memory: the brain, the mind, and the past. Basic Book, New York
30. Schacter DL, Addis DR (2007) Constructive memory: the ghosts of past and future. Nature 445(7123):27
31. Schröder M (1991) Fractals, chaos, power laws: minutes from an infinite paradise. Freeman, San Francisco
32. Schwarz KR (1980) Steve Reich: music as a gradual process, Part I. Perspectives of New Music 19:373–392
33. Schwarz KR (1981) Steve Reich: music as a gradual process, Part II. Perspectives of New Music 20:225–286
34. Snyder M, Uranowitz SW (1978) Reconstructing the past: some cognitive consequences of person perception. J Personal Soc Psychol 36(9):941
35. Vaggione H (2001) Some ontological remarks about music composition processes. Comput Music J 25(1):54–61
36. Varèse E (1971) The liberation of sound. In: Boretz B, Cone E (eds) Perspectives on American composers. Norton, New York, pp 25–33
37. Von Koch H (1993) On a continuous curve without tangent constructible from elementary geometry. In: Edgar GA (ed) Classics on fractals, vol 25. Addison-Wesley, Reading, p 45

Katharina Klement/Transformation and Morphing

Katharina Klement, Daniel Mayer and Gerhard Nierhaus

Katharina Klement grew up in a musical family. Her father played the violin and each sibling learnt an instrument.[1] Klement first learnt recorder, then soon piano, an instrument to which she has remained connected to throughout her career. Klement considered studying piano at the University of Music and Performing Arts Graz but she was reluctant to face the many hours of daily practice with a role that could limit her to a "reproducing" performer. After spending a gap year in the south of Italy, Klement nevertheless decided to study at the University of Music and Performing Arts Graz and passed the entrance exam to study piano performance.

Two years later, Klement moved to Vienna where she began to study instrumental pedagogy, with her main instrument still the piano, at the University of Music and Dramatic Arts Vienna. She began attending the course for Experimental and Electroacoustic Music (ELAK) and studied electroacoustic composition in the class of Dieter Kaufmann.

In particular, the possibility of a prepared piano and by then, the analogue approach to electroacoustic music, for example the cutting of tapes, the multi-track layering of sound material, opened entirely new avenues of composition for Klement. Her early experiences with electronic media and the work in the sound studio changed her compositional perspective: "everything which sounds" can become material and its processing and transformation becomes even a physical act, not dissimilar to some

[1] Biographical introduction and texts from the composer translated from the German by Tamara Friebel.

K. Klement
Institute for Composition and Electroacoustics, University of Music and Performing Arts Vienna, Vienna, Austria
e-mail: klement@mdw.ac.at

D. Mayer · G. Nierhaus (✉)
Institute of Electronic Music and Acoustics, University of Music and Performing Arts Graz, Graz, Austria
e-mail: nierhaus@iem.at

D. Mayer
e-mail: mayer@iem.at

G. Nierhaus (ed.), *Patterns of Intuition*,
DOI 10.1007/978-94-017-9561-6_7

approaches in the fine arts. Alongside the work with electroacoustic music, she took courses such as music theory, counterpoint, harmony and instrumentation which lead to an intense engagement with the traditional techniques of vocal and instrumental composition.

From this time on a series of instrumental, electroacoustic and also mixed style works were created. During her studies she was granted an exchange to attend the course of "Music Technology" at the University of York, to work with the computer music-software CDP[2] and with it came the opportunity to get to know its developer, Trevor Wishart. After her graduation at the University of Music and Dramatic Arts Vienna, Klement taught piano in a music school in Lower Austria and worked alongside as a freelance performer, pianist and composer of instrumental and electroacoustic pieces. One focus during this period was the engagement with "acousmatic music"[3] i.e. music for loudspeakers. The concrete effort "to transform and make sound audible" was the main focus in these years for her.

From 2006 on, Klement returned to ELAK as a teacher, and she currently continues to teach electroacoustic music with a focus on history and aesthetics. She also supervises students in their artistic practice.

Apart from the experimental approach, the traditional compositional craftsmanship is important to Klement. Well-known techniques like diminution, augmentation, inversion, retrograde, permutation, enter again and again into her work, in which musical structures develop often as a result of contrapuntal considerations. The hand-written work, be it sketches or entire scores are an essential part of the composition process. "I am very precise in my elaborations, I contemplate for a long time about things before I write them down. The writing and graphic notation is an essential act of composing. Although the score is of course a 'crutch' for the sound, it is already the first manifestation!"

Improvisation is also a central aspect of her artistic work. Klement calls improvisation and composition "communicating vessels", where it can be seen that her compositional thoughts are often nourished from improvisation experiences and vice versa.

Artistic Approach

Statement

Composing means to first empty myself, to "clear out" for a transparency within, in order to create a climate for inspiration. At the outset I have an idea, which often appears quickly, "dropping down" on me intuitively, without any immediate thoughts

---

[2] *Composer's Desktop Project*, Software package for Computer Music.

[3] "Acousmatique: situation de pure écoute, sans que l'attention puisse dérvier ou se renforcer d'une causalité instrumentale visible ou prévisible". Translated from the French, "acousmatic: the situation of pure hearing, where the attentive capacity is not reinforced through a visible or pre-envisaged instrumental causality", see [2].

accompanying how it might be realised. What follows after is a long journey of considering, calculating, reducing, frequently hearing, amongst many other aspects. Therefore I attempt to organise things and put them into a context of how I imagine them, yet there is also what the material itself demands. Every step of this journey is taken on a high tight rope, leaving below various abysses of doubt and concern about the success of the whole, whilst at the same time being fascinated by the lurking danger of the deep chasm below.

Composing is a transformation. Thoughts are converted into sound or vice versa: sonic material causes thoughtful considerations. These thoughts can be more or less conscious—sometimes it is just a spontaneous intuitive decision to structure something in one way or the other—in other cases they require extensive consideration with numerical calculations.

The act of composing is for me not really tangible, it remains like a distant, unknown, tempting land, which I will never fully understand, but that is what entices me to do it again and again—that I may once manage to circumvent it.

Personal Aesthetics

"Over years she probed the triumvirate of composition, improvisation and electronics in painstaking detail, working to loosen the strict parameters of classification (in relation to form, material and content), until they became one—forming her personal constellation."[4]

During my musical development, which started with the piano, which remains until today my anchor point, three areas have established themselves in my work: the precisely scored composition, the free or structured improvisation and the dealings with the electronic media. That these areas constantly overlap, cross and intersect is on the one hand because of their relatedness and on the other hand because of their often separate existence. Experiences, which are attained by improvising, become integrated into composition and vice versa.

Composition proceeds within different formal regulations and in a shifted temporality. It takes some time until thoughts are fixated and notated, and during this process of determining what matters most is a continuous clearer focus, an aiming towards a centre point, a challenging and detailed execution of an idea.

For example, in an instrumental composition I have tried to achieve a gradual concentration via contrapuntally developed lines and patterns, which make themselves independent. However this continuous process is perpetually interrupted by insertions of varying sound material, so-called "junctions". These provide the change to

[4] "Über Jahre hat sie das Dreigestirn Komposition, Improvisation und Elektronik sondiert und solange an der Auflösung der (auf Form, Material und Inhalt bezogenen) strikten Zuordnungsparameter gearbeitet, bis die drei endlich eins—und so zu ihrem Sternzeichen geworden sind". Burkhard Stangl in the booklet for a Portrait-CD of K. Klement, Edition Zeitton ORF 2008, translation from the German from Lea Rennert.

the process to evolve nonlinearly and discontinuously. I have already experimented with this approach of disruption years earlier, in improvisation, here it appears in full rigour and consequence in a new light or shape.

The work with electronic media requires foremost a different musical definition of material in comparison to an instrumental approach. The possibility to fixate the sound material to permeate, deconstruct and analyse down to the tiniest particles, like atoms, has strongly shaped my dealings with my musical approach. This media has redefined the concept of sound transformation and refocused the treatment of a phenomenon of sounds as a temporal and spectral action. To compose music for loudspeakers continues to be fascinating for me. Being directly exposed to the sound phenomenon in this way I can more or less immediately hear every stage of the sound transformation, I can set up complex systems independent from the practical needs required by human performance, as in the case in an instrumental piece.

One of my instrumental pieces was recorded for an 8-track sound installation, here used as source material, for so called "freeze-transformations". From a multi-layer superposition of the recordings of the various instruments emerged a complex coloured noise which replaced the previously sounding instrument. However, this noise didn't remain constant but underwent transformation through various filters.

The constant interweaving of the three areas of composition, improvisation and electronics in my musical oeuvre puts me in an "in-between-situation" but the meandering is what opens my actual working space.

Formalisation and Intuition

The construction and exploration, yet also the rejection of orders, rules and systems is a fundamental and essential problem in every compositional work for me. How can I find systems, which are in accord with my ideas? How can I find rules for processes, which edit sound material in such a way that it unfolds? Formalisation signifies for me also a certain de-privatising of my own personal ideas.

What do I define as musical material and how do I organise it? The questions couldn't be simpler, yet the answers are all the more complex. The initial igniting force of my work really lies always in the realm of intuition which means that an holistic idea forms the basis of it.

For example in my piece *Jalousie* for saxophone quartet (Fig. 1), I started with the idea to transcribe various recordings taken with an open window, yet with closed shutters. Fragments of these outside recordings serve as an acoustic score, which I translate by ear, first into a drawing and then into notation, a score. In one of these fragments rain could be heard. Finding myself unable to translate this complex texture by ear I applied strict organisational rules within a clearly restricted source of musical material. A contingent of pitches and rhythmical patterns were shifted with augmentation, diminution and inversed according to precise rules. It was my aim to create a "chaotic" texture from an organised set, i.e. to create a texture without

Fig. 1 Score opening of fragment 7 from *Jalousie* (2009)

directly apparent organisation. From an originally intuitive procedure originated a strictly organised musical fragment.

The engagement with qualitative aspects of mathematics where numbers represent not just an amount, a quantity but also a quality as such is another source of my compositional work. Thinking proportionally and correlating numbers and derivations can form the formal basis for a whole piece. In my sound installation *Beton* for instance I have calculated all temporal and spectral relations from the dimensions of a piece of concrete cut off from a construction site. It is a strictly formal work in which intuitive interventions, as in all of my pieces, are also not missing.

I like to spend time moving back and forth within musical data sets (certain pitches, durations, rhythmical fragments, etc.), to repeat them varying, to concatenate, to permutate, to organise them randomly, etc. In doing so I mostly do not use a computer but my head, which constructs such rule-based systems with a good amount of individuality. When, why and how long I stick sometimes strictly to a formalised order and when I free myself from it, I leave up to my intuition, which I trust entirely. Intuition is always radical, as it permits completely new and unexpected events, turns the wheel around, rejects something thoughtlessly and thus not rarely surprises! It is a saving principle which always creates a lively connection.

Evaluation and Self-reflection

Reflection is indispensable for my musical creating, it is an essential part of realising my artistic undertaking.

The creative and contemplating "Me" is constantly present during the work. Nevertheless, it remains impossible for me to be creative at the same time as to be reflecting. Therefore it is important in the process of working to take a step back and to observe my undertaking, to exploit it from a distance, to overview it first in stages and finally, after finishing the piece, to reflect on it as a whole. Only then do the large-scale interconnections become clear to me, which I partly lose when working on the details. I realise often only during contemplating my deviation from the originally planned structure or a particular characteristic or feature of a piece which I have missed before. Pieces often develop their own intrinsic life, they set a course themselves, which becomes clearly visible and audible only during reflection.

The most intensive reflection process is to listen to a piece multiple times. For scored compositions, to listen to the entire acoustic result is often only possible once, during the first performance. A recording of it is indispensible to me, and by re-listening I come to terms with the strengths and weaknesses of a piece and not rarely, the hidden, often unintended connections. This is similar with improvisation—the listening reflection via recordings is for me a crucial part of improvising. By working with the electronic media I can per se permanently reflect by listening during working. Sound structures and transformations can be evaluated pre- and re-listened to directly in the studio, even before the real performance. Reflection on my musical undertaking means, besides listening of course, also a structural, i.e. rational analysis. After finishing a piece I usually reiterate all production steps, I quantify and examine, create a protocol of the whole process. Also a temporal separation of the finished piece is essential for my reflection. Longer gaps, e.g. like several years, have opened up many new insights into my work. I also count comments, statements and feedback from colleagues and interpreters as musical reflexions, which cast a light on my blind spots.

Project Approach: Transformation and Morphing

I have always been concerned with the question of sound transformation. How do I get from material A to material B, from one sound shape to another, respectively? How large are the degrees of relationship, the differences, how are the identifying marks defined? Altogether, where does transformation happen? What is the changeable part, what is the persisting part?

"Music as a temporal medium is based on transformations in time: a musical figure unfolds in time, is repeated, developed, modified. While this is a constitutive feature of all music, for Katharina Klement, it seems to be a central theme in her

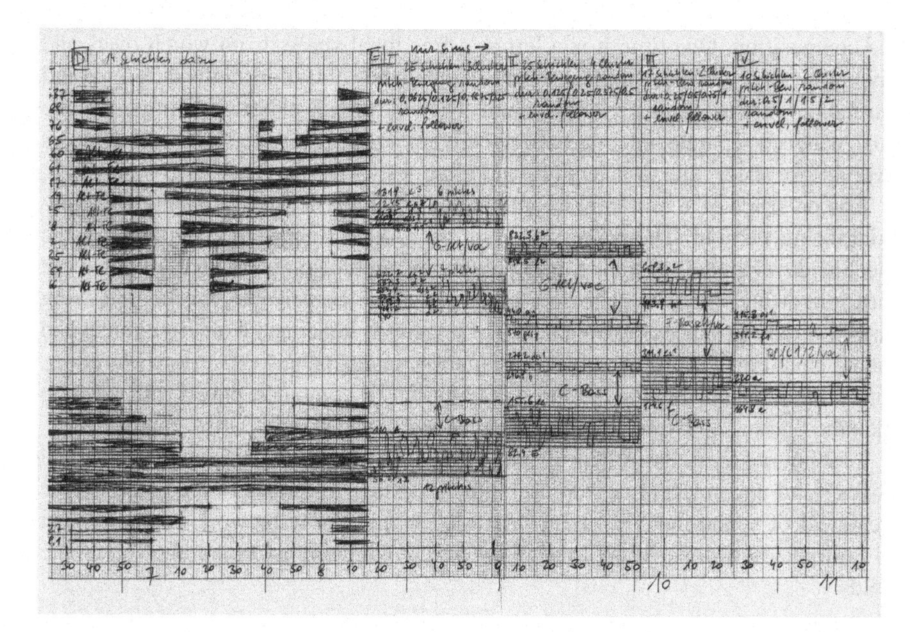

Fig. 2 Sketch for *mihrab*, excerpt from the score

musical work. The composer explores the multiple dimensions of transformations[5] both within and through the medium music."[6]

Transformation has moreover always had a connection with the concept of morphing, which by definition means the continuous metamorphosis of one shape into another. This process includes distortions and overlays, hybrid states between an initial and a final shape. These fragile hybrid states are preferred experimental spaces that I use for my creating, and are the growing ground for new subjects, for example like in biology where new species of a plant are created by hybridisation.

As an example of several works within this context I would like to mention the composition *mihrab* for recorder, (bass) clarinet[7] and electronics (2008/2012), see Fig. 2. The title of the piece is an Arabic word and means an "empty void", which is a place in every mosque, the part which symbolises the remembrance of the presence of the Prophet Mohamed.

Fascinated by the thought that emptiness symbols a "presence" or "attendance", respectively, this void is transformed in multiple ways as a formal characteristic in

---

[5] As a process of continuous change, as communication, as translation, as transmission, as change between the concrete and the abstract.

[6] Ursula Brandstätter in the booklet to the CD *jalousie* with works from Katharina Klement, 2012, from the German translated by Lea Rennert. Ursula Brandstätter, Professor for Music Teaching at the University of Art Berlin, since 2012 Vice-Chancellor of the Anton Bruckner Private University for Music, Dance and Acting in Linz, Author of many books in artistic topics such as [1].

[7] Literally translated from "Bass(klarinette)".

the composition. One time it is a left as a free tonal space, from which over- and undertone series unfold, another time the width of the void is constantly deformed and shifted in order to create varying spectra. The void stands either as an in-permeable zone between separate, unconnected pitch-elements or it vanishes as these elements overlay like sheets due to inversions and rotations.

The presence of a void implies automatically the existence of a divide, of something being twofold. The process of splitting is done right at the beginning of the piece: a development of unison into polyphony, the unison is gradually fanned out into micro-tones until two distinct pitches become perceivable. In the following parts a variety of shapes creating a web of textures is generated from rhythmical and melodic patterns via augmentation, diminution, inversion and repetition. Moreover, tonal contingents, which are ordered randomly by a computer, play a role. Chaotic melodic fluctuations, free of personal decisions, shape several parts.

Last but not least, the void is also interpreted as a temporal cut—these individual zones stand unconnected next to each other. Starting from these experiences in the treatment of sound material, I am interested within the framework of this project to explore a new dimension of morphing as a dynamic transformation of harmonic fields.

Project Expectations

I expect to receive additional facets of reflection on my own creations. In particular I am intrigued in a comparison between calculated computer models and the compositional procedures that I develop and apply without the help of a computer. Can I track down my creating anew, can new pathways be found for my creative process? The topic itself, the oscillation between rationality and intuition in the process of composing is highly interesting for me and I see it as an opportunity to learn more about it in a dialogue with scientists and also with other composers. It happens all too rarely, especially in the field of composition, that one discusses working processes with colleagues. Thus I also expect to get an insight in the creating and thinking of the other participants. I hope that in this collective a discourse happens, which enriches as well as sharpens my individual profile.

Exploring a Compositional Process

Klement: As I am working on several variations for piano in the context of this project, a question arose concerning these variations, and their transitions, where I became interested how these spectra morph into each other. Three sketches shall explain this (Figs. 3, 4 and 5).

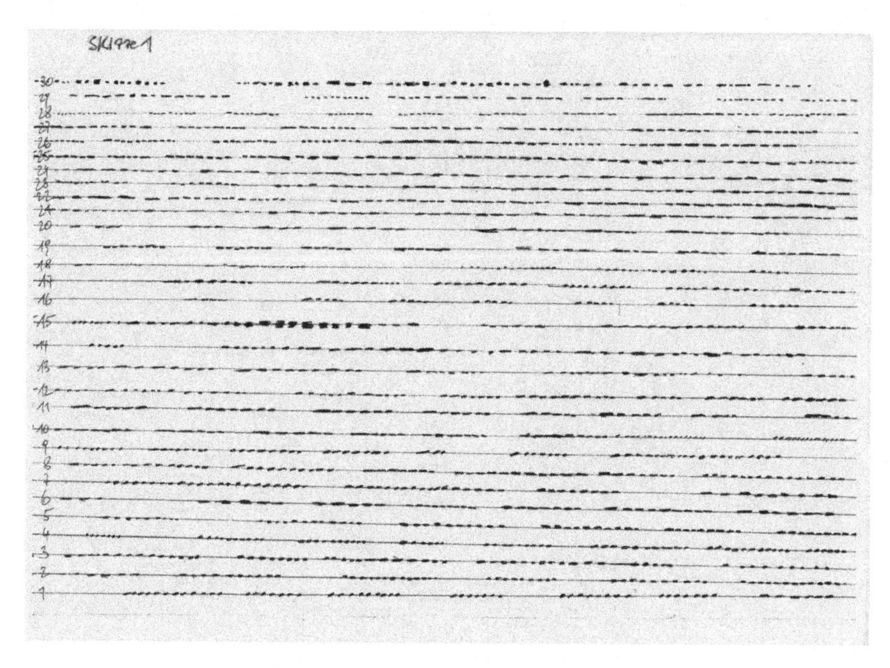

Fig. 3 Sketch 1 from *anständig abgeräumt* (*Anständig abgeräumt* roughly translates to “cleared away properly”) for piano solo (2013)

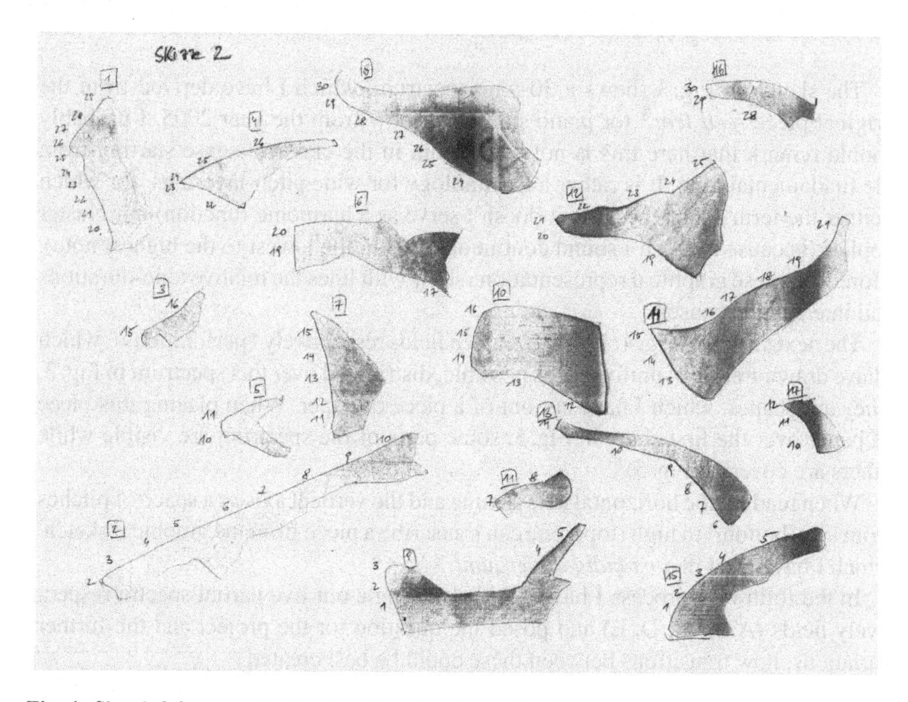

Fig. 4 Sketch 2 from *anständig abgeräumt* for piano solo (2013)

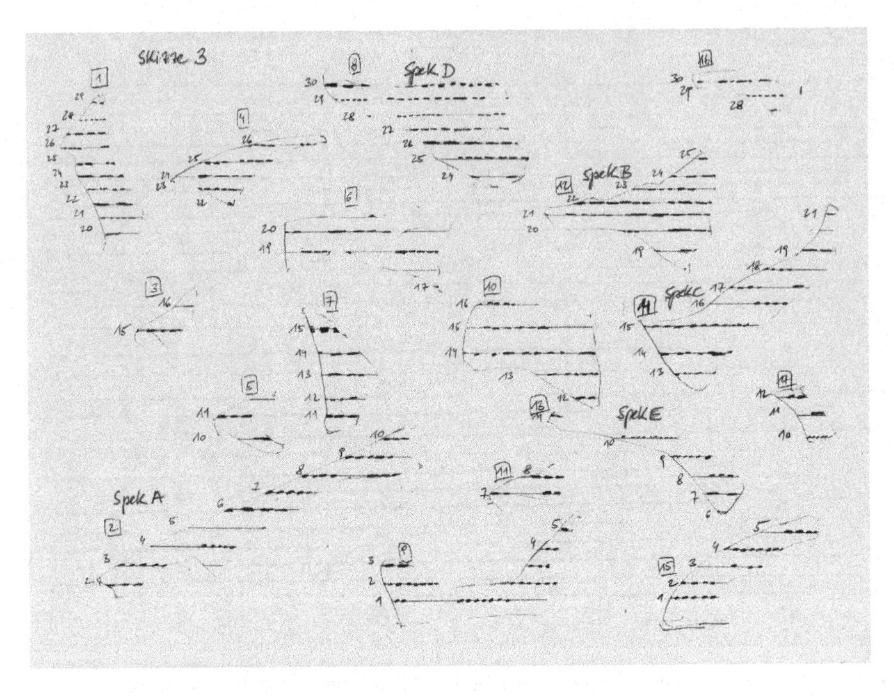

Fig. 5 Sketch 3 from *anständig abgeräumt* for piano solo (2013)

The sketch in Fig. 3 shows a 30-tone spectrum, which I have derived from the original piece *reell leer*[8] for piano solo and e-bow from the year 2005. I probably should remark that here this is not a spectrum in the classical sense starting from the fundamental tone. It is rather a terminology for wide pitch-layerings, for which neither the term chord (because it doesn't serve as a harmonic function) nor cluster applies (because it doesn't sound continuously from the lowest to the highest note). Moreover, these graphical representations show with lines the relative tone-durations and interruptive pauses.

The next sketch in Fig. 4 shows seventeen fields respectively "perforations", which I have drawn freely, as uniformly as possible, distributed over this spectrum in Fig. 3. They are shapes, which I have cut out of a piece of paper. When placing this piece of paper over the first sheet in Fig. 3, some parts of the spectrum are visible while others are covered (Fig. 5).

When reading the horizontal axis as time and the vertical axis as a space of pitches from low (bottom) to high (top), one can transcribe a piece from the graphical sketch, which I have tried in *anständig abgeräumt*.

In the following process I have decided to single out five partial spectra respectively fields (A, B, C, D, E) and posed the question for the project and the further variations, how transitions between these could be best created.

---

[8] *Reell leer* roughly translates to "real empty".

POINT: With regard to Klement's approach, we examined the following transitions and fields of pitches. A further query was whether to regard arrays (ordered collections of items) or sets. We tended towards the former at first, but then switched to sets (with additionally regarding the order of new and removed pitches) because Klement preferred to use the intermediate collections rather freely, as a pool for detailed work in her composition. From now on we will employ the term *morphing* for both kinds of transitions. Having agreed to deal with harmonic morphings we then went back one step and started with a formal description of general morphings of arrays. This seemed appropriate for our use, as transitions in an array lead smoothly from a start to an end.

Let's say we have a word of five digits, *12345*, and want to transform it to a word of the same size, but with different letters, say *abcde*, with one new item per step. A very simple way would be doing it like this (Table 1).

In the above example exchange positions are chosen in random order, positions of items don't change during the morphing. Disappearing items disappear at their initial position, they are not shifted before, new items appear at their final position, they are not shifted thereafter. This doesn't have to be the case in general, start- as well as end-items might be shifted. However for the sake of smooth morphing we stick to the convention that partial words of intermediate arrays do not have a swapped order—they would have to be reswapped later on otherwise. E.g. *e* might appear before *d* in the course of the morphing, but in any case *d* would have to be inserted at a position before *e*. Table 2 shows some variants, where the start- and end-order disappears, respectively, is built up from left or right.

It turns out that such morphings can be described with three sequences of indices, one for the indices of old items to remove, one for the indices of new items to add and one for the indices of new exchange positions. In the first example, where positions of items are not changed, the same index sequence (53124) is used for all

Table 1 Morphing with one new item per step, exchange positions chosen with random order

12345
1234e
12c4e
a2c4e
abc4e
abcde

Table 2 Morphing variants where start- and end-order disappears, respectively, is built from left or right

12345	12345	12345	12345
e1234	2345a	a1234	2345e
de123	345ab	ab123	345de
cde12	45abc	abc12	45cde
bcde1	5abcd	abcd1	5bcde
abcde	abcde	abcde	abcde

Table 3 Description of morphing by index sequences for operations remove, add and exchange

	Remove	Add	Exchange
12345	5	5	5
1234e	3	3	3
12c4e	1	1	1
a2c4e	2	2	2
abc4e	4	4	4
abcde			

Table 4 Description of morphing by index sequences for operations remove, add and exchange

	Remove	Add	Exchange
12345	5	5	1
e1234	4	4	2
de123	3	3	3
cde12	2	2	4
bcde1	1	1	5
abcde			

Table 5 Description of morphing by index sequences for operations remove, add and exchange

	Remove	Add	Exchange	Position of new and old items thereafter
12345	2	1	5	(5) (1 2 3 4)
1345a	5	2	1	(1 5) (2 3 4)
a134b	1	4	2	(1 2 5) (3 4)
ab34d	3	3	4	(1 2 4 5) (3)
ab4cd	4	5	3	(1 2 3 4 5) ()
abcde				

operations (Table 3). Table 4 shows, how shifting is done by an identical remove and add sequence (54321) and reversed exchange sequence (12345).

Table 5 shows an example with three randomly chosen, non-monotone sequences for remove, add and exchange. Items of intermediate arrays might jump from step to step (as *b* does from *a134b* to *ab34d*) if new positions (as indicated in the 5th column) are separated, however start and end order are maintained.

Morphings between sequences of unequal length can be based on morphings between sequences of equal length with additional steps where items are removed or added. These additional steps can be spread as regularly as possible along the morphing in order to get a smooth transition, however this might not be unique: in the example below we have a sequence of exchange steps and remove-only steps of the form (x x ro x x ro x), it could also be (x ro x x ro x x) or (x x ro x ro x x). But still additional remove-only (or add-only) steps can be done quite easily if start and end sequence have no items in common, here bijective morphing is applied to the underlined partial sequences, the first three columns of indices refer to the restricted size (5), see Table 6.

Table 6 Morphing between sequences of unequal length with no items in common

	Remove	Add	Exchange	Remove only
1234567	5	1	1	
12a3456	4	2	2	
12ab345	3	3	3	
				2 of 7
1ab345	2	4	4	
1abc34	1	5	5	
1abcd3				
				1 of 7
abcd3				
abcde				

Things are a bit more complicated with common items in the start and end sequences and/or with multiple occurrences within sequences. In this case bijective embeddings of a maximum size can be calculated by enumeration. In the transition from a twelve-tone row to a whole-tone row, written in pitch classes (Fig. 6), the bold items are the bijectively embedded ones (the ones to be morphed bijectively to the end sequence). Within this embedding four items can be left at place (bold on

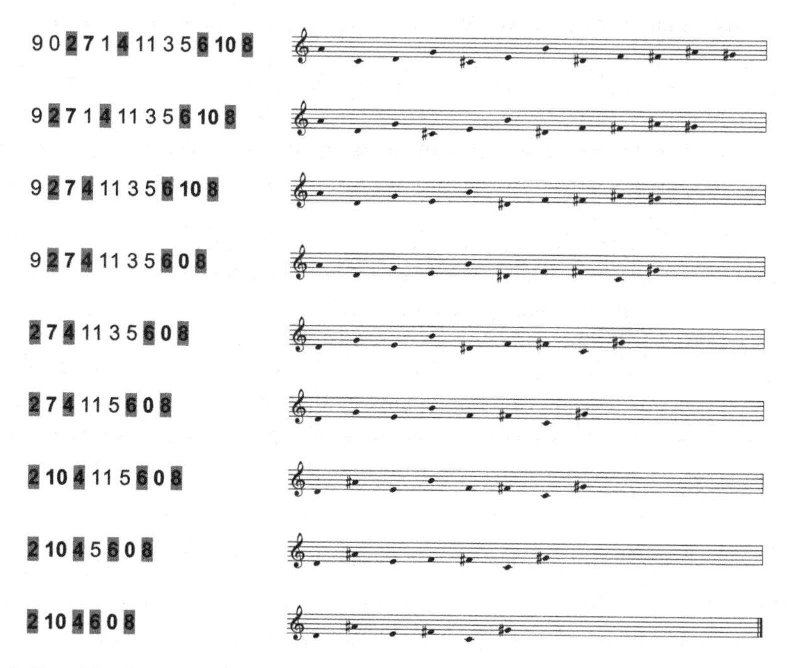

Fig. 6 Transition from a twelve-tone row to a whole tone row

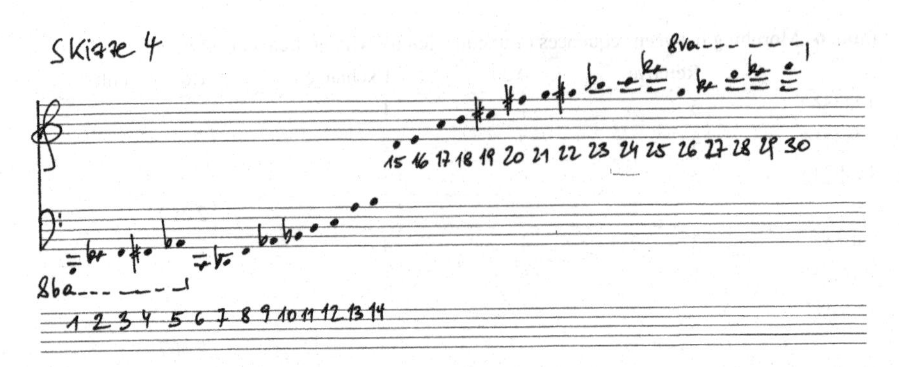

Fig. 7 Complete spectrum

grey background) and two must be exchanged. Here we additionally demanded all intermediate sequences to be rows, i.e. to have no doublings.

Finally, in any case single steps of exchanging, removing/adding might be omitted in order to get faster transitions, that are less smooth, i.e. that contain more than one of the elementary transition steps.

POINT: Could you name concrete spectra/pitches, which them you would like to work in the project?

Klement: The 30-tone spectrum, earlier noted graphically in sketch Fig. 3, appears transcribed into a notation-like score, see Fig. 7. This contingent was the result from the sound material of two patterns and their inversion/mirroring of the original piece *reell leer*.

I have used the following fields respectively "partial spectra" for the work in the project (Table 7).

For all the following spectra shown in the examples the pitches are consecutively notated for the sake of a better readability (Fig. 8), yet they have to be understood as sounding simultaneously.

The fields were selected by me with respect to the presence and absence of congruent tones: e.g. between A and B there are no congruent tones, or between B and C there are two congruent tones (tone 19 and 21).

The position within the complete spectrum is also crucial: very low range (A), high range (B), higher middle range (C), very high range (D), lower middle range (E). My vision was to achieve a continuous gliding from one range to another—which traces are left, which inter-stations are generated? I was less concerned with the beginning- and end-state, but more with what lies between these stages.

Table 7 "Partial spectra" used for the work in the project

Field 2 = A (tone 2–10)
Field 12 = B (tone 19–25)
Field 14 = C (tone 13–21)
Field 8 = D (tone 24–30)
Field 13 = E (tone 6–11)

Fig. 8 Consecutively notated pitches of Klement's original spectra table

POINT: We started with generating some transitions with different parametrical sets of transition lengths (from 4 to 15 transitions) and orders of appearance and disappearance. We produced a large number of them and expected Klement's reply to guide us to the right path, Fig. 9 shows an example for a morphing from A to B in six steps.

POINT: Were the results of these morphs satisfying for you?

Klement: I was looking forward to the results of the first "morphings" with anticipation! Somehow I was imagining with this terminology a temporal-spatial dynamical relation between these fields. But we had agreed to reduce ourselves with only the given concrete pitches. Accordingly it became clear to me after the first results that we here only touch an aspect of morphing. However that didn't make the matter less interesting, because I was forced to examine this aspect of the parameter extremely accurately. It also showed me a way to grab this complex phenomenon of "morphing" head on in terms of these tonal contingents.

In view of my *piano variations* I played through the examples—as far as technically possible, the partial spectra always in rapid repetition, in which gradually a tone is dropped and another is added. By means of this playing technique I could already find with these examples an approximation of my vision and perception of morphing.

After the first results I was also considering the possibility of microtonal deviations or different directions of a transformation process. What would happen when overlaying a process from partial spectrum A to B with one from B to A? It was soon

Fig. 9 Morph from A (lower system) to B (upper system) in six steps

clear that within the limits of this particular project there had to be a clear focus and with it a reduction of these problems in order to avoid generating an unmanageable number of solutions.

POINT: As we are talking about "pools", indeed the order is only a point of representation here, it doesn't play a role in their usage—so we adapted our strategy and went on with longer transitions and neglected inserting positions (Fig. 10).

POINT: Did the results of the last generated material satisfy your expectations more?

Klement: In the example from Fig. 10 I found the disappearance and appearance of the pitches too schematic. I remembered a procedure, which I had already used in my piece *Schiff und Hut*[9] for tenor hammered dulcimer and transducer: there I let new material within an established sound-contingent grow out from the middle, which eventually replaced the old (Fig. 11).

POINT: At the same time we were also modifying the original form of spectra. As Klement tended to interpret morphing results rather freely concerning the octave register, we agreed to bring them to the same register, though we didn't go so far as to regard pure pitch class sets, as Klement insisted that the shape of spectra should remain transposed, neither rotated nor compressed (Fig. 12).

During one meeting Klement once used the wording *flower morphing*. We looked at the possibilities for *flowers* in index tuples $1, \ldots, n$. Strictly speaking we were looking for permutations of size n, where first k integers fill a complete interval for all k between 1 and n and the number of steps towards a certain direction doesn't exceed

[9] *Schiff und Hut* translates to "ship and hat".

Fig. 10 Morph from A to B in 10 steps

a given threshold m. For a given integer m we are speaking of *flowers of type m*. The most simple flower tuple is given by $m = 1$, just a strict zigzag movement. There are two such solutions for each n, e.g. see Table 8. We did a full enumeration of *flowers* for given spectra sizes and $m = 1, \ldots, 4$. Identical pitches of adjacent spectra were taken out of the calculation of morphings, so their purest form occurs with spectra having no pitches in common. Here one result which applies the restrictions given above is shown in Fig. 13, for a morphing from C′ to D′, flower type 2, new pitches of D' appear with order (4356271), old pitches of C′ disappear according to (564372189).

POINT: How did you like these transitions?

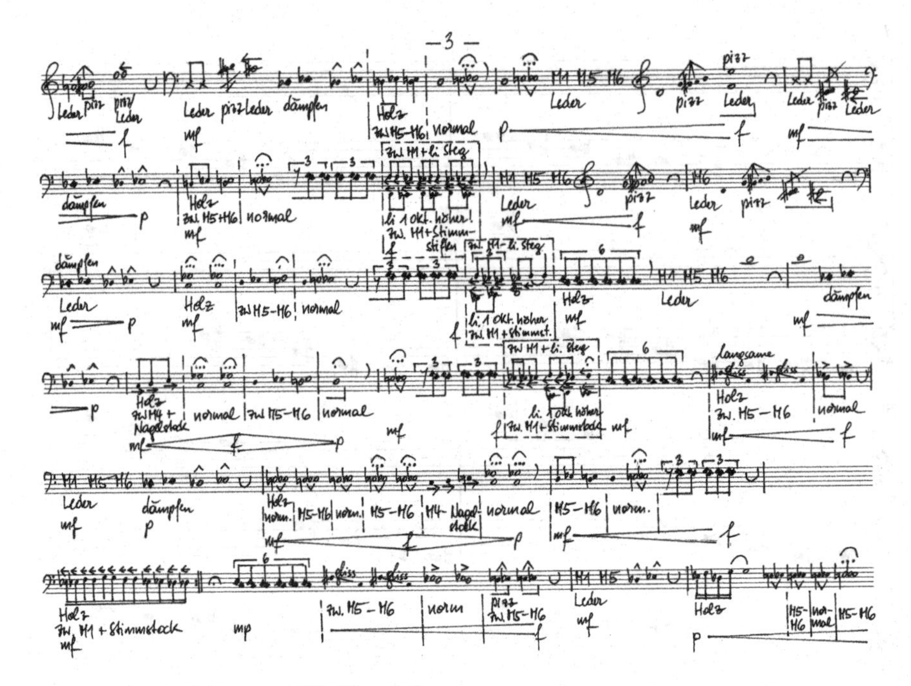

Fig. 11 Page 3 of the score of *Schiff und Hut*

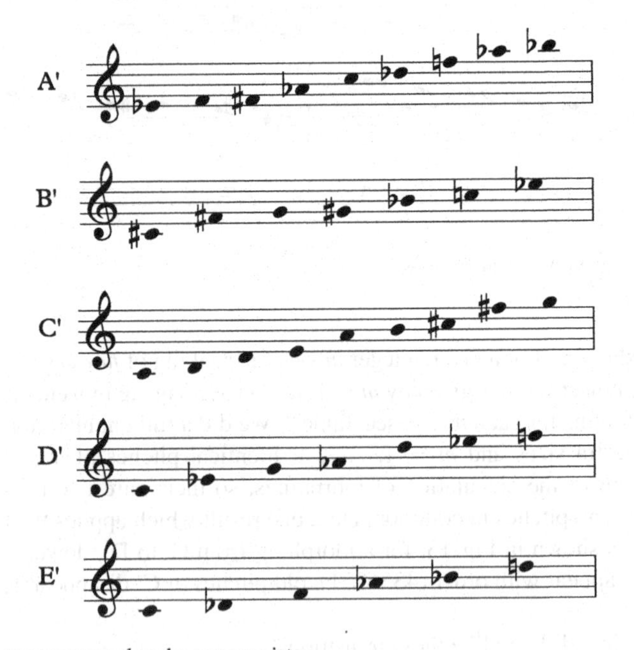

Fig. 12 Spectra transposed to the same register

Table 8 *Flowers* of type $m = 1, 2, 3$ and size $n = 5$

$n = 5, m = 1$:
32415
34251
$n = 5, m = 2$:
23145
23415
32145
32451
34215
34521
43251
43521
$n = 5, m = 3$:
21345
23451
43215
45321

Klement: Some examples I found quite successful, others were less so, which made me try to more precisely formulate my criteria. For instance, tonal relations, outstanding triads or pentatonic I wanted to avoid. Moreover, I realised that for a successful transition the dynamics are also very crucial.

POINT: In the first regard Klement wanted to avoid certain transitions as excluded by rules (1) and (2) which were partially refined later on:

1. Leaps are forbidden for identical pitch classes.
   If a pitch class is contained in a start set and end set of pitches, there must not be any intermediate set of pitches that doesn't contain at least one representative of this pitch class.
2. Adjacent sets of the morphing should not consist of identical pitch class sets (reduction of redundancies).

It should be mentioned that, as often in art and music, a rule-based description has its limits. There were cases in which Klement was bothered by the above characteristics more than in others, however to state them as rules seemed to be a viable way to achieve more suitable morphings. In case of the transition from C′ to D′ a full enumeration brought up only two solutions fulfilling criteria (1) and (2) of flower type 2 (Figs. 14 and 15), from which Klement preferred the second. Here there are no leaps for common pitch classes D and G and there isn't any step with redundancy, i.e. there are no steps with adjacent spectra having identical pitch class sets.

Fig. 13 Morph from C′ (lower system) to D′ (upper system)

However for other morphings compromises and tradeoffs between restrictions had to be made to some extent: while we were able to keep rule (1) for most morphings from A′ to E′, rule (2) could not be fulfilled in any case with the given spectra. In general solutions could be found more easily if more and higher flower types (arbitrary m) were admitted. As Klement preferred lower flower types (with more typical zigzag movement), we suggested taking solutions of the lowest possible flower type with a minimal number of redundancies and dropping redundant intermediate pitch sets.

POINT: What do you think about our latest bunch of flowers and morphings?

Klement: I think this latest example (Fig. 15) is wonderful—here the two spectra blend well into one another and separate themselves very organically. There are also no tonal "edges", a consistently homogeneous and together fascinating process. If I insert here in a composition also volume gradients, articulation and possibly micro tonality, I approach close to a dynamical transformation.

POINT: For the sake of comparison Klement also wrote a "well-suited" morphing with pencil and paper. There in Fig. 16 she strictly follows the flower principle for appearing and disappearing pitches; those, that the two spectra have in common, are

Fig. 14 Morph from C′ (lower system) to D′ (upper system)

notated in parenthesis. But two other criteria, that were implemented in our search, were not strictly followed and for this reason the solution was not given by the machine: There is a leap—pitch class C# disappears in B.2 and reappears in B.5 and additionally two reduction steps (where only old pitches disappear) are performed in sequence (B.5–B.7)—so the actual target spectrum is already reached in B.5. In our implementation reduction steps were spread over the whole morphing as regularly as possible.

So it would indeed be possible to loosen restrictions of our implementation in that sense, i.e. (1) allow leaps that are not too small, so that reentries of pitch classes could again be perceived as new entries and (2) allow irregular appearances of reduction steps. But as we got a majority of results that were judged by Klement between reasonable and very well suited and the loosening of constraints would enlarge

Fig. 15 Morph from C′ (lower system) to D′ (upper system)

the number of possible solutions, thus again delivering a number of non-satisfying results, we decided to refrain from further modeling. The exceptional cases can now mainly be described by guiding principles which the composer applies, a result that turned out from continuously adapting our model.

POINT: What are the next steps in the compositional process?

Klement: I have a composition for string orchestra in front of me, and I am certainly going to incorporate the idea of "morphing" and the insight, which I have gained from this examination. With this instrumentation I can easily also go into the direction of micro-tonality. I will be able to apply the "exchange principle" of contingents needed in this project also to other parameters like rhythm, articulation or sound colour.

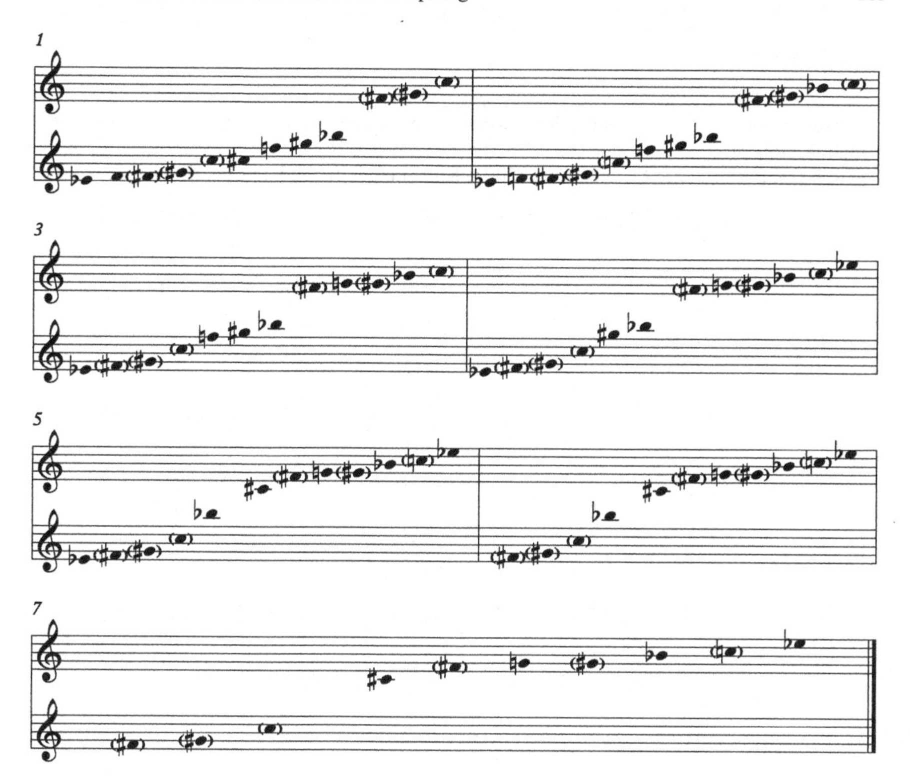

Fig. 16 Morph from A′ (lower system) to B′ (upper system) composed by Klement

Project Review by Katharina Klement

Through the discussions and the "ping-pong procedure" between the programmed results of the project team and the intuitive evaluation I was forced again and again to "X-ray" the topic at hand. The permanent reflection and analysis as to why I prefer some solutions to others has shown me how small the space is where the good examples lie. The precision of my intuition, on which I entirely trust as mentioned in the introduction, has astonished me yet again. It was especially interesting to me at the end of the project where it seemed absurd to make such a computational effort, because I knew exactly how the best solution would be found and I wrote it freely by hand. Would I have known that so well from the beginning of the project? Likely not—the effort to narrow down a topic for so long has certainly paid off. I think that I have thereby acquired a portion of new compositional craftsmanship. I am still too close and cannot judge how far the results have influenced me, but the upcoming composition will show the extent of this effect.

References

1. Brandstätter U (2013) Erkenntnis durch Kunst. Theorie und Praxis der Ästhetischen Transformation. Böhlau
2. Misch I, von Blumröder C, Kersting A (2003) Klangbilder: Technik meines Hörens, vol 4. LIT, Münster

Orestis Toufektsis/Chords in a Black Box

Orestis Toufektsis, Hanns Holger Rutz and Gerhard Nierhaus

Orestis Toufektsis was born in Tashkent in the former Soviet Union.[1] The family originates from Greece but was forced to emigrate to Uzbekistan after the civil war of 1946–49. In his youth he made his first contact with music through learning piano in the local music school. After the return of his family to Greece in 1977 he continued piano lessons at the conservatory in Alexandroupoli. After finishing school he enrolled at the Technical University of Thessaloniki as a student of surveying technologies, mostly to escape the two years of military service and to continue his musical studies. During this time Toufektsis earned his living as a musician in nightclubs and worked as a keyboarder in several jazz- and rock bands.

However, Toufektsis received a crucial impulse for his musical career from his counterpoint studies at the Conservatory of Thessaloniki; he also made a decision at this time to dedicate himself entirely to composition in the future. In counterpoint he was less fascinated by the various techniques of composers like Dufay, Palestrina and Bach, but far more in the inner logic and the corresponding ideas of the equality of multiple layers of pitches, rhythms and different sound colours. This concept liberated his musical intentions and led him in search of his own musical language. Through his teacher Dimitri Papageorgiou, Toufektsis also came in touch with, at this time in Greece, unknown pieces of Scelsi, Nono, Cage, Ligeti and Xenakis.

---

[1] Biographical introduction and texts from the composer translated from the German by Tamara Friebel.

O. Toufektsis
Institute for Composition, Music Theory, Music History and Conducting,
University of Music and Performing Arts Graz, Graz, Austria
e-mail: orestis.toufektsis@kug.ac.at

H.H. Rutz · G. Nierhaus (✉)
Institute of Electronic Music and Acoustics, University of Music and Performing Arts Graz,
Graz, Austria
e-mail: nierhaus@iem.at

H.H. Rutz
e-mail: rutz@iem.at

G. Nierhaus (ed.), *Patterns of Intuition*,
DOI 10.1007/978-94-017-9561-6_8

He finished his study at the Technical University in 1993 and moved to Austria in order to study composition at the University of Music and Performing Arts Graz, first with Hermann Markus Pressl and later with Gerd Kühr. With Pressl, Toufektsis appreciated in particular the questioning of assumingly "axiomatic" concepts: "to assume that that of which one is strongly convinced does essentially not hold".[2] With Kühr, he valued his competence in teaching the craft of composition, but also in particular his ability to refrain from his own aesthetic and to embody the composition sphere and thoughts of his/her student. After finishing his composition study Toufektsis went once again to Greece for a short period; yet a year later he returned as a teacher to the University of Music and Performing Arts Graz, where since then he has taught music theoretical courses.

In his artistic discourse Toufektsis is strongly inspired by methods emanating from science. The design of systems, verifying hypotheses, is a frequent approach for his compositional work, yet not in order to establish an arbitrarily fashioned artistic truth, but primarily to obtain new and unconventional perspectives on the musical material.

Artistic Approach

Statement

For me composing is a part of being human, a necessity,[3] which is something more than the need to express oneself, to "say" something, to communicate or convey something. In my view, art is in general an essential part of our survival strategy because we can obtain experiences by the creation of artworks, which represent snapshots of our thus crystallised knowledge, our intelligence and intuition, which couldn't be reached on a different path.

Music is always created by people for people, yet not only in order to convey a message to each other, but to share an experience, to make it a common good. To share an experience like, for instance, how it would be not to have fear, to sharpen a view on the essential, the substance of a matter, or to practise being aware and in particular to be free. To share such experiences is most substantially artistic and—in my opinion also the only possible—political act, by which music addresses and stimulates the psyche, the spirit, the intelligence and perhaps also can alter an individual. I believe that one cannot escape this political and human component of art, one can only neglect or disregard it.

---

[2] Quote from a discussion with the composer.

[3] Schönberg's statement originally in German which states that "art is not a matter of skill, but of duty (internal necessity and drive)" can hold only together with the important comment that craftsmanship is an indispensable requirement in order to realise this duty.

Personal Aesthetics

Contemplating music requires, in general, a binding logic, which is however mostly culturally and historically conditioned, representing for me only an abbreviated image. I always assume that music is able to be something universal, generally valid where in its own way it can also fulfil more basic human needs. There are indeed "languages", which exhibit an obligatory syntax and are nevertheless culturally and historically not at all or only very slightly conditioned. Such "languages" are found in the sciences and in particular in mathematics.[4] The fundamental aspect of such "languages" seems to contain a high degree of abstraction.

I have tried to view my compositional work in this direction and to find in music a kind of analogy to such "languages". In this respect I view it through the term "absolute music". To find an analogy to such languages means to apply as abstract as possible a composition or design principle, which however does not imply that it need to have been composed exclusively through means of mathematical principles. This route led me increasingly in a direction, which I would essentially describe as an act of "permanent reduction"—reduction of the means, the possibilities of expression, of the material, the formal structures into what seems fundamental and substantial, as free of every kind of cultural and historic epiphenomena as possible, onto what music is: structured time—no message, no statement, as well as nothing spectacular or exciting. I haven't always succeeded in this reduction and it is likely I never will in the future, which is however, certainly not a sign of in whatever way of a "natured" failure. Figure 1 (*EpiTria* for ensemble, 2003) shows an example of such an attempt in "permanent reduction".

In these efforts I deliberately try to abstain from detailed technical and dynamical instructions in order to create a specific type of a musical texture, which provides only the structure—the temporally organised pitches. The "decoding/decryption" of the structural role of each tone or group of tones, respectively, should emerge from the formation and composition of the "tone-material", of the movement and is left to the musician according to defined criteria.

I believe it is no exaggeration if I claimed that most colleagues would agree with the phrase "composing means to permanently make decisions". This doesn't seem to be the primary role to begin with, but what is essential is how these decisions are managed and thus the connected questions: what is the underlying approach of the composition, what do these decisions target, what shall be achieved as the results of the composition work?

I would call my first compositional decision, which stands at the beginning of every composing, as the "artful configuration of the void". As far as I am able to, I try—by drawing an analogy—to start from zero. Composing means a possibility to enter the imaginary space of absolute freedom. This space however is not an arbitrary one because freedom is not felt by arbitrariness. It emerges and is "formed" slowly through the respective rules of the composition, which, importantly do not need to

[4] E.g. the formal languages in the theory of syntax, see [2].

Fig. 1 Excerpt of *EpiTria*

be verifiable in a scientific sense. Yet I also do not consider it secondary, which rules are to be imposed. Due to their specific formulation and application, previously unknown spaces can also open up for new experiences.

The more abstract these imaginary worlds that arise in such free-spaces are, the more differently can they be perceived. However, I do not consider this a problem in itself, it is rather an enrichment. To share an "experience" does not mean to simply communicate something, but to make it experiential in multiple ways.

If I had to describe comprehensively my aesthetic premises, then the following aspects would be important for me: economy of the means, which can lead to a complexity respectively a "multi-layered-ness", which itself allows in return multiple possibilities of associations and interpretations. Thus complexity is not an end in itself, but serves as a freedom of observation, which can be enabled and set free by this "multi-layered-ness". Thereby, I consider the monotony, which can emerge from an application of material oriented towards an economy of means, not as a counter-pole but as a possible form of complexity.

I consider the separation into "extra-musical" and "inner-musical" only as terms, which can serve understanding during a discourse. There is no area for me, which would be extra-musical. Aspects of mathematics, physics, and fine arts can flow into each other or to formulate it differently: everything can be music. Extra-musical, also science-originated sources of inspiration are therefore for me not questionable. The interface between science and art consist as a generalisation via abstraction. The techniques and tools of composition, which allow the realisation of such a sound image, originate from the areas of self-similarity, combinatorics, and from the principle of isorhythmics, which is however not only allied to tone-duration and pitch.

Additionally I have lately looked into possibilities of applying stochastic-based variations to, amongst others, the possibilities of arranging variable self-similar structures. It is very important for me to emphasise that here self-similarity must not be understood as an exact transfer of fractal structures but as a general musical design principle, which can be applied onto both the micro- and macro-level of a composition.

As a further consequence I see combinatorics as a suitable tool to create structures, which, although consisting of a minimal amount of elements, still allow in their combinatorial possibilities complexity and "multi-layeredness" with respect to very relevant musical terms like repetition, development, change, variation, and metamorphosis. For example, the precise repetition of practised, "controlled" movements of the interpreter also produces difference, and this is the decisive point. It is similar to our notion of control, our capacity to decide "freely" and the unpredictable consequences of our actions that arise. As an example for such varying repetitions, Fig. 2 shows an excerpt of *Fraktum 4/EpiEnteka*, for violincello (2006). In this piece the application of self-similar principles was an essential formal principle.

Formalisation and Intuition

During the compositional process, it is important for me to reflect and question the decisions made, with respect to their origins, motives, etc. The answer to the question "why" a certain decision was made can often be simple, or even trivial, yet

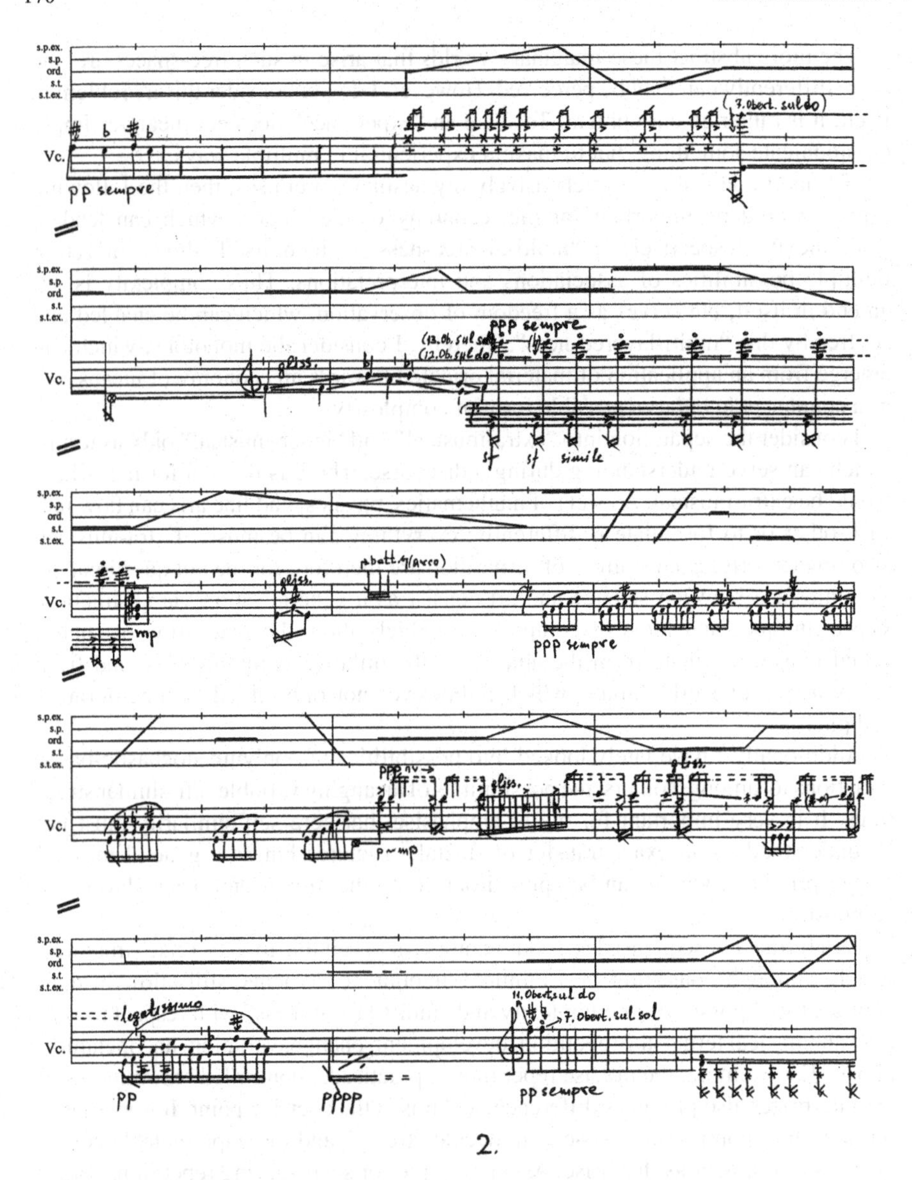

Fig. 2 *eFraktum 4/EpiEnteka* for cello (2006)

also frequently difficult. Therefore there are, on the one hand decisions, which are easily justifiable because they are rational, logical or explicable from a pragmatic, technical viewpoint and on the other hand, decisions which are less easy to justify, since they do not readily disclose their "inner logic" and are more related to the area of intuition.

Intuition and rationality are considered in general, if not as two antipodes, as two terms, which express a certain contradiction. I have repeatedly felt the experience that "irreconcilability" is another term, which is related to intuition and rationality for the awareness of many people. With respect to artistic creation, intuition is mostly associated with terms like inspiration, sentiment, artistic instinct or even taste and in any case something, which cannot be conceived systematically or rationally but thus with the mind. There is talk of "magic moments" or of the "quintessence" of art. Without this necessary "something" art is in danger to be downgraded as "uninspired".

Rationality, in contrast, is usually understood as something that has little to do with an artwork or in any case plays an inferior role. It is associated with construction, structure and "dry" logic, which might be a necessity in a scientific context, however in art it only has relevance in the realm of technical ability. The general idea of how an artwork is created is therefore coined by a "romantic" image that the artist needs, in crucial moments of the creation process, a sort of inspiration. And this inspiration can and must not be based on rational considerations, but should be an ingenious manifestation of what in general is called "talent".

For me there is no discrepancy between the term intuition and rationality. From my own experience I am unable to draw a clear line between these two kinds of mental acts. Though apart from this, is it at all possible or rather even necessary to draw such a line?

With the description "two ways of an intellectual act", an established "common ground" between intuition and rationality could of course conceal a subtlety: oil and water are fluids. The question is what to do with the common ground. Einstein has supposedly comprehended the fundamental principle of his relativity theory for the first time by imagining a person, who falls from a roof or is inside a falling elevator: the flight trajectory of a photon (light impulse), which crosses the falling elevator would appear straight to the occupant and in contrast curved to an observer outside [3, p. 87].

Was that inspiration or the result of a logical, rational thinking about a scientific and physical problem? Is the idea of a twelve-tone series the product of intuition, respectively inspiration, or the logical consequence of the will to finally break with tonality as an act of liberation from its "chains".

An artist does not appear out of nowhere on earth. He/she knows something and this previous knowledge is not only acquired, but also experienced. And it is not only a technical "dry" knowledge. For me, intuition is perhaps nothing more than a thought process, which resorts to this previous knowledge and of which we are simply not aware. Hence "intuition" is quite rationally founded and justifiable. A human being can both "think" intuitively and rationally and on that account I would not like to relinquish any possibility since I can therefore embrace my nature. Are not intuition and rationality simply two sides of the same coin and in fact, perhaps the most important human quality and ability, of mental activity, of thinking? I therefore talk about intuitive and rational thinking. To limit oneself only to intuitive or only to rational thinking amounts to a mental amputation. This holds for me especially when creating an artwork, since I would otherwise rob myself of my own freedom.

If I want to balance a long stick on my finger, I can either rely on my feeling in order to find the point of equilibrium or also measure the stick or mark the middle, where I have to be aware that I can fail or succeed in both cases—the essential is to find the optimal approach.

Evaluation and Self-reflection

Reflection of my work means at first to question the motives of my compositional decisions, to become aware of the mechanisms in the development-and-decision process. For that I usually need a certain temporal distance to when I created the piece. It also means to draw practical, technical conclusions, if, for instance, a compositional strategy bore fruit, or if a certain texture worked out how I had imagined it; or what structuring technique is better suited to generate a certain musical structure etc. In particular it is important, which concrete new findings, insights and experiences I can "take away" with me.

Via the technical, practical layer however, these findings are not easily exemplified. Primarily I am not concerned with the concrete form of a structure, respectively the result in the score, but with how the path came about, with the structural aspect of the implementation. If the structural principle and the implementation is "right", then so are the results.

For the coherence of a structural principle it is in any case important that it stimulates my aural imagination in a way, which allows for unexpected, also not intended sound structures, which nevertheless permit a meaningful musical perspective.

Project Approach: "Harmonic Tendencies"

In the last years I have been increasingly concerned with the question of a possible connection between the structure of the tonal-material and the formal structure. An important aspect, which arises from this problem, is also the nature of the correspondence between harmonic and formal grouping. On the micro-level this aspect of harmonic material manifests itself for me predominately in mostly homophonic chord sequences, which arise from relatively simply describable "rules of voice leading", yet also simultaneously feature an "orientation towards a goal", a kind of "harmonic tendency". The control of this metamorphosis occurs rather on an intuitive level, for which no preconceived systems of rules is applied.

Project Expectations

This project approach touches, for me, a core issue of compositional creation, namely the reflection of the intuitive, but also of the rational decision processes per se.

I consider the questions concerning the decision processes in the creation process, the questions of their function and how they came about in the first place, not only fascinating but also vastly important. Although in this view my expectations are very high, they cannot really be disappointed because any, even the smallest, insight about if and how not fully rationally comprehensible decision processes can be subjected to a formalisation, will be a gain. If I can thereby come closer to understanding how my intuition "works" during the compositional process, that is already a considerable fundamental step.

Exploring a Compositional Process

POINT: We decided to focus in this project on the chord sequences and their "harmonic tendencies" as mentioned above. What is your basic criteria for consideration when composing these chord sequences?

Toufektsis: The chord progressions are generated via a kind of projection from the horizontal to the vertical. The harmony serves to enable linear processes considered from another perspective. In generating the chord I look at the harmony not just as a collection of possible harmonies/chords, but compositionally it is important for me to observe the musical relationships that arise between the sounds, the function that they perform within the formal structure, enabling a particular musical perspective. Although it would be interesting to generate such chord sequences, however, the output parameters would need to be so defined as to allow that certain criteria both in the structures of the chords and their progressions—the musical relationships between their parts—could be defined with particular starting conditions or prohibitive constraints.

Genetic Algorithms

POINT: As Toufektsis composes the "harmonic tendencies" of the chords more on an intuitive level, we decided to consider the application of non-knowledge-based systems, which should converge to more or less optimal solutions based on a human fitness rating. We finally chose to implement a *Genetic Algorithm* (GA) that could potentially produce such harmonic tendencies.

A GA[5] is a stochastic optimisation procedure based on ideas of evolution and genetics. Basically, an originally random set of solutions is improved over a number of iterations by applying a sort of "survival of the fittest". The problem which needs to be solved is formalised as a set of variables, called a 'chromosome'. At any iteration, there are N_{pop} individual variable configurations or 'chromosomes' which make up the current 'population'. When the algorithm begins, the initial

[5] For an accessible introduction to GA, see [4]; also [5].

population is usually generated using random values for the chromosome variables. In a standard GA, a fitness function is then used to evaluate how well each chromosome approximates a given metric. The chromosomes of the population are ranked by their fitness, and a certain percentage of these chromosomes is selected for "survival" while the others are discarded. The N_{keep} selected chromosomes are used to produce N_{pop} "offspring" individuals which form the next population. This 'breeding' typically includes a recombination of two parental chromosomes into a 'crossover', and randomly modifying the selected chromosomes, called 'mutation'.

Often the best previous solutions are kept unmodified in order not to risk degrading them. Any number of iterations are performed until either a formal criterion is fulfilled—such as the fitness reaching a certain value—or until the researcher decides that a solution is good enough.

In our scenario, a chromosome is a sequence of chords, and instead of specifying a computer program for the fitness function, the chromosomes (or chord sequences) are to be *evaluated by the composer*, an approach explained further on. Basically, no constraints should be given for the tendencies, whilst various constraints for the structure of the generated chords, like frame intervals or certain excluded vertical harmonic constellations, could be specified for the generation of the chromosome populations. We asked Toufektsis to help provide some basic conditions for the generation of the chord material.

As a starting point, Toufektsis produced some example sequences by hand, a selection of which is shown in Fig. 3.

Basic Parameters for Sequences

How does one arrive at these sequences? The first step that Toufektsis suggested was to formulate a number of parameters which would constrain the possible chords and the movement of each voice. For the vertical structure, one could for example eliminate octaves or the combination of certain intervals. As well, each voice within a chord would be restricted by its register (minimum and maximum pitch). Using the basic parameters, one can calculate the number of possible chords which conform to these constraints.

The material would further shrink if additional parameters were given for the horizontal structure. For instance, one could forbid the crossing of voices or the occurrence of consecutive fifths. Furthermore, one could define each voice by its register (minimum and maximum pitch), the maximum step size when moving up or down, as well as the minimum and maximum distance between the voices. Still, when an overall form of a sequence is defined, such as having the top voice move upwards while keeping the other voices around a central note or even immobilising them, there are many alternative ways to achieve this form. It was clear that Toufektsis would have an implicit idea about how the voices could actually move and how to decide whether a particular motion was favourable or not.

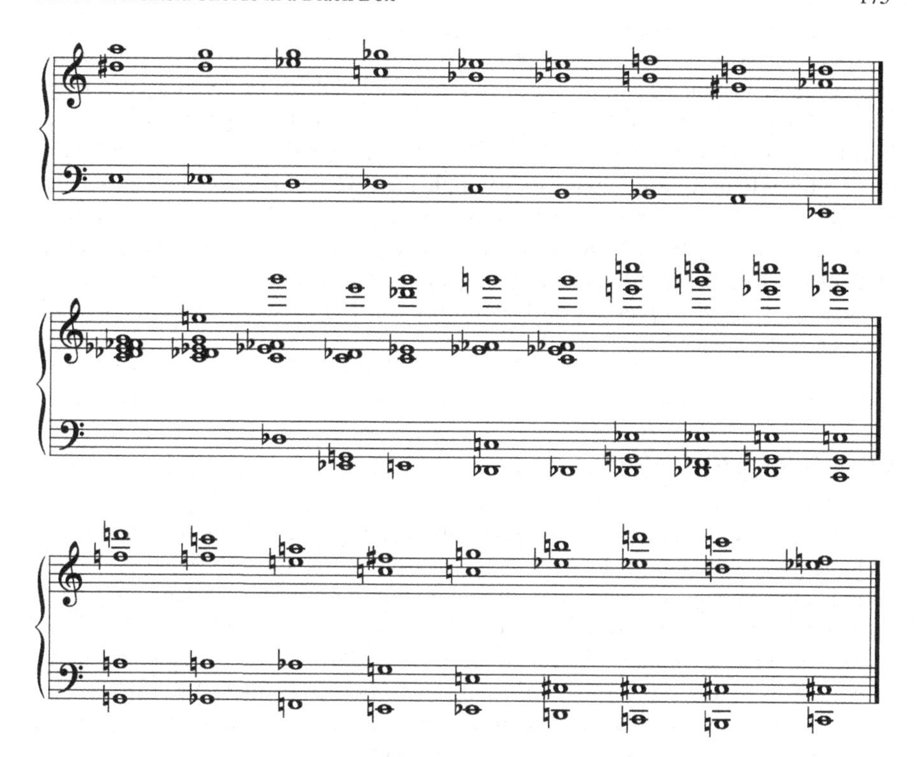

Fig. 3 Manually constructed chord sequences

Interactive Evolutionary Computation

To approach the construction of these sequences, we used a hybrid form of a Genetic Algorithm, where the computer-generated sequences are examined by Toufektsis in an interactive fashion. While the GA is responsible for permutating the material and obeying some basic undisputed rules, the composer takes the role of the evaluation stage of the algorithm. The resulting procedure is thus called Interactive Evolutionary Computation (IEC) [7].

The advantage of using IEC is that a scenario is created where the evaluation criteria cannot be explicitly stated. The user of the system is presented with a list of solutions and has to rate as a "black box", typically using a discrete scale, for example from zero (not good) to five (very good). Other than a computer with a fixed evaluation function, there is no guarantee that a human will not introduce a drift or noise in the implicit criteria which guide his or her evaluation. Another difference is that a human will most likely not arrive at one perfect solution but regard multiple solutions as equally good, so the optimum of the search presents itself rather as an "area" instead of a single point. Again, in our case this is something expected, as there is no reason why two alternative sequences may not be regarded as aesthetically or formally equivalent.

There are however two related problems which cannot be avoided. Unlike the automated algorithm which can easily run through hundreds of iterations and large populations, the manual evaluation is much slower, quickly leading to human fatigue. In IEC scenarios, often only 10 or 20 iterations are performed. The process can also be simplified to minimise fatigue, for example the IEC based jazz melody generator *GenJam* [1] asks the user to incrementally evaluate measure for measure, instead of presenting a complete melodic sequence at once. We have thought about a similar technique, where the chord sequence is built incrementally, however we did not apply it since we wanted to preserve the possibility for Toufektsis to judge the overall contour of the voices.

Constraint Satisfaction

In order to allow Toufektsis to focus on a particular subset of all possible chords and melodic lines, for example by restricting voices to particular ranges and excluding certain intervals, the horizontal and vertical rules as described in the beginning have to be applied to the generation of the initial population and also must be regarded in the breeding phase (mutation and crossover). An elegant method to implement these constraints is to use a solver for constraint satisfaction problems (CSP) [8].

A CSP is defined by a set of variables which are initially unknown except for a given bound for their domain. For example, the yet-to-determine pitch of a note can be represented by a variable p with an integer domain interpreted a MIDI values. We can give some arbitrary initial bounds; we might say that no pitch should be less than 0 or greater than 100, thus $0 \leq p \leq 100$. A chord of M voices likewise is then a vector c of pitches p_i, $i = 0, 1, \ldots, M - 1$. A sequence of N chords becomes a matrix q of integer variables:

$$q_{i,j}, \quad i = 0, 1, \ldots, M - 1, \; j = 0, 1, \ldots, N - 1.$$

Here the column vectors correspond to individual successive chords. We have now placed constraints on the variables. Since crossing of parts is forbidden, we may say:

$$q_{i,j} < q_{i-1,j}, \quad \forall i : 0 < i < M, \; j \text{ fixed}$$

or just looking at the chord level (column vectors):

$$p_i < p_{i-1}, \quad \forall i : 0 < i < M.$$

Here by convention voice indices (rows) are sorted from high to low. So for each adjacent pair of pitches within a chord, a relation $<$ will be used as a constraint to the possible outcomes of the two variables. Similarly, we can establish register bounds, using a constant value on the right hand side of the constraint equation.

Horizontal constraints can be defined in the same manner. For example, to allow voice k to move up maximally by a major second in each step, we specify:

$$q_{k,j} \leq q_{k,j-1} + 2, \quad \forall j : 0 < j < N, \quad k \text{ fixed}.$$

Establishing interval constraints is slightly more involved. For example, avoiding octaves within a chord, can be formalised as:

$$(p_j - p_i) \bmod 12 \neq 0, \quad \forall i, j : 0 \leq i < j < M.$$

This reads as: there may be no combination of any two pitches of a chord which forms an interval of 12 semitones or multiples thereof. Extending this to veto the co-presence of two intervals requires combinations of three pitches and some additional boolean operations on the modulus constraints. For example, to forbid major triads, the combination of major third plus minor third:

$$\begin{aligned}
\Pi_{\text{top}} &: \exists (p_i - p_j) \bmod 3 = 0 \\
\Pi_{\text{bot}} &: \exists (p_j - p_k) \bmod 4 = 0 \\
&\forall 0 \leq i < j < k < M \\
&\neg(\Pi_{\text{top}} \wedge \Pi_{\text{bot}}).
\end{aligned}$$

Property Π_{top} is true if among any combination of three pitches, the top interval is a minor third, property Π_{bot} is true if the bottom interval is a major third. The constraint is that not both properties must be true at the same time.

The "constraints solving system" maintains this set (Z, D, C) consisting of the variable set Z along with their domains D and the set of constraints C applied to these variables. It can then search for a single or a number of solutions for the variables which satisfy the given constraints. When we generate the initial population of chord sequences for the GA, we only want sequences which meet the constraints. For example, if the population size is P, we can ask the solver to give us P solutions for Z and construct the resulting chord sequences from them. If there are less than P solutions, we decrease the population.[6]

Typically, though, the number of solutions in magnitudes is higher than the desired population size. If we used a standard search strategy and a standard selector—the component which explores the possible values in the domain of a variable—and then stop the search after P solutions, we would get sequences which are almost identical, except for the minimum systematic variation. In other words, this subset is a very bad approximation of the overall solution space. The reason is that the standard selectors usually choose the minimum or median of the current domain of a variable for exploration. Luckily, we can use selectors based on a pseudo random number generator, and this way we get a "broad mix" for the initial population.

[6] For an approach to generate chords solely based on the application of constraints, see Nachtmann's project in this book. In his case, a full enumeration of all possible results is used.

The same applies to the breeding stage of the GA. In crossover, when two sequences S_1 and S_2 are taken and recombined such that the beginning of S_1 is concatenated with the ending of S_2 and the beginning of S_2 with the ending of S_1, we want to make sure that the part writing rules are not violated. In mutation, when moving the pitch of individual voices of selected chords up or down, we want to make sure that neither the part writing rules nor the harmonic rules are violated. Therefore, we use the solver again to reject modifications which violate constraints in crossover, or to calculate the mutations directly as randomised solutions.

Running Evaluations

Figure 4 shows the scheme of the algorithm. In a non-interactive scenario, the manual step "rate each chromosome" is replaced by an automatic evaluation based on a given fitness function. The second difference to a standard GA is the addition of the

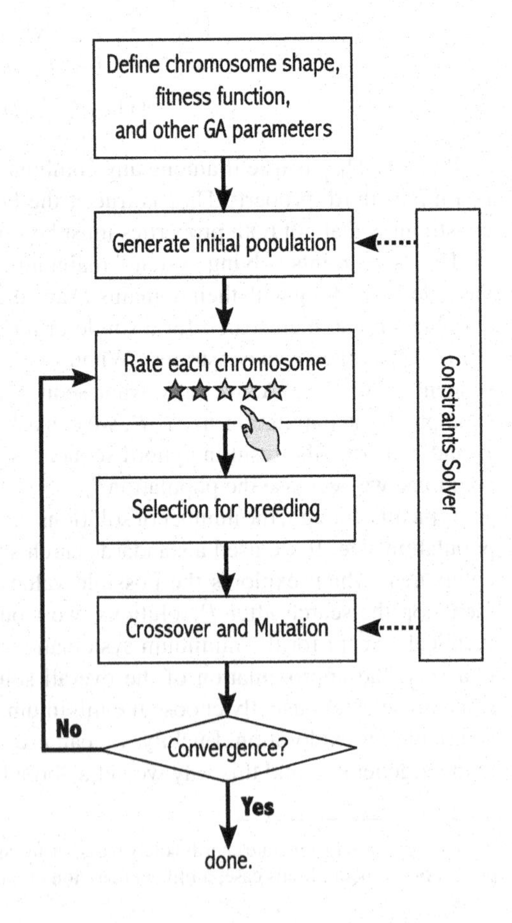

Fig. 4 Schema of the Interactive Genetic Algorithm

constraints solver directing the generation of the initial population and the breeding stage. In our implementation, the GA parameters can be edited on a graphical user interface. Here one specifies the number of voices, the length of each chord sequence, and the total number of sequences or population size. For each voice, one may specify the constraints explained in the previous section, such as allowed registers, allowed or forbidden pitch steps. Global constraints can be added to avoid particular harmonic constellations between the voices.

The main interface element is a table showing the current population, each row of which corresponds to a chord sequence. With the beginning of each experiment, a new fresh population is generated, typically using a size of 100. On the right hand side of each sequence in the table, the rating can then be made via mouse or keyboard, using a discrete scale of six steps (0–5). Initially the rating of each sequence is zero, allowing Toufektsis to skip over sequences which he does not want to consider at all. After the rating has been done, one presses a confirmation button which then proceeds to the selection and breeding stage, updating the table with the new population of the next iteration.

The GA settings as well as individual iterations can be saved on disk for the research team to examine at a later point.

Breeding is done through a combination of mutation and crossover. In mutation, a customisable percentage of the chords of a sequence is affected. Within an affected chord, a number of voices are selected and the respective pitches are moved up or down randomly within a given maximum range. A single crossover selects a position with two parent chord sequences, cuts the sequences at that point, and then produces two child sequences by exchanging the tails. The constraints solver verifies that none of the melodic or harmonic rules are violated by either mutation or crossover.

For our experiments we aimed to generate chord sequences of a certain harmonic tendency to be evaluated by Toufektsis. We wanted to keep the evaluation criteria somehow fuzzy to be rated more on an intuitive level and thus provoking surprising results.

To rate the principal functionality of the algorithm we started our experiments by indicating simple targets, which could also be easily solved by an algorithmic fitness function.

At first Toufektsis should rate the generated chord sequences in regard to a minimum distance between the first and the second voice. By rating longer chord chains Toufektsis experienced rather quickly the fatigue mentioned above. In addition, the procedure was experienced as quite boring as the fulfillment of the criteria was not challenging and optimal solutions could be easily found by pencil and paper. It turned out that satisfying results in a reasonable time should be achieved by reducing the number of chords to five.

In our next experiment single melodies were generated where Toufektsis should rate them in regard to an overall optimal upward direction. To counter the downsides of the user rating in the last experiment and set up a more challenging task for Toufektsis, we indicated a length of 13 notes within a frame interval of a minor seventh (F_4–$E\flat_5$), thus an optimal solution by a chromatic tone row could not be achieved within the given constraints. Figure 5 shows eight melodic lines from different generations that were rated 1–3 possible stars out of 5.

The previous experiments clearly showed some of the drawbacks when applying human fitness ratings to well-defined specifications, that can also be evaluated through an algorithmic fitness function. Nevertheless it should be mentioned that even if an algorithmic fitness function can be defined which meets all the necessary constraints, the solution space, according to the nature of the GA, is still incomplete,

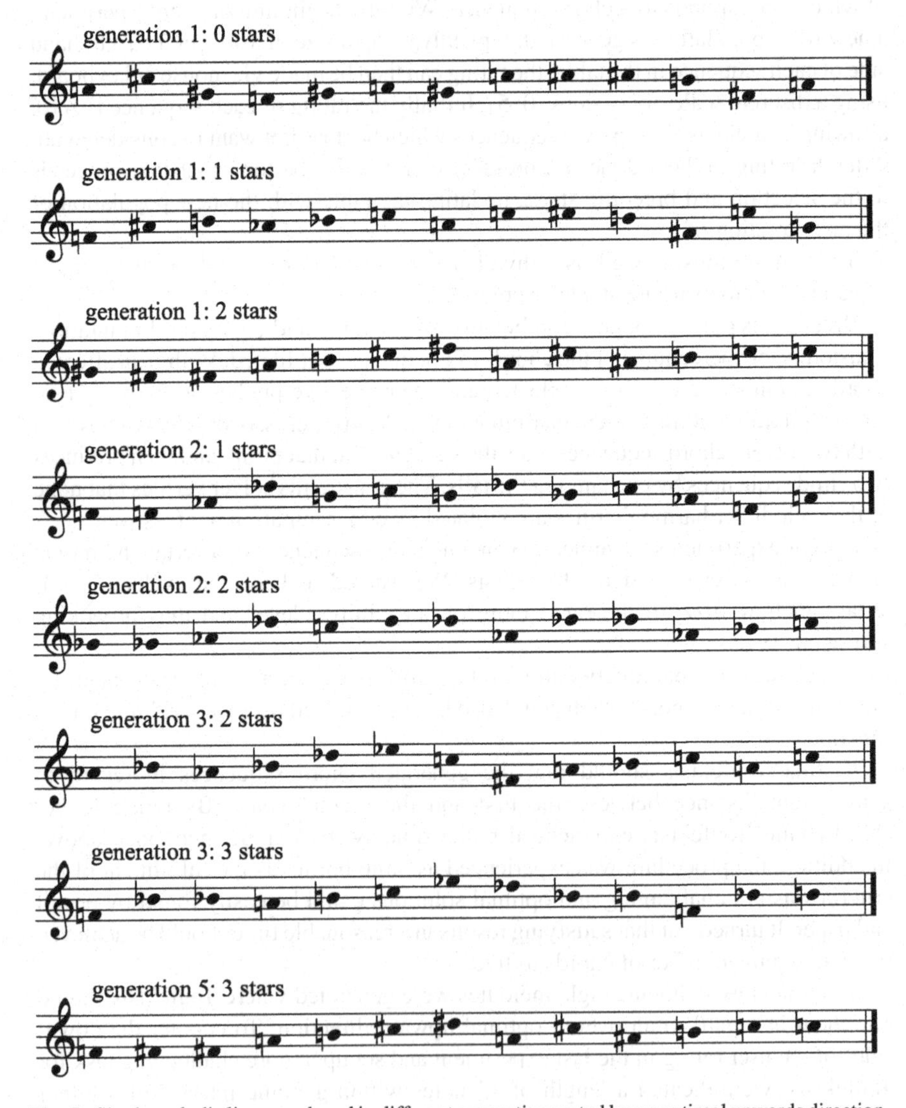

Fig. 5 Single melodic lines produced in different generations rated by an optimal upwards direction

and as a result, it is not always possible to reach an optimal solution for the musical task.[7]

Our next experiments involved the generation of 4-part chord sequences with a general upward tendency in the voice leading, using 12 chords, where each individual voice should be generated within the following frame intervals:

- voice 1: G_4–$E\flat_7$ (MIDI pitches 55–99)
- voice 2: G_4–$E\flat_7$ (MIDI pitches 55–99)
- voice 3: C_3–$E\flat_7$ (MIDI pitches 48–99)
- voice 4: C_2–$E\flat_7$ (MIDI pitches 36–99)

The same output chord is given as a starting condition for all generations. At this point the evaluation criteria was less strictly held than in the previous experiments. Toufektsis was to consider the upwards movement, but also to judge the sequences according to their potential use in his compositional processes.

In this experiment there was a maximum rating of 3 stars, but the number of these solutions increased fivefold in comparison of generation 1 with 8. Figure 6 shows two solutions rated 3 stars from the 1st and the 8th generation.

POINT: These examples show a structurally very different path, what was the reason to evaluate both solutions with the same rating?

Toufektsis: Starting with the same two chords, the sequences depict a very clear upward movement. The first example shows a more homophonic character, which in a negative sense especially displays the transition from chord 4–5. Overall, however,

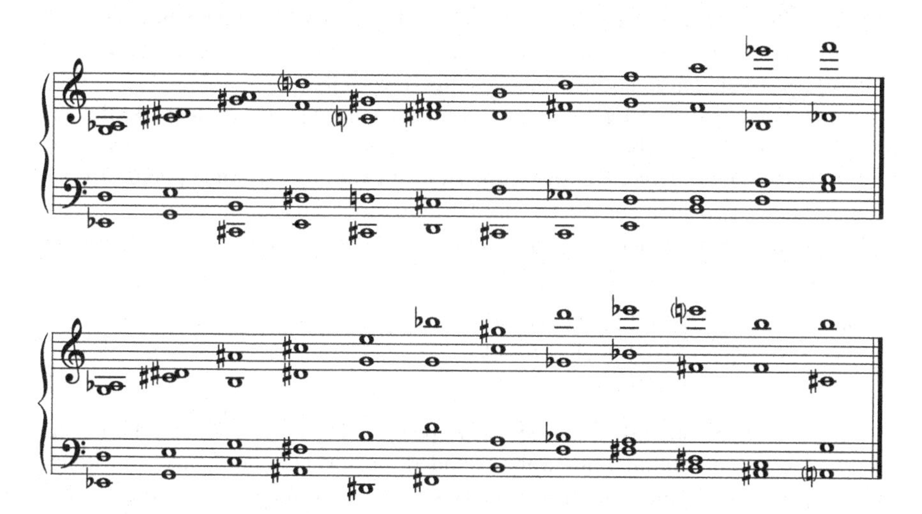

Fig. 6 Solutions rated three stars out of five, from generations 1 and 8

[7] If there exists sufficient problem-specific knowledge in a given musical domain, then rule-based systems are in most cases superior to a Genetic Algorithm as shown by Phon-Amnuaisuk and Wiggins who harmonised a given soprano voice with both approaches, see [6].

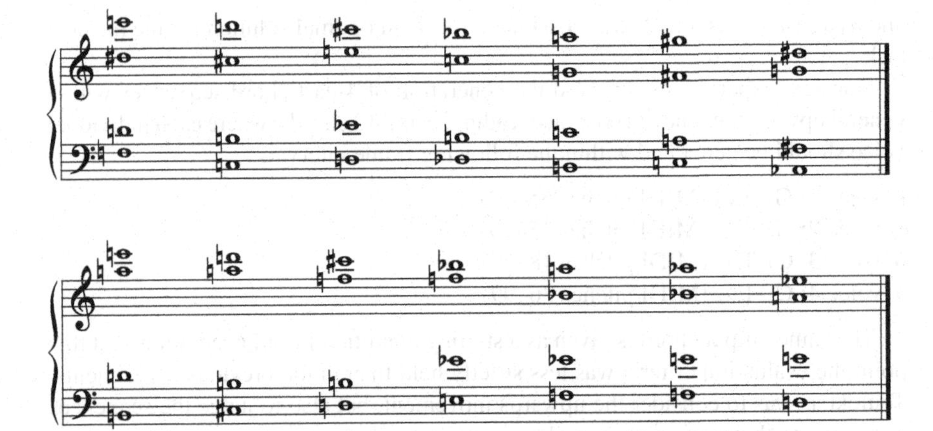

Fig. 7 Chord progressions with a fixed soprano voice composed by Toufektsis

the upward movement is felt more acutely in example 1 than in 2. The somewhat indistinct upward movement in the second example compensates itself for me, to a certain extent through the movement of the voices and turning points creating a more interesting overall sequence than in example 1.

POINT: For our next experiments we moved away from clearly defined initial conditions and wanted to compare composed chord sequences with generated ones. The idea behind these experiments was to allow Toufektsis to evaluate the likelihood that he could have composed them. It was to bring him further insight into his own preferred interval combinations, voice leading flows, etc.

In his piano piece, *Diminutionen*,[8] composed during the project, Toufektsis intended to work with chord sequences with a given soprano voice (E_6, D_6, $C\sharp_6$, $B\flat_5$, A_5, $G\sharp_5$, $D\sharp_5$), "to look at the sequence from various harmonic perspectives". Figure 7 shows two such composed chord progressions from Toufektsis.

For the following generations the soprano voice was accordingly given and for the other voices the frame intervals were as follows:

- voice 2: $F\sharp_4$–F_5 (MIDI pitches 66–77)
- voice 3: G_3–$G\sharp_4$ (MIDI pitches 55–68)
- voice 4: $A\flat_2$–F_3 (MIDI pitches 44–53)

All generations should be based on a specific chord. As further criteria, the following interval combinations were excluded: no seventh chords and no tritones or repeated notes in the bass. The solutions were evaluated by Toufektsis with a maximum of 4 stars, two of these generations are shown in Fig. 8.

POINT: How do you judge these chord sequences?

[8] *Diminutionen* translates to "Diminutions".

Fig. 8 Two solutions from the 5th generation, using chord material from *Diminutionen*, rated with 4 stars

Toufektsis: The first chord sequence basically shows interesting interval combinations in the individual chords, spread well within the respective frame intervals. The voice leading has for me, on the whole, also succeeded. Chord number 3 is particularly appealing because it is made up of a chromatic cluster, like the opening chords in the first composed sequence, see Fig. 7. The bass line is also successful, but I would not compose a parallel fourth leap in the bass and soprano, as it appears in the last chord, nor would I compose tritones between the bass and tenor voices.

The second sequence also shows interesting interval combinations with an appealing voice leading. The 4th chord has a particular character because of the unusual distance between the bass/tenor and alto/soprano voices, but is still very well integrated into the context of the other chords due to the favourable voice leading. For me, also very successful is the transition from chord 2–3: three voices lead downwards, one upwards; also appealing is the D in the soprano voice of chord 2 which then reappears in the bass voice in chord 3—there is a strong likelihood that these solutions could have been composed by me.

POINT: In our final experiment, the same soprano voice was given, we applied the same exclusion of certain intervallic combinations and the same requirements were used for the voice leading in the bass, but in this example no starting chord was specified. Toufektsis was to assess the originality of the solutions and also to what extent it was likely that he could have composed such chord sequences. In this experiment some sequences were rated by 5 stars, of which two solutions are shown in Fig. 9.

POINT: In addition to composed chord sequences Toufektsis used generated chord sequences from the final experiment in his piano piece *Diminutionen*. What was the reason to evaluate the two illustrated sequences with 5 stars?

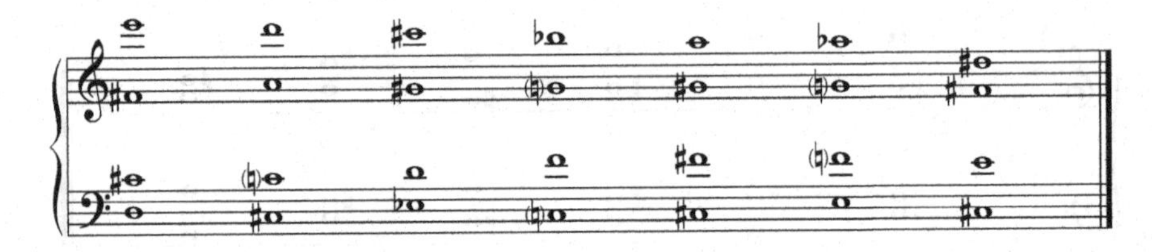

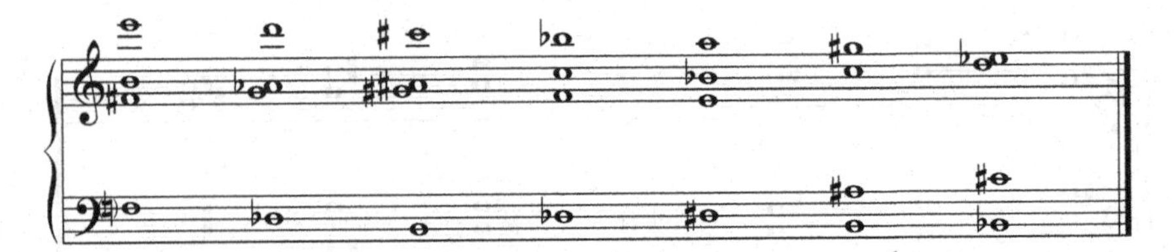

Fig. 9 Generated chord sequences in the final experiment, which were assessed with 5 stars by Toufektsis

Toufektsis: In principle I find both chord sequences very successful. The voice leading and the intervallic combinations are quite appealing in both solutions. The minor ninth chord with the fifth in the bass (chord No. 5 in example 1) I would probably not have composed, but it still works in the context of the other harmonic constellations. I find example 1 particularly attractive because there are a limited amount of pitches in each voice and the melodic intervals make up a maximum of a minor third. In the second example I find the unconventional voice registers together with the conclusive interval combinations very successful.

POINT: These and other chord sequences you have used in very different manners, but the chords appear in very different rhythmic configurations?

Toufektsis: In this piece the harmonic material is used both in homophonic and rhythmically complex layers. The structure arises from the combinatorial arrangement of three elements:

1. Chord progressions.
2. Fast monophonic succession of notes derived from the harmonic material.
3. The sound of prepared piano keys in the upper register.

In principle, seven durations in different rhythmic proportions are dynamically processed, described most appropriately with acceleration and deceleration. In addition, varied repeated notes in several places are accentuated by the prepared piano set up. The chord sequences used in various passages also blend into each other or are simultaneously used.

Figure 10 shows a homophonic refinement of the composed chord sequence from Fig. 7. The first chord sequence is processed from bar 6 until 11, the processing of chord sequence 2 begins in bar 13.

Fig. 10 Treatment of composed chord sequences in *Diminutionen*

Fig. 11 Processed generated chord sequences in *Diminutionen*

A rhythmically complex variant is shown in Fig. 11. Here the two 5-star rated chord sequences are processed. The first chord progression begins in bar 37 with C♯ in the left hand and ends in bar 38 in the first beat with C♯, D♯, E, F♯. The second chord progression begins in bar 38 in the second beat with E, F, F♯, B, where there is already an overlay of the second chord (D♭, G, A♭, D in the bottom sequence of Fig. 9) and ends in bar 39 in beat 2 with a two-tone chord (D, D♯), see the last chord (B♭, C♯, D, E♭) in Fig. 9.

Project Review by Orestis Toufektsis

A central aspect of my work as a composer is the creation of coherency between material and form. In my previous work, I have always used algorithmic principles, especially to structure formal aspects of my compositions—the material originating as a consequence of dealing with the formal structure.

In this project, however, the generation of harmonic material was the focus of our investigation and therefore a rare opportunity arose for me to work with the emanated material, and through my active confrontation in working with the qualities of the chords, further formal structures were explored and developed.

What was revealing in this investigation on the work with the Genetic Algorithms, was the openly designed human fitness function, which enabled me, starting from simple constraints, to incorporate increasingly complex considerations in my evaluation criteria. In the last experiments I rated, for example, not only a general tendency of the harmonic chord sequences, but also the functional relationships between the elements of the harmonic progressions.

An algorithmic fitness function would certainly not have led to these results: firstly, because it would have needed a very complex set of rules in order to incorporate even a small amount of my preferences in a meaningful way, and secondly because if so, the solutions would not have been surprising as the ones which were generated through my open evaluation criteria.

The chord sequences that were highly rated by me, apart from their "syntactic correctness" [9] also had another interesting effect on my work—occasionally surprising results arose, which would not have been created as easily by "pencil and paper".

Detailed examination of the results of the Genetic Algorithms opened up a new dimension of reflection: through the necessity of coherently evaluating the generations, I became aware of the structure of numerous constraints, which I had already intuitively applied in my compositional work.

---

[9] **Point:** The correctness of the solutions in relation to the given constraints.

References

1. Biles J (1994) GenJam: a genetic algorithm for generating jazz solos. In: Proceedings of the 19th international computer music conference (ICMC). Aarhus, pp 131–137
2. Chomsky N (1957) Syntactic structures. Mouton and Co, The Hague
3. Davies P, Gribbin J (1992) The matter myth: dramatic discoveries that challenge our understanding of physical reality. Orion Productions, New York
4. Goldberg D (1989) Genetic algorithms in optimization, search and machine learning. Addison-Wesley, Boston
5. Haupt RL, Haupt SE (2004) Practical genetic algorithms. Wiley, Hoboken
6. Phon-Amnuaisuk S, Wiggins G (1999) The four-part harmonisation problem: a comparison between genetic algorithms and a rule-based system. In: Proceedings of the AISB'99 symposium on musical creativity. Society for the Study of Artificial Intelligence and Simulation of Behaviour. Edinburgh, pp 28–34
7. Takagi H (2001) Interactive evolutionary computation: fusion of the capabilities of EC optimization and human evaluation. Proc IEEE 89(9):1275–1296
8. Tsang E (1993) Foundations of constraint satisfaction. Academic Press, London

Alexander Stankovski/Mirrors Within Mirrors

Alexander Stankovski, Daniel Mayer and Gerhard Nierhaus

Alexander Stankovski was born in Munich and grew up in Vienna.[1] He had piano lessons from the age of six years and music theory lessons from the age of 12. When he was 16 he attended an analysis course with Karlheinz Füssl, where Beethoven's Piano Sonatas were analysed from the perspective of the Second Viennese School. The concentrated atmosphere and stimulating discussions gave Stankovski his first opportunity to speak precisely about music, where form could be analytically described, and therefore in part, its meaning and reception could be better understood. Shortly after, he began to study composition and music theory at the University of Music and Performing Arts Vienna. His main memories from this period of study are of a conservative, and sometimes authoritarian mode of academia. Memorable rays of hope for him were at the electronic music department, ELAK, where instead he found an environment which fostered chaotic-production, where Stankovski created his first composition for tape. It was also unforgettable for him, when he heard the first concerts of the ensemble known today as Klangforum Wien (when it was still under the name, Société de l'Art Acoustique) at a time when composers such as Sciarrino, Grisey, Lachenmann, Furrer or Nono, were still completely unknown in Vienna.

[1] Biographical introduction and texts from the composer translated from the German by Tamara Friebel.

A. Stankovski
Institute for Composition, Music Theory, Music History and Conducting,
University of Music and Performing Arts Graz, Graz, Austria
e-mail: alexander.stankovski@kug.ac.at

D. Mayer · G. Nierhaus (✉)
Institute of Electronic Music and Acoustics, University of Music
and Performing Arts Graz, Graz, Austria
e-mail: nierhaus@iem.at

D. Mayer
e-mail: mayer@iem.at

G. Nierhaus (ed.), *Patterns of Intuition*,
DOI 10.1007/978-94-017-9561-6_9

After graduating, Stankovski went to Frankfurt to study with Hans Zender in his newly founded composition class, where he also met personalities like Isabel Mundry and Hans-Peter Kyburz. They all inspired his future: on the one hand he learnt a reflective analysis of tradition, which is not just simply taken for granted and continued but instead is based on the experience and broaching the issues within its historical distance, and on the other, he absorbed the development of rational, compositional strategies that could be formulated, forming something like the "syntax" of a possible music language.

However, perhaps the most important influence during this time in Frankfurt came from literature and painting, where from Fernando Pessoa and Gerhard Richter he learnt about the recognisable splits and schisms of the creative personality in diverse artistic media: the oeuvre of an artist is not a succession of separate "creative periods", but a conscious contrasting, so to say, a contrapuntal juxtaposition of sequences of works that abruptly oppose each other; although between these oppositions, subcutaneous connections can exist. As in Pessoa's heteronyms, where multiple imaginary characters can be created by one writer to write in different styles, fictional poets with their own biographies and different aesthetics, for example, seen in Richter's harsh coexistence of the most diverse painting techniques, where there is no continuous style or personal signature, but the person appears as a "common denominator" of the differing, conflicting expressions. The Belgian poet, Henri Michaux, 10 years younger than Fernando Pessoa, had already held this stance for a long time. He wrote in 1937: "Il y a pas un moi. Il n'est pas dix moi. Il n'est pas de moi. MOI n'est qu'une position d'equilibre (Une entre milles autres continuellement possibles et toujours prêtes.)".[2]

These thoughts are also present in the work of Stankovski. The continuity and quality of his work doesn't show itself in an intended, readily recognisable personal style, but in the continuous leading of new and differing working compositional modes. In 1996 Stankovski returned to Austria and worked for a few years as an assistant in the composition class of Michael Jarrell at the University of Music and Dramatic Arts Vienna.

Since 1998, alongside his composing career, he teaches counterpoint, music theory and musical analysis at the University of Music and Performing Arts Graz.

At the moment, besides the string quartet he is composing for this project, he is also working on two multimedia pieces that demand very different aesthetics and compositional techniques. The first is an opera project using an old Chinese ballad and the second is a melodrama for speaker and instrumental ensemble, using a text from the Austrian author Xaver Bayer.

[2] "There is no I. There are not 10 I's. There is no I. I is only a position of equilibrium (which is only one among a thousand others, with unending possible variations, always ready for delivery on demand)." Quote [7, p. 217] translated from the French by Alexander Stankovski.

Artistic Approach

Statement

I believe in the meaning of art, in which each artist in his or her work must find and invent a means of expression, which is independent of its use and worth.

I believe in intuition, which through the artist as a person enables a vision of something not yet in existence to emerge, becoming reality uniquely from him or her.

I believe in a communication between the composer and musician, the musician and composer and between the composer and listeners.

I believe in being a self-critic, where a view of one's own work is as if they had an outsider's perspective.

I believe in chance, where unexpected results can arise, even with the most detailed planning.

Composing means for me, that decisions are made, "lines are drawn" and constraints are envisaged. I am unable to compose without a selected and defined scope of constraints. The definitions themselves, the containment of my possible decisions, can change from piece to piece and even within that, from movement to movement, from layer to layer or from section to section. I'm interested in the juxtaposition of differently defined regions. It is not about the mediation of opposites, rather it is about the representation and experience of incommensurability.

Personal Aesthetics

I have no personal, recognisable style and I also do not aspire towards one. I attempt, on the other hand, to put out as many different artistic goals from piece to piece as possible, as far as it appears achievable within my means. On the other hand, I often come back to already posed queries and thoughts. Various differing work groups and series are formed, intentionally, where it could appear to have been produced by different composers.

Pieces with implicit or explicit reference to works from past epochs, which will in turn become their own structural foundation, where the association to the original text of "komponierter Interpretation" (Hans Zender),[3] reaches its own full re-forming of the musical material. I have directly referred to compositions of Arnold Schönberg, Johannes Brahms, Anton Webern, Girolamo Frescobaldi and Claude Debussy in a row of pieces and in each case have reworked them in very different ways. There is also, alongside, a reference to one's own tradition, i.e. fragments from earlier works can become the basis for new compositions.

---

[3] "Komponierter Interpretation" refers to "composed interpretation", see Hans Zender: *Schuberts Winterreise—Eine komponierte Interpretation* für Tenor und kleines Orchester (1993).

Pieces, which are conceived as monodic lines and respectively as contrapuntal networks of multiple lines. One of the applied techniques at this juncture is an imprecise mirroring of material, in order to bring forward a self-referential virtual, unending continuity.

Pieces, where non-musical "objets trouvés" are used (for example, sounds of nature) and an attempt to most accurately transcribe these sounds for instrumental music, which implies to refrain mostly from an immanent musical logic, replacing it with a given "extra-musical" sound shape.

Pieces with a spontaneous approach, without premade conscious defined rules: being thrown back on one's own subjectivity without diversion of one's own decisions made through a self-inflicted resistance.

Pieces especially written for radio, with a focus on text.

Pieces, where the central compositional strategy is based on the reduction of the available means.

Pieces with mixed approaches, which consist of stylistically and technically very different parts. The resulting tension should simultaneously create the impression of incommensurability and the interrelation of individual parts of a composition.

It can be seen, that out of these anytime-expandable-categories, my compositional work should reflect our present time, with its abundance of artistic possibilities, but also at the same time should place itself as something new against the virulent questions about the definition and meaning of the artistic subject which has been a theme since the end of the 19th century.

Formalisation and Intuition

At the beginning of the compositional process a number of things can exist: a formal idea, the involved instruments, a text, etc. The imagination of a piece at the outset is indeed undefined with respect to details, but it can however have a very strong conception, that already over a long time, sometimes over years, has stayed in the thought process before it becomes a reached goal. The way to this point, the compositional technique, must be invented and found during the compositional process.

Starting from a general conception of a piece, I arrive, via the formulation of rules, to the realisation of these. The rules here are not ends in themselves, but are preliminary signposts, which after the musical results that they lead to, are judged and can accordingly, if necessary, be changed. The deviation of the rules can also lead to their abolishment; in extreme cases the rules serve only as a beginning point in order to dismantle them.

Composition can refrain from formalisms only with difficulty, although this was necessary in certain moments in music history during which especially interesting music was created—e.g. during the so-called free atonal phase (around 1908–1923) of the Second Viennese School.

Composition must always be more than a act of formalisation. Formalisation is only the first step, then a second must follow: a critical debate with the rule-generated, and the detailed post-editing from within (transcribing data), or if applicable, also its destruction from outside (overwriting data).

The decision, when, if and to which degree the transcription or overwriting occurs, can in my judgement, not be met on the level of formalisation, but through an instance which attempts to receive the vision of the relation between technical means and their musical effects.

Evaluation and Self-reflection

In judging the quality of one's work and thus its meaning, opens a wide range of self-delusion and even self-deception. I think that a composer cannot decide alone if a piece has turned out well or not: the quality of the piece reveals itself only in the process of how it is dealt with. A piece turns out as what it is, by communicating with musicians and listeners (and the composer is also one of them). This is the only way to release the potential that is inherent in the piece, and to open the possibility for it to act in which way, whatever way it should. The self-judging of the composer has to go beyond aesthetics, compositional techniques and subjective private matters and has to consider the effect on others, foremost on the performers. Otherwise there is the danger that "reflection" degrades to an academic ritual of navel-gazing.

Project Approach: The Mirroring Technique

One of multiple techniques, which I have developed in the course of time, to realise a specific artistic goal, is what I call the "mirroring technique" (Spiegeltechnik).[4] This is in contrast to traditional mirror techniques, for example the canon by inversion ("Spiegelkanon", literally, mirror canon in German), the inversion of a fugue theme or a retrograde inversion construction, which can be found in the work of composers like Guillaume de Machaut, Johann Sebastian Bach or Anton Webern. Musical material is not directly worked with, but the intervals and durations are carried over into numerical values, which are ordered in a retrograde, where the inverted number either remains the same or becomes varied by a certain value. The resulting number series is then translated back into traditional notation.

The goal is a kind of genetic code that is based on a clearly defined initial condition, allowing in every moment a tangible musical connection, but offers nevertheless

[4] Purely on a technical level, this distinguishes between spatial mirroring, the reversal of the direction of the movement of intervals (inversion) and the reversal of the temporal sequence (retrograde). While in the first case the rhythm remains unchanged, in the second part there is "mirroring". What is here called "mirroring technique" is primarily concerned with the temporal mirroring of the musical material, but in a broader sense refers to other applied techniques.

sufficient room for unexpected development. An example of this is my *Courante* for solo violin—a piece that belongs to the above mentioned second category, that means it is constructed as a monodic line: a rhythmic and intervallic initial cell (a shorter plus a longer value) becomes symmetrically mirrored around an axis (denoted by a dotted line), the cell and its inverse mirrored a second time and so on. A melodic flow is created which repeats the initial material over and over but it also transforms it constantly into a different shape or form. The intervals and rhythmical values are inverted independently from one another, where the mirrored values can show minor deviations, in a way that keeps the information of the beginning present but broken in a constantly changing way (Fig. 1).

The starting material consists rhythmically of the proportions 1:4, which is mirrored as 4:1 (bars 1 and 2). The proportions remain identical, but their assignment and therefore durations are changed: a 16th quintuplet (5 notes per quarter note) becomes a 16th quadruplet (4 notes per quarter note). In the following mirror (bar 3) the allocation is changed as well as the value of the first number pair: 1:4/4:1 becomes the proportion 2:5:4:1, measured in the 16th sextuplet (6 notes per quarter note). From these results in the following two mirrors (from bar 4 until bar 8 inclusive) 1:5:4:2:2:4:4:1/1:4*:4:1:2:3:5:1:1:4:4:1:1:4:3:1, with changing allocations (sextuplets, quintuplets and quadruplets). The values marked with a star (*) are split in a pendulum movement made from identical notes.

Fig. 1 *Courante* for violin solo, first section

With respect to the value of the intervals 1:1 (bar 1: two ascending quarter-tones between G and G#) to 2:1 (descending semi-tone + ascending quarter-tone) mirrored, the value 1:1/2:1 on their part again to 2:1:1:1/2:1*:2:1*:2*:1*:1:0*, at which those values marked with a star become played on the next highest string, thus become transposed up a fifth. Four different dynamic levels are used: *pp*, *mf*, *f* and *ff*. These four levels are assigned 1:4, to produce the following pattern (bars 1–8): 4:2/1:4/3:1:2:4/4:2:1:3:3:1:2:4/4:2*:1:3:3:1:2:4*(:3:2):1:3:3:1*(:2:4). From the * begins crescendo or decrescendo, which quasi absorbs the values following the brackets.

Rhythm, intervals and dynamic are encoded as a series of numbers that are inverted with minimal deviations (+1 or −1) and in fact without a directed tendency of this deviation. In addition, the following rules are applied:

- The allocation of rhythmical proportions with precise durations is variable within narrow boundaries.
- Longer durations may be broken down into shorter but equal durations.
- The direction of movement of the intervals is not determined.
- Intervals can be transposed through a change of the string (from bar 4) or harmonic fingering (from bar 10). A punctual dynamic with a sharp contrast from note to note becomes here and there smeared and fused in sporadic local developments, also with help of the articulation, which on such positions often moves from *detaché* (one bow length per tone) to *legato* (one bow length for multiple tones) respectively merges to *glissando* (unbroken connection of two tones).

Project Expectations

The thing that interests me about the work in this project, is at first the development and refining of the "mirroring techniques". As shown in the score examples it was necessary to have several additional rules besides the mirror itself, in order to create a musically satisfying result. To what extent can the additional rules create a "feedback" in the mirrors? Formalising aspects of my mirror technique might not necessarily lead to an acceleration of the compositional process, but I'm happy to invest, especially in our era of perverted economical thinking, in the luxury of this time-intensive and "apparently" ineffective mode of working. However, maybe new possibilities for the extension of one's own composition strategies originate right through the automisation of the process.

Exploring a Compositional Process

POINT: We see Stankovski's use of mirrors in his compositional work as continuing a historical debate: one and many, unity and variety, unity within variety (Einheit in der Mannigfaltigkeit), identity and negation, difference and repetition, several forms of an often bespoken pair of terms in philosophy since ancient times, which has also been influential in music history, though in different interpretations.

The thought of "One and Many" is seen as a basic principle by Plato, appearing in several dialogues in several forms, e.g. in *Phaedrus* [9]. As is typical for Greek philosophy, aesthetical and ethical questions are interwoven; for Plato a good life is an ordered life that integrates or subdues its plurality. But the "Many" must be ordered as an all-embracing principle and this also concerns the individuals of the state as well as the elements of a work of art [8]. Plato's critical thoughts on music are often cited, this mainly regards music not compliant with his general demands of order [10].

In his *Monadology* [11] Leibniz describes "Einheit in der Mannigfaltigkeit" as characteristic of the monads, the ensued points of the universe in his metaphysical view. In a note[5] he also identifies harmony as "Einheit in der Mannigfaltigkeit", the idea of this relation had deep impact on the music philosophy of the classical era [6].

From the beginning of the 19th century romanticism and subjectivity became a matter of philosophical debate. Hegel develops a concept of the duality of unity and plurality based on perception: unitary perceptible things do not exist without a plurality of properties [4]. The idea of a pre-stabilised or over-individualised harmony, still alive in Leibniz' thinking of "Einheit in der Mannigfaltigkeit", vanishes with Hegel. For him music is "subjektive Innerlichkeit" (subjective inwardness) [5]. Besides the plurality of perceptible things, Hegel's dialectical process of thesis, antithesis and synthesis creates a varied identity and emphasises the teleological aspect of unity and plurality.

Rejecting Plato and Hegel, Gilles Deleuze describes difference and repetition as "leading and undirected forces" [3]. This is a critique about identity and representation, a plea for the otherness and to relish the use of these concepts. As Deleuze's concept transcends classical ideas of balance as well as romantic ideas of a subject expressing itself, even denying the existence of a stable subject at all, he has become philosophically attractive to contemporary artists. Reciprocally much of his work is referring to art, in his works on cinema Deleuze differentiates between a unified view on the world connected with traditional ways of storytelling [1] and the predominance of discontinuity and missing order [2], a distinction that might well be adapted to music too.

POINT: Your use of iterated and varied mirroring leads to structures that let the dualism of identity and variety appear in several forms. What are the aesthetical reasons determining the choice of using them, do you feel obliged to any of the philosophically enrooted interpretations of identity and plurality above, or others?

Stankovski: First of all I would like to emphasise the differences between the varying discourses. I'm primarily concerned with queries of a musical nature, rather than philosophical. I am suspicious to identify music and philosophy with each other because this identification limits a potentially open scope of experiential understanding, which through precise ideas certain standards were derived, within which this scope was exactly designed for. It may be useful to refer musical and philosophical concepts to each other, especially when composers explicitly gain inspiration for their creative work from philosophy, or find elements of their artistic activity from philosophical texts.

---

[5] In a draft of a letter from Eckhard from May 1677 Leibniz denotes "Harmonia autem est unitas in multitudine".

Of the above-mentioned positions I acknowledge that my compositional interest, not surprisingly, again lies best in Deleuze's thoughts, which in turn, reflect the fundamental uncertainty of contemporary European culture. The deliberate destruction of the subject, certain in itself, seems to me to be the common theme.

I am explicitly concerned with the question of the identity of the creative personality, since my encounter, as previously mentioned, with the works of Fernando Pessoa and Gerhard Richter. Earlier I was also fascinated with Stravinsky, not only because of his impressive and perfectly crafted musical works but also because of the diversity of his stylistic interface, which raised queries about the criteria one uses to consider an oeuvre as a whole.

The mirroring technique can also be seen as a response to these particular queries. The focus lies in the foundation of an associated context, directly between very different, unpredictable musical events, through variation of a common idea. Having said that, the mirroring technique is only one part of my compositional work. I also use completely other techniques, at times in sharp contrast with each other—as a complementary reaction to the same query, but here with a focus on the diversity.

POINT: To sum up the results of some of your procedures: the beginning and ending in full measures as well as in parts show varied identity, an iterated application which leads to self-similar structures. Is self-similarity a guiding principle for you? Do you see it related or independent and in addition to principles of identity and variety?

Stankovski: The term "self-similarity" is for me too much related to very defined mathematical structures, from which I have limited precision as a mathematical layperson. What interests me in the mirror technique is a personal actualisation of the musical principle of variation. I place my work rather in relation to the musical tradition than to mathematical concepts.

Some of the musical phenomena, which emerge from the mirror technique, could be called "self-similar" in an extended meaning: for instance if the initial cells reappear in the course of a series of mirrors again almost unchanged. What is important for me is however not a greater principle, but the construction of a coherent musical speech.

POINT: In order to approach Stankovski's use of mirror principles we provided an algorithm that can perform iterated mirroring with arbitrary sequences of operations op_i and depth parameters d_i. In this way we generalised the procedure he works with, which is not restricted to musical parameters, but it isn't restricted to numbers either. We needed an operation or a sequence of operations that was defined for all of its possible results. The operations are not functions in a mathematical sense, as they might contain non-deterministic elements. For Stankovski operations were defined for numbers and worked as deviations.

Let's say we start with an axiomatic tuple of items,

$$x_0 = (x_{0,0} \dots x_{0,n_0})$$

an operation op_0 is applied to each element of the mirrored start tuple and we get:

$$x_0^* = (op_0(x_{0,n_0-1}) \ldots\ op_0(x_{0,0})).$$

It is not relevant for the explanation of the principle if the last element of the starting tuple is mirrored or not, we omitted it in this case. The depth parameter d_0 determines what amount of the mirrored tuple is actually taken. Let [ ] denote the rounded integer, then the size of the mirrored tuple is

$$j_0 = [d_0 * n_0]$$

i.e. only the first j_0 elements of x_0^* are used and x_1, the overall result of the first mirroring is the concatenation of the starting tuple x_0 and the shortened tuple x_0^*:

$$x_1 = (x_{0,0} \ldots x_{0,n_0}\ op_0(x_{0,n_0}) \ldots\ op_0(x_{0,n_0-j_0+1})).$$

The procedure is applied to x_1 and so forth. The amount of change done by the last operation determines the similarity of the start and end points. As a simple example with numbers let's start with a tuple

$$x_0 = (3\ 6\ 9\ 12\ 15)$$

with a non-varying operation that randomly adds 1 or -1 and a non-varying depth $d = 1$ a possible result could be:

$$x_1 = (3\ 6\ 9\ 12\ 15\ 13\ 8\ 7\ 2)$$
$$x_2 = (3\ 6\ 9\ 12\ 15\ 13\ 8\ 7\ 2\ 8\ 9\ 14\ 14\ 11\ 8\ 7\ 4)$$
$$\ldots$$

with $d = 1/2$ a possible result could be

$$x_1 = (3\ 6\ 9\ 12\ 15\ 13\ 8)$$
$$x_2 = (3\ 6\ 9\ 12\ 15\ 13\ 8\ 7\ 14\ 16\ 11)$$
$$\ldots$$

It is interesting to regard overall developments of the iteration process. For example in Fig. 2 with six iterations of a non-varying operation that randomly adds 1, 2 or 3 we started with tuple $x_0 = (0\ 1\ 2\ 3\ 4\ 5\ 6\ 7\ 8\ 9)$ and took full depth $d = 1$ in all iteration steps (Fig. 2).

We see a self-similar structure, a large bow form consisting of smaller bow forms with increasing and finally decreasing deviations. In this case the deviation operation, adding random values within non-varying bounds, is independent from the mirrored values and hence from the starting sequence, i.e. if we only regard the deviations in Fig. 2, or equivalently take a starting sequence of zeros, we get Fig. 3.

So in Fig. 2 the deviation sequence of Fig. 3 is just added to the repeatedly mirrored start sequence.

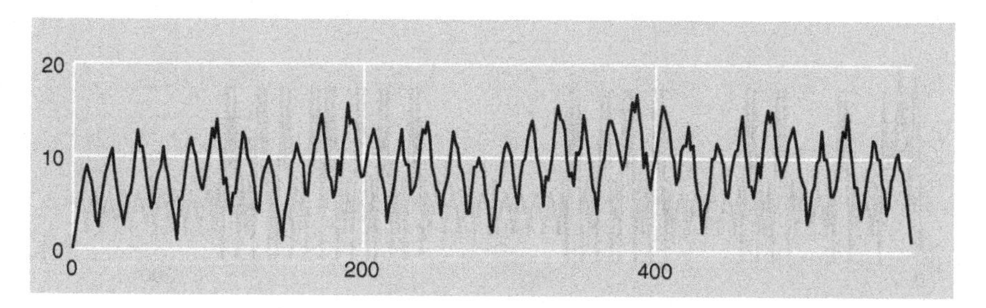

Fig. 2 Iterated mirroring with random addition of numbers 1–3, starting with numbers 0, ..., 9

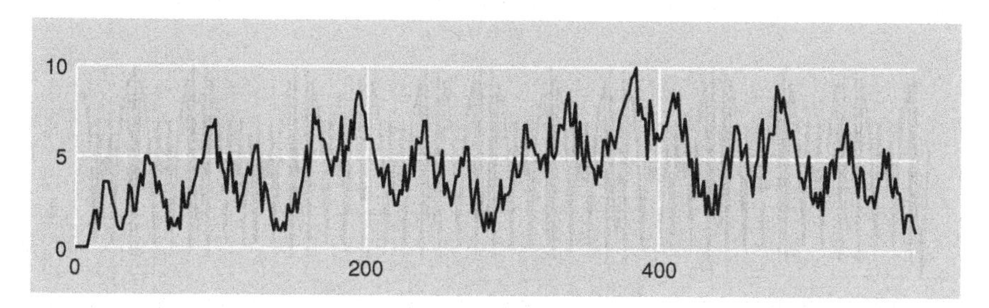

Fig. 3 Regarding only iterated mirroring and additioning of Fig. 2

With shortened mirroring, typical patterns also occur, partial sequences with mirrored shape of increasing length enfold in combination with a global tendency. Now we chose again a zero sequence at the start and a deviation operation, adding random values within non-varying positive bounds between 1 and 3. A mirror depth $d = 0.3$ and 18 iterations result in a graph shown in Fig. 4.

Regarding only bare shortened mirroring with identity operations we observe typical behaviour depending on constant depth d, independent from the start sequence. For $d < 0.5$ we end up with an oscillation between two states of increasing respective lengths, see an example of this in Fig. 5.

This, again independent from the start sequence, doesn't seem to happen for $d > 0.5$ (Fig. 6).

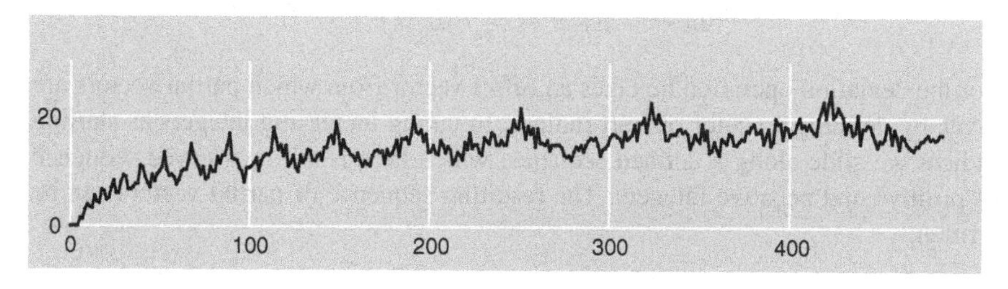

Fig. 4 Iterated shortened mirroring (d = 0.3) with random addition of numbers 1–3, starting with a zero sequence

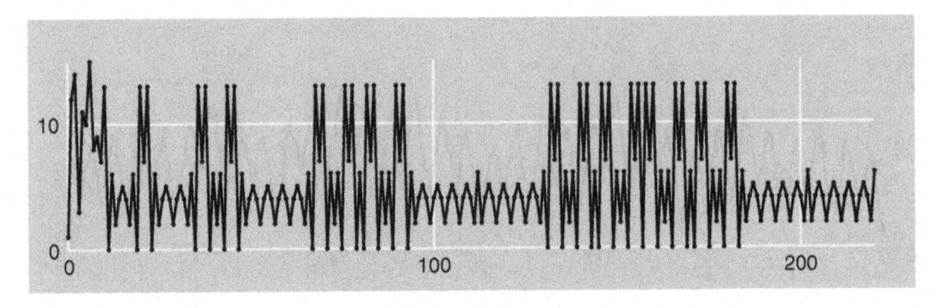

Fig. 5 Shortened mirroring with identity operations, d = 0.4

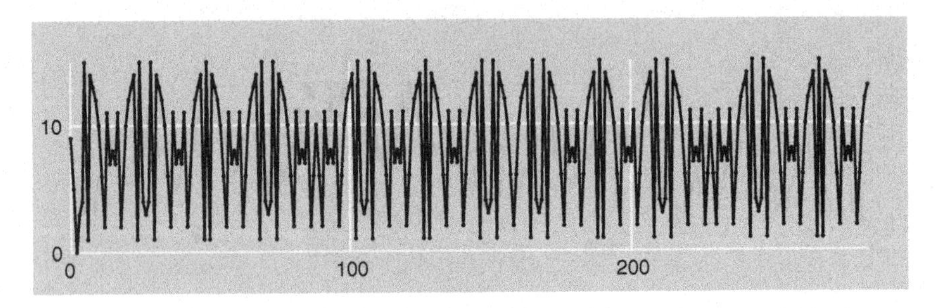

Fig. 6 Shortened mirroring with identity operations, d = 0.8

In his string quartet *A House of Mirrors III* Stankovski explores the generalised mirroring algorithm with specific characteristics. Stankovski uses several such processes to generate interval and rhythmic data, which he also subsequently adapts. Let's regard the first one which determines intervallic data for all instruments.

For a starting sequence of one element

$$x_0 = (7)$$

mirroring depths are varied, he defines them in absolute lengths (hence notated as $\underline{d}$), here just with increasing integers:

$$\underline{d}_0 = 1,\ \underline{d}_1 = 2, \ldots,\ \underline{d}_i = i + 1.$$

For the deviation operation he takes an offset vector from which partial vectors are taken by defining a vector of start indices. In taking increasing integers as starting indices we slide along a defined sequence which here is the interleaved sequence of positive and negative integers. The resulting sequence of partial vectors can be written:

$$O_0 = (0)$$
$$O_1 = (1\ {-1})$$

$$O_2 = (-1\ 2\ -2)$$
$$O_3 = (2\ -2\ 3\ -3)$$
$$\ldots$$

20 iterations give the following sequence, see Fig. 7.

As the order of increasing depths is linear, the order of increasing mirrored sequences is quadratic, hence relative depths decrease, below 0.5 quite rapidly. Again, oscillation between two states can be clearly observed.

As Stankovski uses the values as step values, deciding the directional changes from step to step, the development of absolute values is relevant (Fig. 8):

7 7 8 6 5 10 5 7 8 8 3 1 11 5 11 1 4 8 9 7 6 2 5 10 3 14 3 10 5 1 6 8 9 9 4 2 9 13 7 1 15 3 15 1 7 13 8 2 5 9 10 8 7 1 4 18...

For each instrument of the string quartet Stankovski occasionally added seconds and quarter-tone sharps and flats, the first violin starts with the interval sequence (see also score, Fig. 9):

7 7 8 6 5 10 4.5 6.5 8 8 2.5

In the first part all instruments play only intervals, adjacent intervals usually have one pitch in common, so that the interval sequence is somewhat folded. Rhythmic data is determined by a similar mirroring procedure which is not included here in this report. Finally the above interval sequence translates to the violin part (Fig. 10).

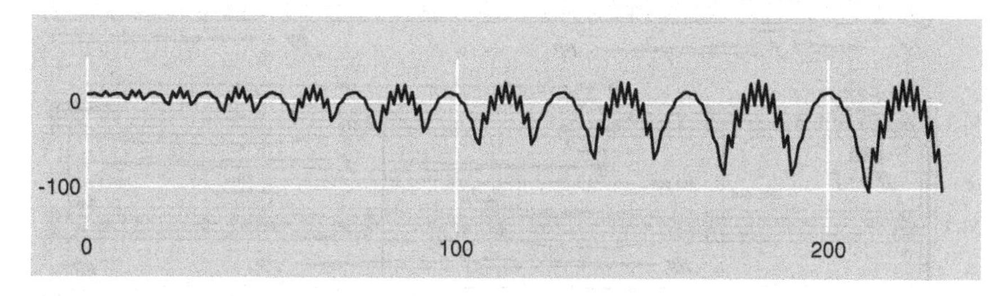

Fig. 7 First raw mirroring sequence for intervallic data of *A House of Mirrors III*

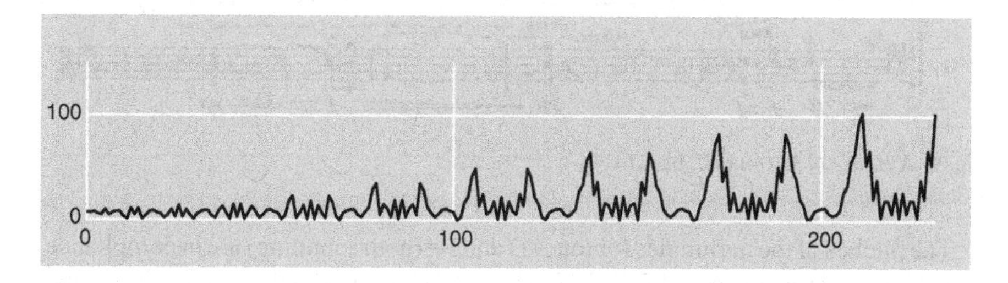

Fig. 8 Absolute values of sequence in Fig. 7

A House of Mirrors III

Alexander Stankovski (2010/2014)

Fig. 9 *A House of Mirrors III*, bars 1–10

The pitches of the harmonics for tones D and F# (both sounding) are in compliance with the partial sequence (−4.5 10 3 13.5 1.5 10 −5 −1 6 7 8) taken from the original (5 10 3 14 3 10 5 1 6 8 9 9).

Fig. 10 *A House of Mirrors III*, 1. violin, bars 20, 21

POINT: Sometimes you make quarter-tone deviations from the original interval sequence and sometimes diatonic deviations (mostly up to a maximum of a second), what are the reasons for these deviations?

Stankovski: On the one hand I wanted to use quarter-tones; on the other hand I had to probe each quarter-tone for its musical meaningfulness in order to avoid the risk of an "indifferent" microtonality. I had certain instrument-dependent and musical context-dependent criteria for the use of quarter-tones, which I believed should not be left up to the algorithm. For example it was therefore important for me to relate a quarter-tone to a simultaneously or immediately previously or afterwards sounding tone, either as a melodic deviation or as harmonic roughening.

Moreover it was part of the compositional idea to increase in the first part of the piece more and more the room for manual deviations in order to allow for further significant subjective disturbances of the perspective of the originally planned mirroring.

POINT: Experiments generalising the mirror principles you used, resulted with typical patterns. One is the oscillation of two states with increasing respective lengths, in this example from the beginning of the piece likewise with increasing values. Were there points you considered when choosing this type of procedure? How does it comply with your aesthetical preferences, i.e. concerning identity, variance and escalation?

Stankovski: I didn't give much thought in advance to repeating numerical patterns, but made at first very simple general musical considerations. It was clear to me that the original material might return several times but that its recognisability would in addition be strongly affected by the separated treatment of pitch and rhythm on the one hand and on the other hand by the continuous inversion. The continuous mirroring of the rhythmical "basic cell" (short-long) yielded very rapidly a polarity between long held tones and fast passages of single instruments (Fig. 9). The relation between both poles is slowly inverted: at the beginning long continuous tones dominate, interrupted by scattered chords, whereas at the end of the first part there are only gestures left, which are interrupted by rests.

POINT: Stankovski added additional rules: he filtered out rhythmical or interval values above a certain threshold. Applying a threshold value of 30 to the sequence of Fig. 2 lead to a development towards an almost periodic fluctuation (Fig. 11).

The values actually taken in the piece come from three generations of iterated mirroring with different depth and deviation inputs and, after every generation, filtering out zeros and values above a threshold (dependant on the instrument) plus adding

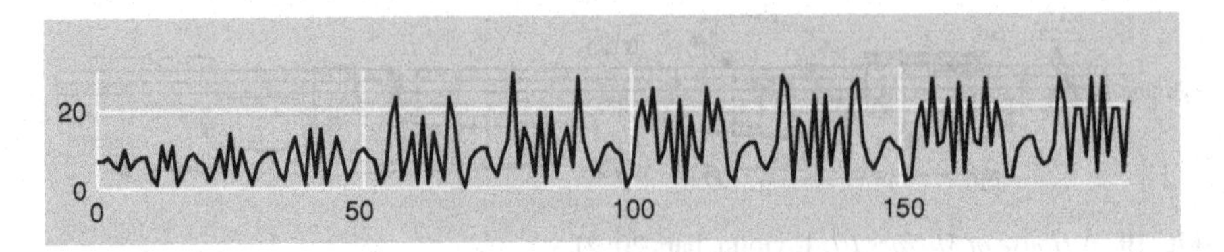

Fig. 11 Sequence of Fig. 7, filtering out values greater than 30

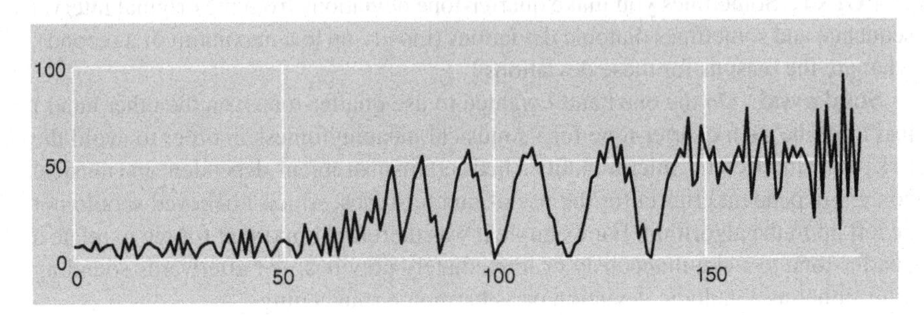

Fig. 12 *A House of Mirrors III*: concatenation of three mirroring sequences for intervallic data, absolute values, no filtering of values, no inflections

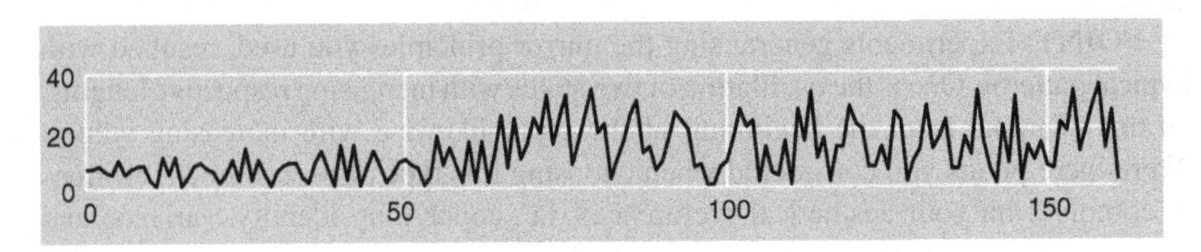

Fig. 13 Sequence of Fig. 12 with filtering, no inflections

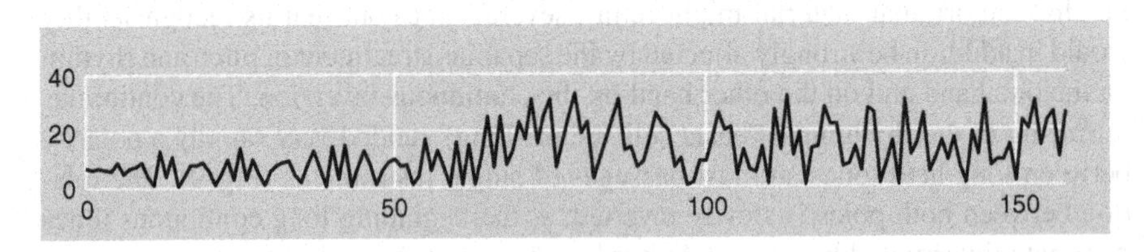

Fig. 14 Sequence of Fig. 12 with filtering and inflections (sequence of interval values used for the viola part)

microtonal inflections. That way the microtonal inflections also have an influence on the mirroring performed after them. Here one can see a comparison between iterated mirroring without filtering and inflections (Fig. 12), iterated mirroring with filtering and without inflections (Fig. 13) and finally mirroring with filtering and inflections

(Fig. 14), always reduced to absolute values. Figure 14 shows the sequence of interval values used in the piece for the viola part.

POINT: In the first part of *A House of Mirrors III* you performed three generations of iterated mirroring. In each generation you used different mirroring depths and deviation operations. What were the reasons for altering the starting conditions?

Stankovski: Parallel to the above-mentioned shift from surface to gesture runs the separation of the four instruments via the individualisation of the mirrors. The first mirror indeed separates already between intervals and rhythmical values, yet a set of mirrors holds for all four instruments. In the second mirror I separate the quartet in two pairs of voices each with different initial data, also for intervals and rhythms, to be used by the algorithm. In the third mirror each instrument finally has its individual initial data. My idea was—independently of the applied mirror technique—the transition of a homogeneous texture to a clearly audible separation of the voices. In the second part there is a development of isolated sound islands towards more integrated episodes.

POINT: What is the role of the mirroring in the second part of *A House of Mirrors III*?

Stankovski: The second part of the piece is also based on mirroring techniques. Rhythmically, the second part is based on the retrograde result from the events in the first part. The entries of all instruments are combined in a sum of rhythms and can be read in reverse order, in which, as in the first part, increasing deviations are possible (and necessary!) both at the micro level of specific rhythmic values and at the macro level of tempo. Therefore, the speed of the second part, in contrast to the constant tempo of the first part, is unstable and fluctuating.

The pitch of the second part goes back exclusively to the cello part, which in the end of the first part remains solely alone (Fig. 15). However, the pitches are multiplied through vertical mirroring so that chords of symmetry, or respectively, balanced intervals arise. These symmetries are broken in places through individualised mirroring within the single parts, so that apparent motivic imitation can arise (Fig. 16, bar 89). The mirroring axis between the first and second part is in the middle of the pause in bars 85, 86. The durations are arranged in retrograde, where pauses can be replaced by sounds, and vice versa, the pitch of the cello in bars 85–78 is the basis for the symmetrical chords in bars 87–89. In bars 91, 92 the first noise material replaces the pitches.

As the number of entry points from the additive rhythms is much larger than the number of pitches of the cello, empty periods arise, which become occupied by noise material, so that the way back to the intervallic "original position" becomes displaced through an increasing isolation—although structurally it goes back the same way, musically there is no way back.

Fig. 15 *A House of Mirrors III*, bars 77–86

Fig. 16 *A House of Mirrors III*, bars 87–94

Project Review by Alexander Stankovski

The "balance" of the outcome of the project has been ambivalent—on one hand it has brought me a refinement of my originally fairly rigid use of the mirroring technique, in particular by addressing the additional parameters of the mirroring depth, which I will certainly take into account when I apply this technique elsewhere. On the other hand, the compositional process in *A House of Mirrors III* was much more tedious than expected. Not only was the manual process for each value in the first part of the piece very time consuming and the actual process of notating not inspiring, I came in this part to a dead end, which I only overcame by changing my compositional strategy. So the principle of mirroring has unexpectedly changed from a technical tool into a psychological reaction, as a break with the prior technique that was used.

But that does not mean that I am not satisfied with the music that was created through the engagement with the automatically generated data. On the contrary, I would not have been able to write the first part of the quartet so coherently with such creative exploration without the rigid structural framework. When the function of the framework had been met in the first part, I had to respond in a completely different way in the second part, so as not to remain a servant to an abstract principle.

Whereby my summary is as follows: the development of personalised compositional techniques, however they might be defined, is essential, but this continues to remain a means to an end. Consequently, when it becomes necessary that they need to be modified, ruptured or replaced by completely different ones: only in this manner do they become the expression of a personality.

References

1. Deleuze G (1986) Cinema 1: the movement-image (trans: Tomlinson H, Habberjam B). University of Minnesota Press, Minneapolis
2. Deleuze G (1989) Cinema 2: the time-image (trans: Tomlinson H, Galeta R). University of Minnesota Press, Minneapolis
3. Deleuze G (1994) Difference and Repetition (trans: Patton PR). Columbia University Press, New York
4. Hegel GWF (1986) Phänomenologie des Geistes. Werke 3/20. Suhrkamp, Frankfurt am Main
5. Hegel GWF (1970) Vorlesungen über die Ästhetik III. Werke 15/20. Suhrkamp, Frankfurt am Main
6. Leisinger U (1994) Leibniz-Reflexe in der deutschen Musiktheorie des 18. Jahrhunderts. Pommersfeldener Beiträge, vol 7. Königshausen & Neumann
7. Michaux H (1963) Henri Michaux. Plume, précédéde Lointain intérieur. Gallimard, Paris
8. Plato (1997) Gorgias. In: Cooper JM, Hutchinson DS et al (eds) Plato: complete works (trans: Zeyl DJ). Hackett Publishing, Indianapolis, pp 791–869
9. Plato (1997) Phaedrus. In: Cooper JM, Hutchinson DS et al (eds) Plato: complete works (trans: Nehamas A, Woodruff P). Hackett Publishing, Indianapolis, pp 506–556
10. Plato (1997) Republic. In: Cooper JM, Hutchinson DS et al (eds) Plato: complete works (trans: Grube GMA, Reeve CDC (Rev.)). Hackett Publishing, Indianapolis, pp 971–1223
11. Ross GM (1984) Leibniz. Oxford University Press, Oxford

Matthias Sköld/A Topography of Personal Preferences

Matthias Sköld, Hanns Holger Rutz and Gerhard Nierhaus

Matthias Sköld grew up in Stockholm in the 1980s in a family of musicians. His father was a cellist at the opera of Stockholm, his mother was a singer and also a violin teacher. Also his larger circle of relatives consists mostly of musicians. The career wish of Sköld was already early settled on becoming an orchestra musician, right in the line of the family tradition. In the beginning he took clarinet lessons, then violin, piano, and saxophone, yet true enthusiasm emerged only from e-guitar, an instrument, which interestingly is played by many Swedish composers.

At this time Sköld received also important stimuli from regularly choir singing, which was extensively practiced at his secondary school with music focus. After high school Sköld studied first English and looked at this time intensively into the subject of literature theory in particular with phenomenology and poststructuralism. At the same time he applied for a two-year course in radio journalism. The training contained cutting, editing, and arranging of analog and digital audio in a professional studio-environment, an ideal addition/complement to his private experiments with different computer software and various synthesisers. After finishing his training Sköld started to work as a radio-reporter, yet at the same time the performing and composing of music became more and more important for him, which brought him finally to the Gotland School of Composition, where he studied under Sven-David Sandström and Per Mårtensson. Inspired by Per Mårtensson, Sköld engaged himself at this time extensively with various methods and software to analyse and synthesise sound.

M. Sköld
Department of Composition, Conducting and Music Theory,
Royal College of Music, Stockholm, Sweden
e-mail: mattias.skold@kmh.se

H.H. Rutz · G. Nierhaus (✉)
Institute of Electronic Music and Acoustics, University of Music
and Performing Arts Graz, Graz, Austria
e-mail: nierhaus@iem.at

H.H. Rutz
e-mail: rutz@iem.at

G. Nierhaus (ed.), *Patterns of Intuition*,
DOI 10.1007/978-94-017-9561-6_10

His composition studies coincided with the birth of social media in Sweden which might explain why the sense of community was very strong in his composers' generation here, a country with few large cities, far apart. Moreover laptops were still really not affordable, which was great for composition education since everyone had to sit together in the computer room, providing plenty of opportunity for spontaneous aesthetic discussions and coffee.

After his graduation 2000 Sköld continued his studies with Pär Lindgren and Bill Brunson at the Royal College of Music (KMH) in Stockholm. Besides "traditional" compositions also a series of electroacoustic piece were produced. At this time Sköld became more and more involved with the experimental music scene in Stockholm, particularly with the Fylkingen society,[1] where he was member of the board and also the president for some years. At first he separated his electroacoustic pieces from the experimental work with live-electronics at Fylkingen—they were two completely separate entities. The former were of an electroacoustic tradition, often in multi-channel, while the latter was playful, noisy, glitchy etc. With time, he became more and more detached from the traditional electroacoustic idiom, focusing more and more on the laptop as a live-instrument. At the same time, encouraged by the inspiring choir conducting professor Anders Eby, he composed a lot of pieces for the choral community in Stockholm. Another passion during his college years was working with percussion. A very influential personality was the percussionist Pontus Langendorf, who became a close friend and later on premiered also all of his compositions for percussion. Beside the live-electronic music and the emerging interest in choir and percussion Sköld wrote chamber music, orchestral pieces, solo pieces, also organised large-scale collaborative projects, performed and recorded with great jazz musicians. These days he has many areas of interest. One being closing the gap between traditional electroacoustic music and live-electronics, strongly believing these areas will grow closer together over the next few years. Another passion is teaching at the University, meeting with students. And discussing great music constantly reminds him of why he wanted to be a composer in the first place.

Artistic Approach

Statement

We are now truly past the era of modernism and postmodernism, though many of their followers are still around—"art music" continues to be amongst the most slowly evolving art forms. The possibilities of new technology of communication and expression leave little reason for new generations to continue the principal battle of modernity, that against tradition. And in any case, the last hundred years of modernity invalidates any continued aesthetic revolution. There is simply no all-encompassing

[1] An influential society for experimental music and arts, founded 1938 in Stockholm.

agenda to rebel against anymore. Through the internet you can easily find artists and critics who agree or disagree with an aesthetic position regardless of its content. The obvious way forward is to continue the tradition of western "art music" in much the same way as composers have done over the last millennium, incorporating new ideas, technology and media as they arise. We need to constantly remind culture administrators that classical composers like Mozart and Beethoven didn't reach their positions in our culture by accepting music as it appeared in their time, even though their ideas for renewal may have been less revolutionary than Luigi Russolo's ideas in the beginning of the last century. I strongly believe that the composer plays an important role in society, and this means a big responsibility to our culture as a whole.

Personal Aesthetics

I consider my own work very much a part of the tradition of western art music. However, this tradition is obviously very diverse; the old concept of learning the composition style of your professor simply won't give aspiring artists the perspective needed to decode 21st century contemporary music. Today, ideas like serialism, post-minimalism and neo-romanticism exist side by side and one can use one or the other simply as musical material rather than as aesthetic positions. I love the sparse instrumentation of Shostakovich, the dense harmonies of Messiaen, the extreme dynamics of Boulez, the carefully constructed timbres of Denis Smalley, the counterpoint of Bach and the rhythms of Autechre, and in listening to these artists' music their musical structures become part of my musical vocabulary. I often work with contrasting musical structures where their purpose is to mediate meaning rather than the origin of their stylistic behaviour. In other words, the techniques don't appear as style markers, which is why I don't consider my work postmodern, but rather postpostmodern, or digimodern to use Alan Kirby's term [3]. Because of the multitude of styles and ideas brought forward during the last century, and the ease with which young artists now remix and re-contextualise existing techniques and musical material, what was traditionally considered postmodern is almost implicit in music creation today.

Rather than attempting to develop 20th century ideas like serialism and indeterminacy further, I consider it my task to explore the implications and meaning of using these ideas in specific contexts. How can they function in a musical narrative? How do they relate to one another? When too much emphasis is put on following a too homogeneous aesthetic path, like that of serialism, the music starts to behave like a sub-genre, and western art music should never be a sub-genre within our music culture. A composer should consider the music of our culture as a whole; interesting musical ideas are developed and explored all over the world in all parts of society, even though the aesthetic discussions may reside in the academies. To sum up, I believe it is my duty to pick up the pieces from the explosion of ideas from the last century and make sense of them, put them in context. Which particular ideas I explore in relation to my composition work will depend on the context and the subject matter. In my sacred choral music I have, beside the classics, been much inspired by Swedish composers Sven-Erik Bäck and Ingvar Lidholm and their work with twelve-tone music in a choral context.

In electronic music inspiration comes not only from my contemporaries, most of which are more or less concerned with live-electronic music and interactive music programming languages, but also from visual and conceptual artists. My orchestral work shows influences from the twelve-tone composers and Shostakovich. Integrating twelve-tone ideas with other more or less traditional ideas is of course nothing new, Alban Berg's violin concerto is maybe the most famous example of this. Twelve-tone technique could never be the true liberation of tonal hierarchies, but provided a highly interesting alternative to traditional harmonic progressions. Sven-Erik Bäck demonstrated this beautifully in his twelve-tone hymn *Du som gick före oss* from 1959, number 74, in the Swedish Church's book of hymns.

Formalisation and Intuition

I think of formalised structures as the grammar of the musical language—they are the condition for everything said, yet say little by themselves. The formalised structure of music is at the same time its conception and its description, which is why musical analysis is so important for composers' education. Once you have described how the sounds of a piece relate to one another, even if they appear at random, you have described their formalised structure. Therefore composing in itself implies some level of formalised methods—if you don't apply them, you're simply not composing. Naturally, I include rules of traditional harmony in both classical music and jazz in my definition of the musical structure. Only, the improvising jazz musician is so familiar with the rule system that there is no need for making calculations on paper beforehand. The less experience you have of a system, the more important it is to be strict with your methods. Once you get to know the rules better you can start bending and breaking them to fully explore their boundaries. These boundaries are usually the place where musical rules come alive.

Working with music programming software has made possible a more tactile approach to musical structures. By constructing a computer model of the music you can explore various aspects of a set of musical rules without the need to create elaborate scores. I think of it as similar to how 3D-artists work as opposed to traditional 2D-artists; instead of drawing a 2D-representation of an instance of your imagined structure, you create a 3D-model that can be viewed from different angles and distances. With this in mind, my first task is to decide when a model is finished, then from what angles I want to view it. At this point, I am not breaking any rules. I am rather testing their boundaries, their relevance for the overall musical idea. For some pieces the structural laws will be open enough to make any crimes against them unnecessary, but in most cases the structures are not constructed to be perfect, but provide raw material for further rule-based or non-rule-based composition work.

In working with structures, I am constantly guided by intuition. For me, musical intuition is the possibility to perceptually measure, weigh and balance musical material in heard and unheard musical structures. Hearing is the key word here; we tend to rely heavily on our eyes when studying and working with music, but we must remind ourselves that there is no given correlation between the experience of seeing and hearing, or as Murray Schafer puts it: "I have never seen a sound." [6]. At the same time, visualising music is what has made the elaborate structures of western classical music possible. What a composer needs then is not only an intuition for heard musical structures but also an intuition for how a given visual abstraction relates to auditory perception, in other words the capability to imagine sound that is not there. No matter how mathematical or logical a set of musical rules might be, there are always crucial musical decisions, some of which may seem insignificant on a structural level, that may change the whole universe of the musical piece. Such impact from structural changes can rarely be calculated but must be understood in terms of making musical intuition the condition for musically relevant decisions.

Reflection and Evaluation of My Works

I constantly evaluate what I do. This is at the core of all art forms, particularly in the era of conceptualist art when craft is no longer the focus, then all you are left with is your ability to evaluate your ideas. I do however disagree with Sol LeWitt's notion that the artistic idea is the machine that drives the artistic output [5]. In that sense I believe music to be closer to poetry than to visual art, since music, like poetry, unfolds in and through time. An understanding of how time works in music is central to a composer's ability to evaluate his/her work. The more I learn of music, the more I realise how transient it is, its unwillingness to be defined. And still the sounds keep communicating, expressing, exploring, making artistic research at the same time impossible and completely possible. I evaluate music not based on any logic but rather based on what I know of music in terms of experiencing it on different levels, from different aspects.

My judgment of my music stems from the knowledge of how extraordinary the experience of music can be. Through listening to music, studying music and making music I am still learning day by day how my craft relates to musical experience, while being aware that I can never truly define one or the other without constraining the possibilities of musical expression. What puts me as a composer apart from a music listener is that to evaluate a structural component of music. I have to be able to imagine its possibilities as placed in various contexts, as tweaked to fit a certain musical idea. Most people can distinguish music they like from music they don't like, but I believe it is this type of musical imagination that not only informs improvisers in their "instant" composition but also composers as they work with discreet music components while not losing sight on its purpose in the main framework of the piece. The difficulty lies in the necessity to be aware of the structure on different hierarchical levels at the same time, while imagining music that is not yet there. Furthermore, there is the aspect of interpretation which comes with working with notated music; another important discussion that I will pursue at a later stage.

What Is the Actual Focus of Your Composing?

I used to be more focused on processes where the process was the piece; that was how I first started working with musical structures with computers. In those pieces, the rules were usually clear from the first bars of the music and you could hear them operate throughout the piece. These rules were easily discerned by any listener which was a conscious choice on my part. I wanted the structure to be experienced in a very tangible way. These days I have a more semiotic approach which I think is at the heart of the term *composition*, meaning that the act of composing implies placing different musical entities in relation to one another on a vertical and/or horizontal level.

Practically for me this means separating the structuring processes from the ordering process. I usually work with a landscape-A4 drawing or data table as a superstructure, while collecting material for this overall form in a MIDI sequencer. The sequencer's arranger window acts as the empty score sheet, where different music materials meet. Then I bring in ideas generated in various music programming software patches as I need them. Some pieces may be completely generated from a particular patch while other pieces are composited of various structures with different levels of complexity. Structural ideas may be as simple as traditional counterpoint but formalised in a programming environment, while others may be functions of serial or stochastic principles with different levels of hierarchy. Some programmed structures produce material for several pieces while others only work in the one piece.

Examples from Concrete Works

My Saxophone quartet, *Ups and Downs*, was basically made with one quite simple programming patch that was designed specifically for this piece. This was at a time when I was consistently working with very simple and easily discernible ideas. I wanted the structural idea of a piece to be obvious from start to finish. In this particular piece I worked with gradually expanding arpeggios in four layers. I used the computer to build a model for the behaviour of one instrument over a given period of time. Then it was easy to test the behaviour of the four instruments together and evaluate different settings for the expansion of the overall arpeggio structure. In this particular work, the timing of the structure was crucial, which it often is with regard to process music. How long can you listen to a particular musical algorithmic process without losing interest? Once I was happy with the settings and the durations of each take, I recorded the music as MIDI and edited the result in a notation software to make the score readable. This is something you have to do, particularly when you work with constantly changing and irregular rhythm patterns. Figure 1 shows one saxophone part in the beginning of the piece.

In *B–A–C–H* for flute, violin, cello and piano I used a similar method. But in this case I was working with a gradually accelerating structure derived from a well-known Bach choral. In both cases, the idea came first, then a computer model of the idea, and last the production of a score.

Fig. 1 One saxophone part from the beginning of *Ups and Downs*, bars 1–22

These days I usually work with more general structures, aiming to develop a musical language that can be used in various contexts. This work has so far been focused on counterpoint and independent voices that relate to one another in different ways. The first piece where this approach was used was in my *Requiem* from 2007. In the Introitus movement I let a twelve-tone row form the centre of the four-voice structure while building one voice above, and two voices below the centre as seen in Fig. 2.

For the final version, the pitch structure was basically kept as originally constructed, while the phrasing and division of notes into words were done manually. Figure 3 shows the resulting edited score for the same passage.

Project Expectations

The more music I make, the more intuitive I allow myself to become in my approach and this has to do with the fact that as I get more and more familiar with my structural ideas, I don't have the same need to model them. I can, as it were, draw the 2D-image directly. But there is always a danger with intuitive composition. While

Fig. 2 Twelve-tone row in *Requiem*

Fig. 3 Edited score from *Requiem*, bars 1–9

intuitive adjustments can be vital for a particular work's playability, they can also compromise the structure to the point where it starts to lose its value. This is very much a danger for performers of new music as well; if you "sell" a piece to an audience in the wrong way, they may leave the concert hall invigorated but completely unaware of its ideas. I am here somewhat echoing Adorno's critique of lip-smacking euphony distracting the listener from experiencing the overall musical structure [1]. With this in mind, I think it makes sense to make proper inquiries into one's own intuitive processes, and surely this is something that most composers do on a regular basis. But this particular project takes this inquiry one step further by allowing the participating composers' intuitive decisions to be monitored externally and with the aid of computer models. As a result we are not only exploring to what degree we are compromising structural processes but also to what degree our intuitive decisions form other subconsciously constructed patterns.

Exploring a Compositional Process

POINT: Sköld programmed a so-called patch, a generative structure written in a computer music software, to produce polyphonic sequences. It has certain controls for the harmonic and part-writing rules, and Sköld may change the parameters while the output is being produced. Figure 4 shows an example of such a four-part movement in a piano plot. However, we treated this patch as a "black box" and instead analysed the sequences merely judging from its output. By doing so, we wanted to avoid seeing the outcomes too narrowly through one particular pairs of glasses.

The output produced by the patch is not always considered "perfect" by Sköld, so he usually edits it. What we did in the first step, is to look at a few outcomes from a harmonic point of view, filtering out the chords, and compare the raw output with the edited version. Figure 5 shows the distribution of intervals between the voices of these chords. As can be seen—even versus odd columns in the figure—weighting their occurrence with the note durations does not significantly alter the result, therefore we continued to disregard such weighting.

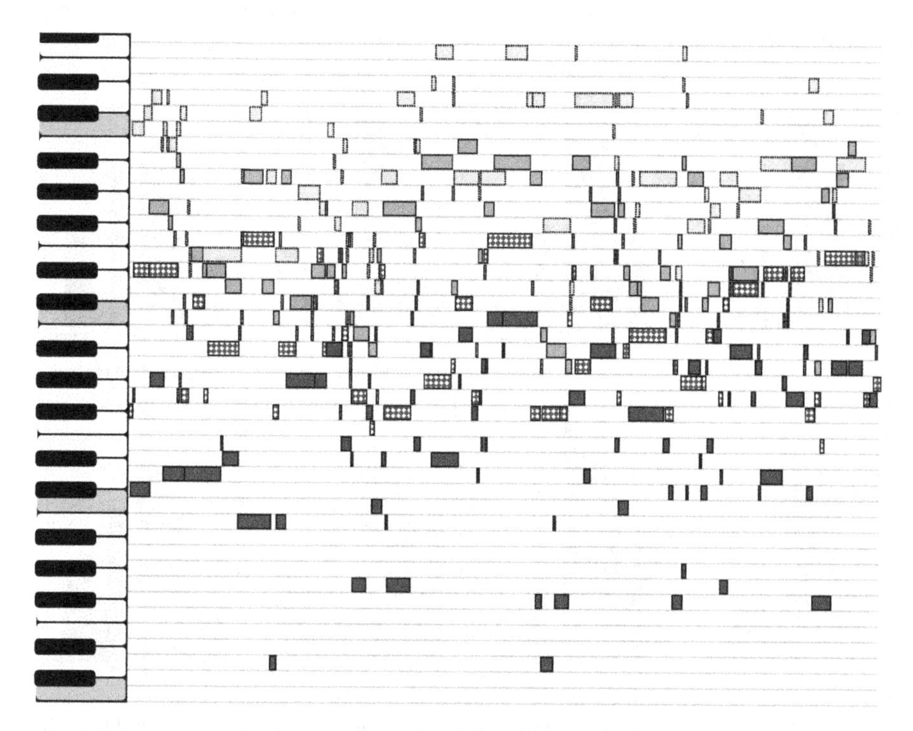

Fig. 4 Four-part movement produced by a generative patch from Sköld

During the editing, particularly octaves were removed and also a few occurrences of minor sixths. In exchange, more minor thirds appear. The overall picture shows a dominance of perfect fourths and with frequency decreasing towards the smaller intervals, and only the perfect fifth appearing significantly in the greater intervals.

The next idea was to refine the view by looking at simultaneous occurences of interval pairs to see if certain combinations are particularly prominent or missing. This is shown in Fig. 6, again distinguishing between raw and edited version and neighbouring intervals versus all intervals.

Consecutively, Sköld provided more files, also indicating whether he found them rather "boring", or on the contrary if they were "promising", meaning that they are "considered for further editing but may turn out to be very different sounding when used in an actual piece".

Figure 7 shows two such sequences in comparison, the boring one on the left and the interesting one on the right. It can be observed that the promising file is richer in terms of a broad distribution of interval constellations, with particular focus on minor seconds, and the co-appearances of two intervals of the same size is unlikely (the diagonal is significantly brighter). The boring file contains a lot of fourths and fifths combined with octaves and major seconds.

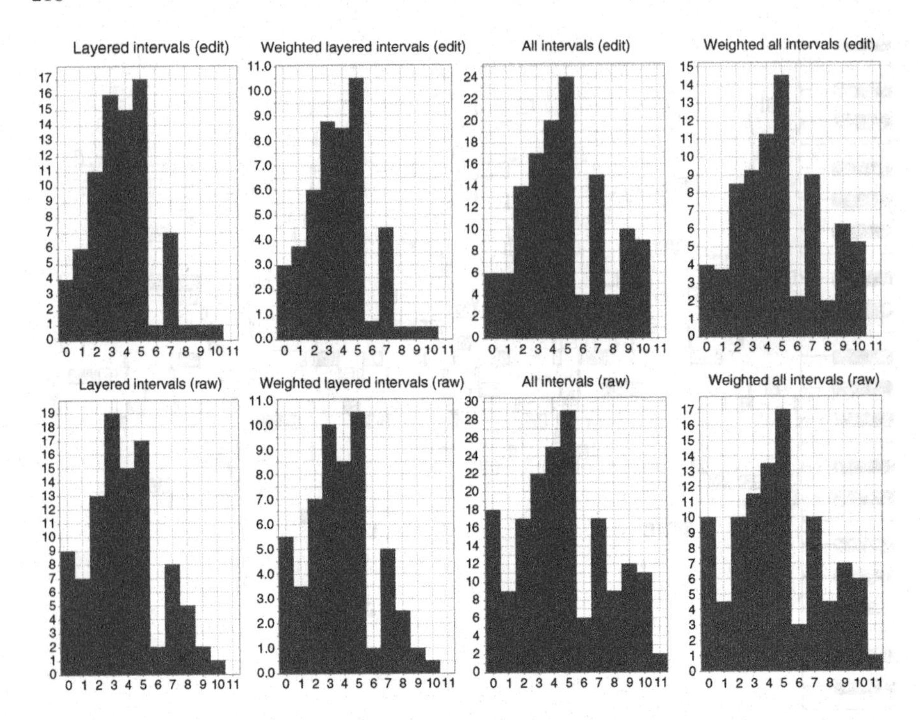

Fig. 5 Interval distribution in a sample sequence. The raw output is shown on the *bottom*, the edited version on the *top*. The *left* counts only neighbouring intervals, whereas the *right* half counts all intervals. The *odd* columns simply count the intervals as they occur in the file, the *even* columns show these intervals weighted by the respective durations of the notes in which they appear

Sköld: It is interesting to see how these charts differ regarding corresponding intervals and the distribution of intervals and even though these findings were made with relatively little material I nevertheless find the conclusions interesting especially since the difference is so obvious. We'll see if this holds when we add more material.

POINT: When selecting other files, however, these statements do not generally hold. In general, the broadness of distribution seems to correlate with the distinction into the two categories, but for each of the particular interval constellations, counter-examples can be found. We were looking for a different angle. Sköld was interested in finding out whether the pieces drift in some form over time. The next analysis we thus conducted, was using a sliding "window" over the sequences, and calculating a measure for each of these windows. The window size was mostly 16 s with a step factor of 1/8, i.e. adjacent windows overlap by 87.5 %. For each of these overlapping time frames, the computer could filter out the relevant notes and measure the contents of the window—e.g., pitches or durations—according to a given metric, producing thus a function of that metric over time.

The measures we used were the rhythmic functions of Vladimir Ladma [4]: *mobility*, *tension*, *entropy*. Durations are derived from the offset between two note offsets within each voice.

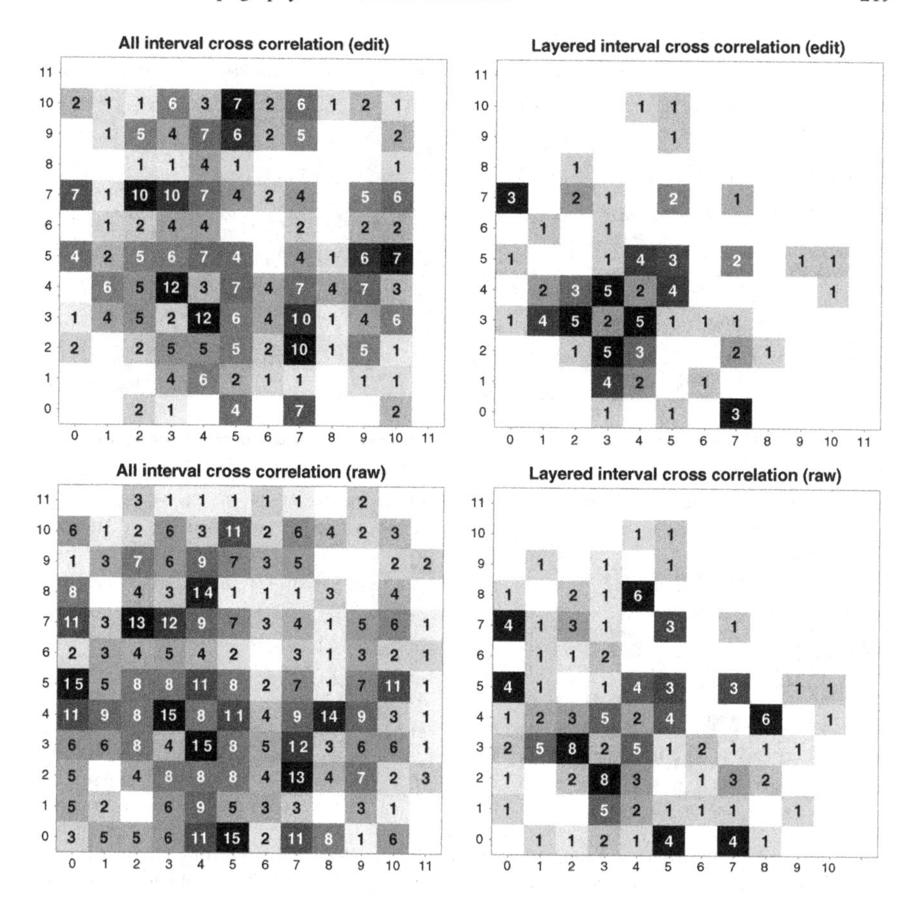

Fig. 6 Distribution of interval co-occurrences in a sample sequence. The *matrix cells* denote the frequency of the interval on the horizontal and the interval on the vertical axis occurring together within one harmonic constellation. The raw output is shown on the *bottom*, the edit version on the *top*. The *left side* shows all interval relations, the *right side* only counts adjacent intervals

The entropy curves for four selected studies are depicted in Fig. 8. The average entropy of the interesting files is higher than that of the boring ones, although study No. 26 seems to become more diverse over time.

Mobility in overall does not change that much, so the pieces do not seem to be significantly accelerating or decelerating over time. The boring files tend to be slower than the interesting ones, and the interesting file No. 5 has the greatest variability in speed. The rhythmic tension (deviation from durational means) is quite low, only one of the "boring" files seems to be spiking at times. However, it must be said, that tension is also somehow anti-proportional to tempo, so slower parts tend to produce higher tension values.

Sköld: The findings make sense in light of how these files were produced. Not much thought has been going into creating rhythmic variations on a microlevel, but I

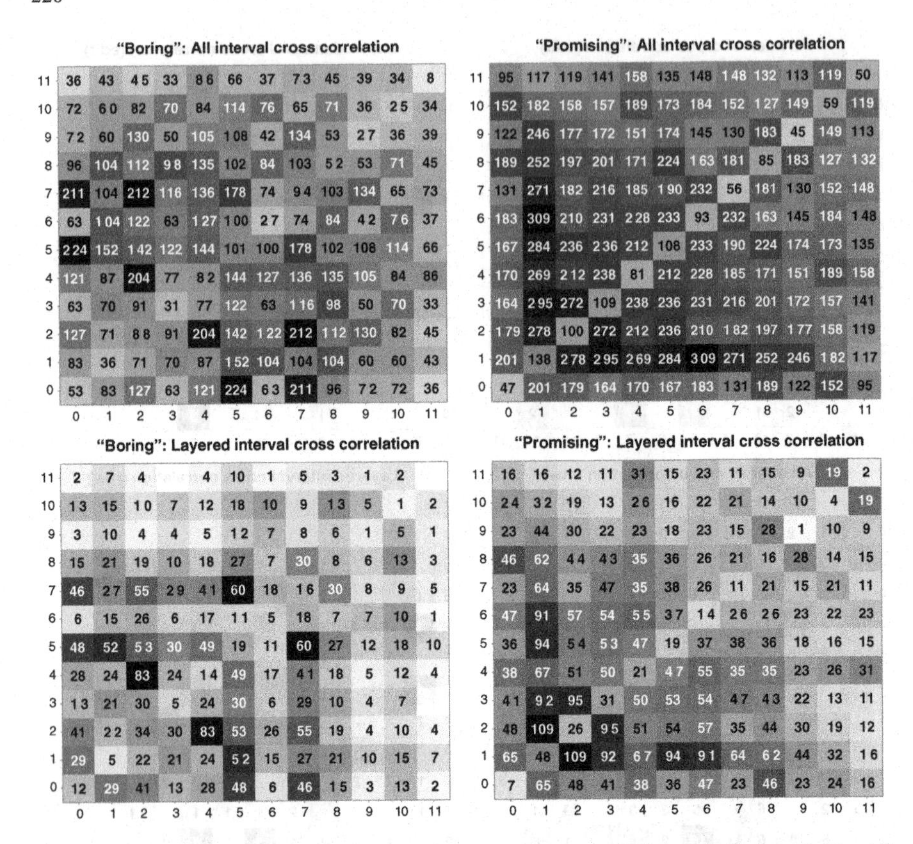

Fig. 7 Interval co-occurrences compared between a boring (*left*) and interesting (*right*) file. All intervals are included in the *top* figures, whereas in the *bottom* figure only neighbouring intervals are accounted for. *Numbers* indicate absolute frequencies

don't rule out the significance of these findings (that the interesting one had slightly more varied rhythm and somewhat higher tempo) bearing in mind how important even small varieties in tempo and rhythm can be for the interpretation of a piece, so I think making combined mobility, tension and entropy charts makes sense. But it should not be the main focus for the continuing work—maybe a tool one might use when two files that seem harmonically similar still fall into different sides of the boring/promising line.

POINT: We thus shifted our attention back to the harmonic structure, looking at the variance within all interval classes, as they change over time. In Fig. 9 only vertical intervals are observed, segmenting the sequences into harmonic fields or "chords". The variance within all interval classes is taken as it changes over time. It seems to exhibit a motion at more or less the same rate, where the variance grows and shrinks. The interesting files have a higher average interval variance than the boring files, although the difference is not strong, with the exception perhaps of No. 8, and No. 7 spiking once.

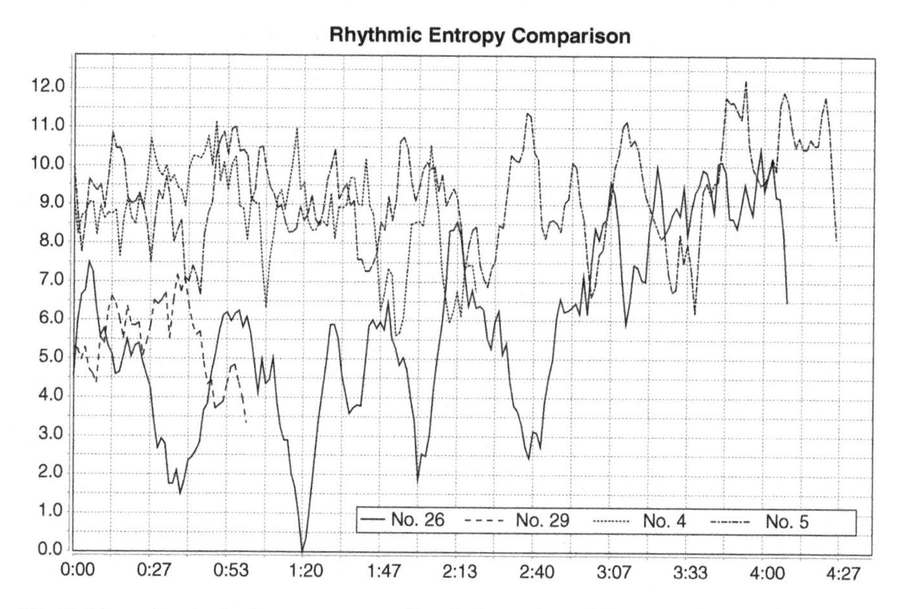

Fig. 8 Measuring rhythmic entropy across time for four selected studies. Time is given in minutes and seconds. Studies Nos. 26 and 29 were marked "boring", whereas study No. 5 was "promising"

Another possibility is to look at the melodic development of the individual voices. This was done in Figs. 10 and 11. Again a sliding time window is applied. The first figure calculates the variance of the different interval steps found. The second figure shows how inside these windows the pitches of each voice vary, given as a range of semitones (ambitus).

In terms of the variance, the boring and interesting files largely differ. The horizontal motion of the boring files is much smaller in mean, and narrower in width. The two files marked as promising (Nos. 5 and 10), have particular strong variance in the horizontal pitch steps, and also the variance moves quite strongly across time. The ambitus is significantly larger and has greater variability for the interesting files. The boring files barely use more than one octave of space within a 16 s time window, whereas the interesting files are centred around two octaves of ambitus.

Sköld: I was surprised to see that there was a measurable difference already at interval class variance level. By systematically investigating occurrences and varieties of simultaneous (Figs. 8 and 9) and successive (Figs. 10 and 11) chords and intervals we can already see a clear pattern. Judging from these results, explicitly working with more variety in terms of different intervals and interval pairs over time would provide a more interesting musical output. Though this makes perfect sense, we need to investigate the material further before I start make changes to my patch based on these findings. Would it be possible to evaluate my MIDI structures as a whole, and possibly learn to predict whether a given material is interesting or not?

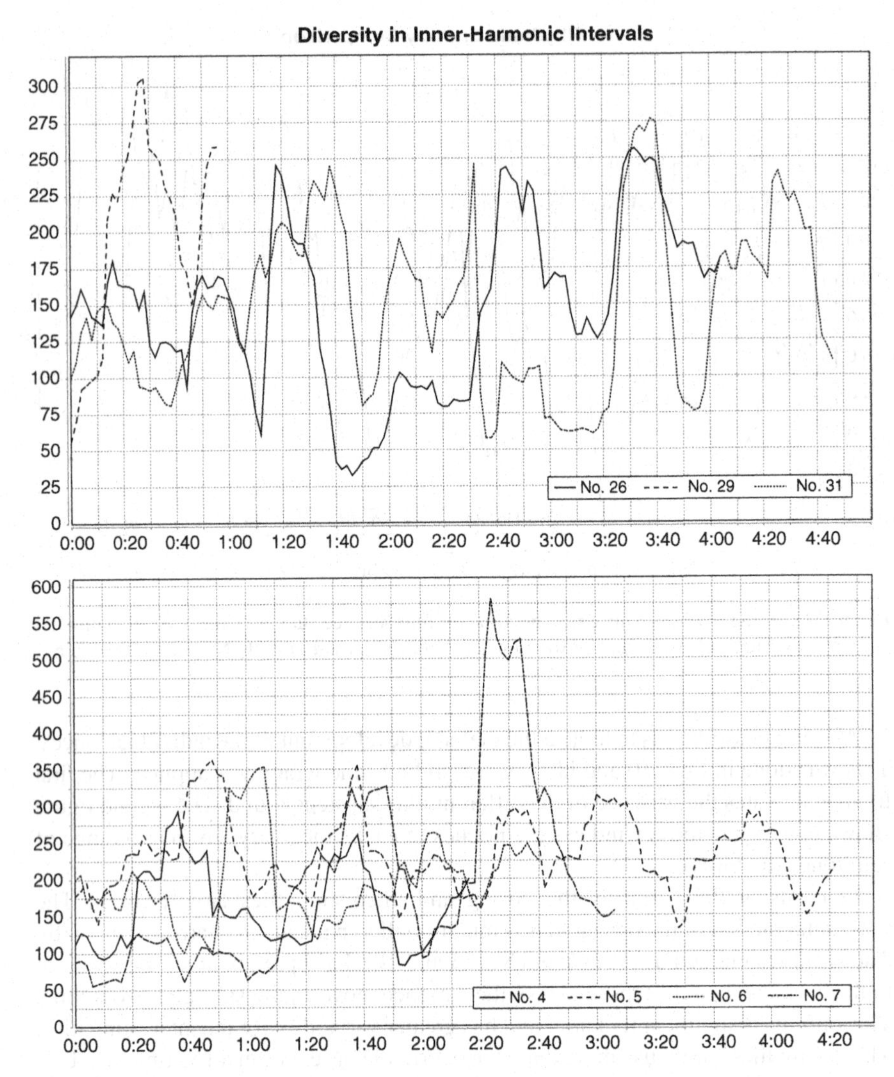

Fig. 9 (Mathematical) variance over time in the number of intervals found in vertical structures. The *top* shows three boring files, the *bottom* shows three average files and one promising one (study No. 5)

Classification

Having analysed different aspects of the uninteresting versus interesting files, we wanted to formalise our findings further by finding out which are the crucial factors and to which extent they play a role. Perhaps some of the magnitudes calculated are insignificant for the discrimination of interestingness, perhaps one or two factors

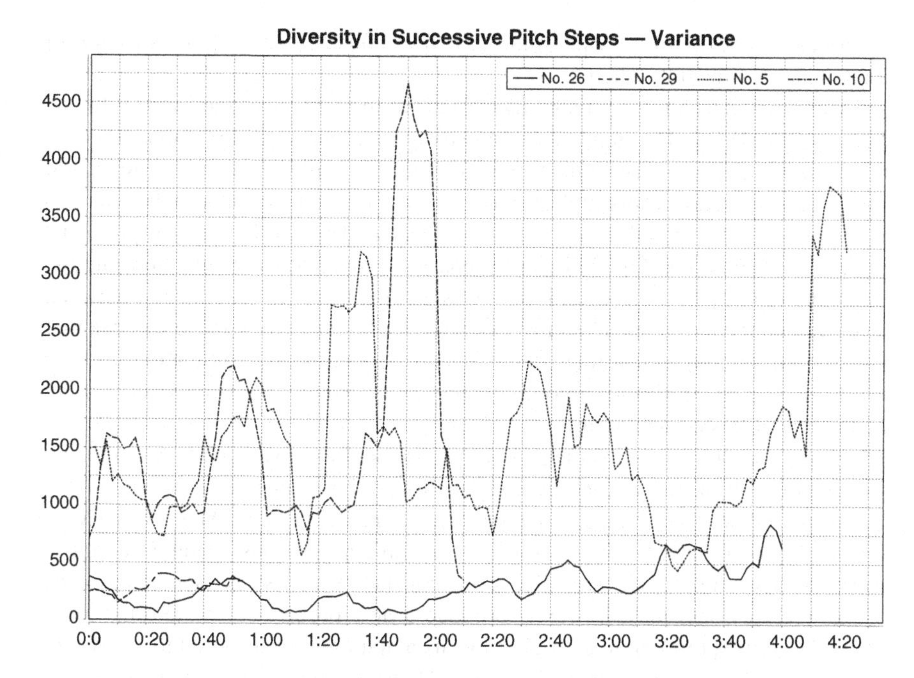

Fig. 10 Development of variance in the number of horizontal intervals over time. For each sliding time window, the mathematical variance of the intervals across all four voices is calculated

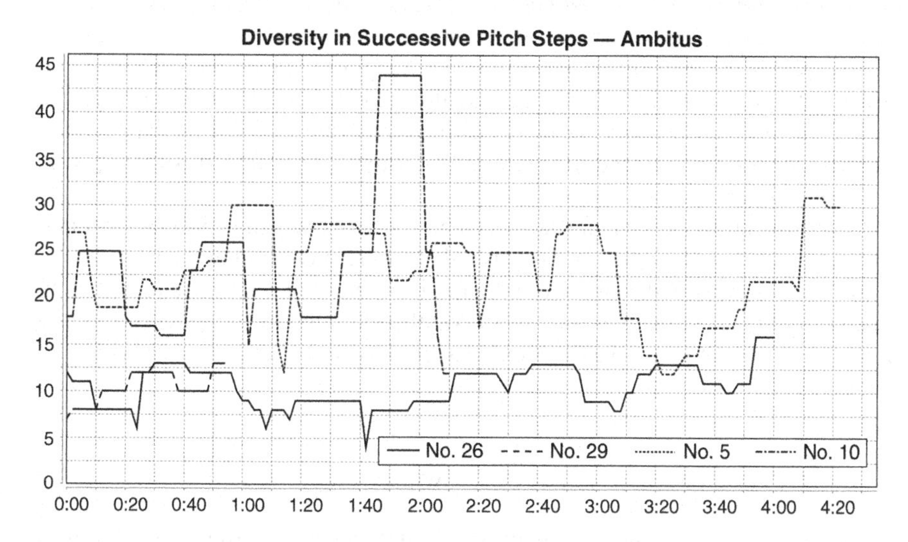

Fig. 11 Development of pitch step ambitus over time. For each sliding time window, the range between the minimum and maximum of the intervals across all four voices is calculated

are sufficient to explain it. A possible way to establish these weights is to use an automated classification scheme.

A well-known technique for solving classification problems is the *Support Vector Machine* (SVM) [7]. The items to be classified are represented by feature vectors, and the task is to find a discriminant function that optimally separates the items into individual categories. The simplest classifier is a linear function which separates the hyperplane in which the data points are located. This works well for problems in which the data points can be linearly separated. Imagine we would calculate the tempo and tonality of a piece, and there was an unambiguous relationship between these two and the judgment about their value—say, an interesting piece was one that was fast and without a clear tonal centre. Then using a linear combination of these two features, a categorisation into interesting and boring could simply rely on the coefficients of this combination.

In many real-life scenarios this is not the case, and an advantage of SVM is that it allows the use of non-linear functions such as polynomials or Gaussian curves. As a result, SVM is described as having a high accuracy even when dealing with high-dimensional data, finding classifiers that have no obvious "geometric" interpretation. The tuning of the so-called *kernel* of the SVM (its representation of the separating functions) may, however, produce a new problem of overfitting: With enough flexibility, one can fit even the most irregular data set such that its points are separated according to the category at hand. However such a function then entirely depends on the particular samples of the dataset and may perform much worse with new unknown data points.

To employ an SVM, the user has to make many decisions [2]: In which form is the input data represented, what kernel to use, and how to tune the chosen kernel? For example, the support vectors—the sample points that are closest to the surface that separates the categories—determine the margin by which the categories are separated. One parameter typically specifies the "penality" of misclassifying a sample point. The lower the penality, the more marginal errors are accepted, perhaps gaining more freedom in defining the curvature and robustness. Therefore, this penality C is co-dependent on the allowed bendiness γ.

A specific problem is an unbalanced dataset, where many more samples fall into one category than the other. In our case, there were more boring than interesting files. The algorithm will now have a higher probability of wrongly classifying an interesting file as boring than vice versa. We therefore follow the suggestion [2] to introduce a weighted misclassification penality based on the relative frequency of the samples:

$$\frac{C_i}{C_b} = \frac{|\mathbf{w}_b|}{|\mathbf{w}_i|}$$

Here, category i would be "interesting" and category b would be "boring". Because the number of boring files, $|\mathbf{w}_b|$, is larger than the number of interesting files, $|\mathbf{w}_i|$, the weighted penality C_i for wrongly identifying an interesting file as belonging to the category "boring" becomes higher.

The other crucial step is the selection of features that make up the vectors to be classified. We developed multiple feature vectors which we tested for their suitability to explain the classification of studies into promising and boring ones. We started with small vectors combining the individual measures taken in the first part, such as mean and variance of the horizontal ambitus over lapped time windows or the chord variation. We also took the variance within the table of vertical interval correlations into account.

None of these vectors yielded particular high success rates in categorising the studies. We then used a different strategy of a brute force combinatory process. The idea is to use a larger feature vector and let the computer determine through an exhaustive search which subset of these elements is actually relevant. For example, we used the cross-frequency tables for intervals again, but reduced them to interval classes to limit the vector size. This vector is made up of 28 components—the seven interval classes on the diagonale, plus the upper-left triangle, or $(N(N-1))/2$ where $N = 7$. In order to be able to compare these vectors in between different studies, the tables are normalised to represent the relative frequencies of the cooccurrences of interval classes.

Then, to find out which of the table cells are best suited for classification, we calculated all combinations of subsets of these vectors. One would begin with a vector size of two, iterating over all possible combinations of selecting two of the 28 table cells. For each selection, we generate all possible sets. The number of combinations is given by:

$$\frac{N!}{(N-M)!\,M!}$$

Here N is the size of the complete vector (28), and M is the size of the subset. For $M = 2$ there are 378 possible vectors, for $M = 3$ the number becomes 3,276. With a sub-vector of size seven, the number of combinations exceeds one million, and the calculation time becomes too high, so we restricted ourselves to finding vectors of up to six elements.

We use the *leave-one-out* approach to train our model. That is to say, in order to avoid overfitting, we train the model with $k-1$ samples and verify the result by predicting the left out sample. The total body of exemplars comprises of 46 studies. For each of the sub-vector combinations found in the previous step, we use $k = 46$ iterations in which we train a models based on 45 studies and using the left out study as the target to predict. We define the robustness of the prediction by the minimum of the rate of the correctly identified promising and the rate of the correctly identified boring studies.

The prediction exceeds the 50 % margin at a vector size of three, the best selection here being the interval-class pairs (0, 2) (1, 2) (4, 4)[2] predicting the class with 65 % accuracy. We have traced the increase in accuracy up to a vector size of six where it

[2] The co-presence of unison/octave and major second, minor and major second, and two major thirds.

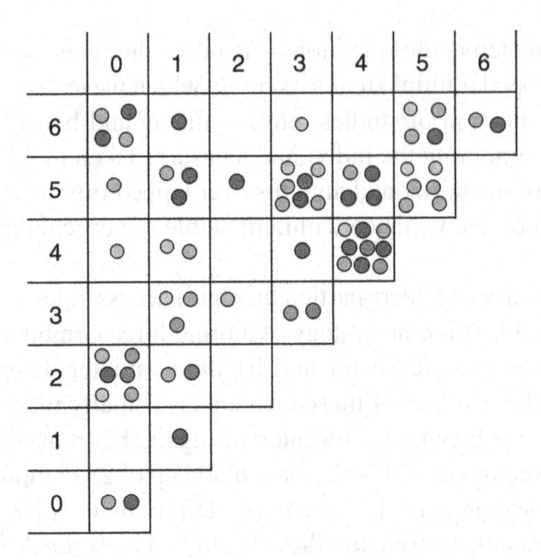

Fig. 12 Selection of elements for feature sub-vectors during exhaustive search. *Row* and *column* indices correspond to interval classes (semitones) within vertical structures

peaks at 76 %. This is illustrated in Fig. 12. Whenever the prediction becomes better, we add a circle to each of the table cells represented by the current vector. Higher accuracy is reflected by darker colours. Although the selection of cells changes over time, one can see specific cells which are never chosen (empty) to explain the classification, while others get chosen frequently such as the co-presence of interval-classes (0, 2), (3, 5) and (4, 4).

So far, the accuracy has not been overwhelming, and we decided to repeat the experiment with a different super-vector accounting for horizontal structure. It is made up of the normalised pitch class histograms of each of the four voices, resulting in a vector of size 48. The corresponding diagram is Fig. 13, showing the selected sub-elements whenever the accuracy improved. The lightest circles reflect a prediction accuracy of 59 %, the darkest circles reflect an accuracy of 93 %.

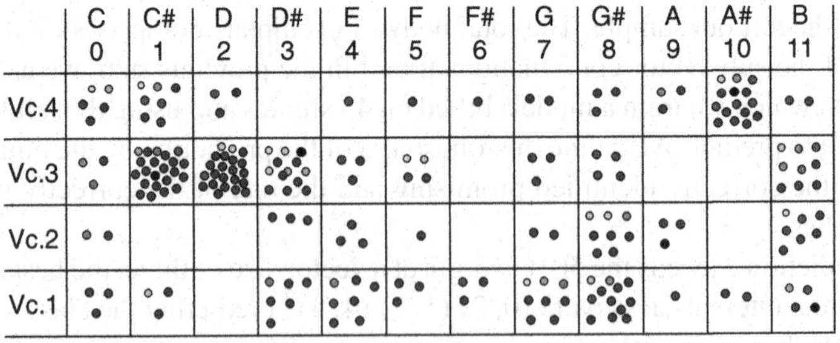

C C# D D# E F F# G G# A A# B
0 1 2 3 4 5 6 7 8 9 10 11
Vc.4
Vc.3
Vc.2
Vc.1

Fig. 13 Selection of elements for feature sub-vectors during exhaustive search. Total vector consists of pitch classes for each of the four voices

Here all of the following four vectors of size 2 already gave us an accuracy of over 70 %: ($D\sharp_3$ $G\sharp_2$), ($D\sharp_3$ $C\sharp_1$), ($D\sharp_3$ $G\sharp_1$), (B_3 $G\sharp_2$).[3] The best accuracy of 93 % was achieved with a vector of size 5 (we stopped at this size as the search space became too large for a vector of size 6): ($A\sharp_4$ $C\sharp_3$ $D\sharp_3$ A_2 B_1). This includes two of the most frequently encountered cells, the A♯ in the fourth voice and the C♯ in the third voice. It is interesting to note that generally the third voice seems to be most prominent in the suggested vectors, whereas the second voice is chosen the least.

To verify our results, we wanted to put the algorithm to a "blind test". Using a combined vector from the best preceding vertical and horizontal results, we did not retrain our model but used it to predict labels of further 34 files provided by Sköld. He had previously marked them as either boring or promising, but this was not revealed to our algorithm. We then compared the algorithmic evaluation of these new files with Sköld's decisions. The accuracy of this blind prediction based on the previous model was much lower than the one obtained in the leave-one-out phase. Within the group of files marked as boring my Sköld, the algorithm agreed to 64 %, however within the group of files marked as promising, the congruency with the algorithmic prediction was as low as 33 %. The total score was 56 %.

Sköld: With the aid of the SVM technique the computer was able to recognise the promising files from a number of MIDI file collections with remarkable accuracy. Still, when putting it to the test with a batch of new unmarked files, it proved hard to predict what I found promising or not. This could either be because there are important factors that inform the appeal of a structure that are not yet part of the model, or because unlike a computer model, I can sometimes be unpredictable. Therefore, thinking of possible future evaluation of my music, I believe we would need to further problematise the evaluation of the structures themselves while also establishing some basic rules for how my intuitive assessments of the MIDI files are made.

Looking at the structures themselves one could start looking at whether the files marked as promising show similarities with traditional, non-algorithmic music in terms of consonance-dissonance, harmonic cohesion, cadence-like behavior etc. And concerning my own intuitive assessment, one could lay down some rules for the process of marking the promising and boring files. One such rule could be that I mark the same batch of unmarked files on three different occasions to provide a higher level of consistency to this procedure. One could also raise the bar for how clearly and consistently promising a file must be to earn this label.

POINT: There may be indeed different explanations for the rather disappointing robustness in the "blind" run. It could be that the exhaustiveness of the search invalidated our robustness protection designed with the leave-one-out procedure, leading again to a form of overfitting. It could also be that the categorisation of the first batch of files depended on features less prioritised in your decisions for the second batch. In other words, we might have only gained an incomplete picture of the features. Lastly, since the criterion of being "interesting" is very unspecific, it might depend

[3] We use the indices here to denote the voice number and not the octave.

on additional external factors that are not stable but change with the context within which you evaluated the files.

Nevertheless, two things should be pointed out. First, we retried the prediction with slightly different selections of features (not just the best-ranked ones according to the exhaustive search). None of these yielded better results in terms of the overall accuracy, however there were quite a few studies for which the agreement between Sköld and algorithm remained stable despite the exchange of vector elements. This indicates that at least a subspace of "interestingness" was indeed covered by our algorithm. Second and more importantly, the SVM approach was not considered an end in itself—instead, it was an auxiliary devise to shed light on the possible ways of analysing the material and elucidating the reflexion process of the composer.

As at this point our analyses came to an end, we were curious to know about the functioning of Sköld's patch and especially his evaluation strategy concerning "boring" and "interesting" material.

Sköld: My decisions so far regarding the appeal of a file have been purely intuitive, in the sense that I have tried to hear the structures as music and not as music material. It is similar to browsing an online music store, listening for something interesting to buy. The patch I have used for this project is generating musical structures based on different series of intervals that relate to a core melody not unlike a cantus firmus, and this melody is shared among the four voices. There are some random elements involved and also slow automated parameter changes. Everything that comes out of the patch has the integrity of my structural ideas, but still only a few examples will have the magical appeal of what I consider good music.

POINT: As you composed a piano piece in the course of the project, can you tell us about the further compositional steps in the making of this piece?

Sköld: For this piece, I mainly used three of the promising structures with different characteristics. These files provided harmonic progressions and outlined the movements of each voice in the counterpoint structures. I then carved out the phrases I wanted, adjusting the harmonic content as I went along. For some passages I made dramatic changes, as in the middle part where I applied an arpeggio pattern to the harmonic content. At this stage I also applied stylistic behavior, like the baroque-like ornaments as in Fig. 14. Often, my notation looks more complex than it sounds, because of the different voices that need to be notated in different layers. An important aspect for me is the question of time signatures. It is crucial for me to get the division of time just right and I tend to have a lot of time signature changes to ensure that each phrase gets the correct underlying pulse.

Fig. 14 Ornamental detail from the piano piece

Project Review by Mattias Sköld

Looking back, this project had many interesting aspects; the intuitive side of composition is such a significant part of a composer's work. My initial goal was to locate and study tendencies in my own artistic process, and it did not come as a surprise to me that some of these decisions could be revealed using a Support Vector Machine. Nevertheless I was amazed by the efficiency of the model. We can conclude that there are statistically noticeable aspects of a structure that in my ears makes it a "better" choice than another structure. However, it remains to be seen if my judgement is shared among the composer community leading to the conclusions that there is a common idea of what constitutes a good structure, or if the computer has indeed managed to define a decision process that is only mine.

Regardless of this, there are both general and more specific ways in which the results from this project will enrich my further composition endeavors. I think of the different computer analyses performed in this context as a number of spotlights illuminating my work from different angles, making my work cast shadows in new ways. Besides helping me make more informed and more deliberate intuitive decisions in the future when evaluating my musical structures, there are very concrete ways in which the findings of this project could be integrated with my work since it is already a computer model in itself.

References

1. Adorno TW (1985) On the Fetish-character in music and the regression of listening. In: Arato A, Gebhardt E (eds) The essential Frankfurt school reader. Continuum, New York, pp 270–299
2. Ben-Hur A, Weston J (2010) A user's guide to Support Vector Machines. In: Carugo O, Eisenhaber F (eds) Data mining techniques for the life sciences. Springer, New York, pp 223–239
3. Kirby A (2009) Digimodernism: how new technologies dismantle the postmodern and reconfigure our culture. Continuum, New York
4. Ladma V (2004) Rhythmical structures. 29 January 2004. http://vladimir_ladma.sweb.cz/english/music/articles/links/mrhythm.htm. Visited on 20 Nov 2014
5. LeWitt S (1967) Paragraphs on conceptual art. Artforum 5(10):79–83
6. Murray Schafer R (2009) I have never seen a sound. Can Acoust 37(3):32–34
7. Vapnik V, Golowich SE, Smola A (1997) Support Vector Method for function approximation, regression estimation, and signal processing. In: Mozer MC, Jordan MI, Petsche T (eds) Advances in neural information processing systems (NIPS), vol 9. MIT Press, Cambridge, pp 281–287

Djuro Zivkovic/Difference Tones

Djuro Zivkovic, Daniel Mayer and Gerhard Nierhaus

Djuro Zivkovic was born in Belgrade, Serbia and comes from a non-musical family, however his parents inspired in him a love for art. His father adored painting and folk music, and his mother was a ballerina in her youth. When he began primary school he started to play the violin, an instrument to which he has remained connected throughout his career.

By the age of 12 he discovered Baroque music and was fascinated by composers such as Vivaldi, Handel and Bach and even started to compose music in the style of that period. When Zivkovic attended secondary school, besides his compositional attempts, he also began to improvise with friends and organised concerts for their performances. At this time, skipping almost the entire classic and romantic periods, he was strongly impressed by the music of Shostakovich, who opened for him the doorway to the music of the 20th century and also by Prokofiev, whose first two violin concertos he chose as pieces for the entry exam at the music academy in Belgrade, which he successfully passed at the age of 18.

Two years later, alongside his performance study, he began to study composition with Vlastimir Trajkovic, whose enormous knowledge, not only in the area of music, was a great inspiration for Zivkovic. The composers are still friends and from time to time enjoy long conversations concerning a broad range of topics.

Zivkovic later continued his studies at the Royal College of Music in Stockholm (KMH), where many of his compositions were performed by professional ensembles.

D. Zivkovic
Department of Composition, Conducting and Music Theory,
Royal College of Music, Stockholm, Sweden
e-mail: djuro.zivkovic@kmh.se

D. Mayer · G. Nierhaus (✉)
Institute of Electronic Music and Acoustics,
University of Music and Performing Arts Graz, Graz, Austria
e-mail: nierhaus@iem.at

D. Mayer
e-mail: mayer@iem.at

G. Nierhaus (ed.), *Patterns of Intuition*,
DOI 10.1007/978-94-017-9561-6_11

During this time he also had the chance to meet composers like Penderecki, Stockhausen, Magnus Lindberg, Esa-Pekka Salonen, James Dillon and Bent Sørensen. The latter made an especially strong impression on Zivkovic through his personality, music and teaching. After finishing his studies Zivkovic went on to work at the KMH where he taught composition and various aspects of music theory, though recently he made the decision to concentrate solely on his composition in the years to come.

Besides his compositional work, he also performs and improvises as a professional violinist and pianist. Zivkovic has developed a broad range of compositional techniques, from sophisticated polyrhythmic structures to microtonal techniques and an innovative approach to working with difference tones, which was explored in the course of this project.

His music is frequently performed by renowned contemporary music ensembles and, alongside other awards, Zivkovic received the Swedish Grammy award in 2010 and was given the 2014 University of Louisville Grawemeyer Award for his piece *On the Guarding of the Heart*.

Artistic Approach

Statement

"Mapping the truth", expressed in its totality through music is what composition means for me. Composing is therefore a way to express an understanding towards a "wholeness" of truth, and this notion contributes to how I view myself in my "humanness", as a spiritual and intellectual being.

I see composing as a way to search, with curiosity, for a kind of "mind-map" of the total essence of truth, which has both perceptible and noetic qualities. By noetic[1] I mean not only a pure mental and intellectual activity, but also one which draws from a spiritual and religious perspective where the Orthodox language of the "Fathers", is not "what is perceived by the mind, but what is perceived by the heart". The "heart" is not the biological organ, rather "heart" means the internal psychosomatic space where the kingdom of God is found. This "space" is part of one's ontology. And "nous/mind" is not the brain as organ. The vigilance of the mind, nepsis (νῆψις), combined with the purity of the heart creates the combination of mind/heart that performs the "noetic" spiritual function.

I find this principle very close to what in ancient Greece they called "the music of the spheres". For me personally, it is a two-part idea: the first idea is a philosophical direction crystallized by Archyta, who grounded the science of acoustics where through numerals we are able to observe the "correctness" of music[2]; the second

[1] Which comes from "nous" (νόος)—intellect.

[2] For example, as strings vibrate in certain frequencies where the result of it has a specific effect on the listener, rhythms in a certain number of beats or accents gives a result which can be perceived as correct or incorrect rhythm according to lyrics, dance or emotional conditions of the listener.

idea, the older Pythagorean *akousmatikoi* order, is less about the correctness of music, but more about the metaphysical and ethical quality of their ideas and the unity between harmonies: astronomy, music and humanity. Music is therefore for me not just "correct sound" in numbers, or sounds that can or cannot please other ears, rather it has the power to *change and develop* human psyche and ethos on the cosmogonic[3] level, as a spiritual transfiguration.

In this sense I keep in mind Plato, who had from arithmetic and geometry distilled a deep doctrine of the art as constituting human goodness and contemplative wisdom, through the second Prometheus—Pythagoras [5, Sect. 16c].

The creation of music for me is an ultimate human creation that takes all kind of effort. It has always been something very deep, on the level of looking for the truth in layers: emotional, intellectual or, above all, spiritual. I would describe it as a balloon. There is a membrane, it appears just to be a surface: an intellect, numbers, forms and outlook, yet on the inside of the membrane we find a strong, dense and enormous potential which is the spirit.

Personal Aesthetics

When I first started to take composition seriously I drew from folk music in my pieces, which was well received among my teachers and the audience. I was still young and at the beginning of my studies in music but I immediately realized that composing, or any art creation, must go beyond the limits what is accepted to be expected. For example, when I composed a piece in a "folk music style", many people expected that it would be similar to Bartok, or some modern arrangement of Balkan rhythms, etc. but instead I took a single stringed folk instrument as inspiration and made a piece for classical instruments. The music, which I wrote out of that inspiration, was so direct, so plain, in a way it could be said it was hard to listen to. I wanted it to be played in a rude, but not ugly manner and I did not ask that the performers play it as the classical musicians would, with nice vibrato or a pleasant sound.

Let us take an example of inspiration that I have used. Imagine a village and a recently dead person lying on his bed. His family sits around him, with the so-called "crying mothers",[4] women who cry and talk at the same time. I tried to make music out of this sound, or rather just from a second, a snapshot of this sound. I wanted to make music that goes so directly into the place where you sense the "crying mothers" sounds, towards this high intensity, without any disturbance, without the emotional relief. Some reactions of the audience were that my music was too heavy, stressful to listen to it and too intense.

[3] Here I have in mind not only the origin of a human, but also the whole set consisting of the human pre-existence and as well both its ontology and eschatology, where music serves as an ultimate medium to re-establish the true, divine source of human nature and the whole environment.

[4] *narikače* (*Serbian*).

Another example is the inspiration I have received from the music of the Orthodox Church. I was not interested in composing "spiritual" music that was very close to the original but I wanted to create music out of the very short ornaments in church singing, which are usually left unnoticed by the average listener. If you apply those kind of ornaments to your own music, whatever it is, it will sound related to "spiritual orthodox" but at the same time that small information of the ornament is very transparent, almost invisible.

I've also composed music that is inspired by church bells. That approach is not new, but I don't use any computer analysis of the bell's sound for later re-synthesis, neither do I use a real tubular bell. In the beginning of my piece *On the Guarding of the Heart* there is a sound inspired by a church bell but to achieve that special sound or impression I tried to extract the most important information from the bell sound and put it in 14 instruments in the first two minutes of the piece. I have listened to church bells for years since I was child and after some time I could dissolve the bell sound in numerous different ways in my imagination: diverse pitches, unusual attacks, lingering sounds, complex harmonies. By "dissolving" these sounds and digesting them acoustically, when my inner "sound machine" was ready, I arranged everything for orchestra, altering the envisioned bell sound by stretching its resonating time-span, not by computer processing software but just with the power of imagination.

Figure 1 shows a section of my piece *On the Guarding of the Heart* where I applied such procedures to "paint" the sound of bells through the orchestra.

In all these processes I push myself to the edge to render my musical ideas in a simple but nevertheless condensed and complex form. When I reach the point that my "inner need" is ready to notate that idea, I have to be very careful how to transcribe the music on paper. For me the ambition is to write as simple as possible but not to compromise the expression of musical ideas.

I believe that foundation for the equation "the most simple notation = the maximum of expression" is in strong connection and association to my performing of music as a professional violinist, as well in my interest in folk music and improvisation. When I talk here about improvisation I use it in two different ways, one is analytical and the other is synthetical.

It is my normal procedure, which I find it very useful, to record my improvisations and to analyse them. In some cases a huge portion of a piece is created entirely by notated improvisation. I have faced many times, reading various analysis of my music written by others, stating that it is organized in many different ways, and in some cases there are discoveries of structures which I hadn't done consciously. Furthermore, the analysis shows a certain level of logical concept of the diachronic or chronological development. For instance, when I notate a recorded improvisation I can notice that the acceleration of a rhythm or the dynamic nuances show some shapes close-to the golden section ratios; that imitation in different instruments is almost a total canon; or formal parts are exactly mirrored. The fact is, I have actually improvised all instruments and recorded them at different times, without any conceptual preparation or making a plan in advance. I don't have any speculative clarification for how it occurs. In a quixotic manner I would say: while the soul vibrates, it vibrates in coordination

On the Guarding of the Heart
instrumental cantata
(2011)
Djuro Zivkovic (*1975)
Echoing. ♩ = 58

Fig. 1 *On the Guarding of the Heart* for chamber orchestra (2011), bars 1–3

with some pre-defined powers, similar to pre-defined powers in a genetic code or chemical reaction. But of course, not all my music is done in that way.

An example of an early work, inspired by folk music, *Duo* begins with two violins performing heterophonic melodies in the entire first part (Fig. 2). Although there are numerous chromatic movements, it doesn't allude to chromatics in the late romantic sense.

At the end of my almost exclusively dedicated "folk music period", in a piece for string orchestra, *Serenade*, I used numerous layers to achieve a total, super-dense sound of melodies. Every melody was put in a different pitch area where each of them was additionally split with heterophonic techniques. The result I would call "layered-polyphony" (Fig. 3).

After *Serenade* my desire became strong to develop my music on the level of harmonic progressions, not just on the melodic and rhythmic level, but also my

Fig. 2 *Duo* for two violins (1997), bars 7–11

intention became to develop music in its "totality". *Serenade* is entirely polyphonic and there was an absence of harmonic development and this piece led me to a crucial upcoming question: "how should I continue this?".

In my pursuit for harmonic improvement, I was seeking something sustainable, something that has both fundamentals in the historical harmonic foundations and in acoustics, which would give me a good feeling of security in my own composing and which could be developed further as a system my compositional approaches might build on.

At the beginning of that investigation I simply listened to a row of chords and tried to alter one after one in such a manner that they "transform" into each other in different ways. My ear and inner feeling was the only tool in that process. The first piece I did in that approach was *Eclat de larme* for chamber ensemble, 2005. In all pieces after *Eclat*... I have been increasingly using these harmonic progressions and the constant use of it has resulted with the further development of that method.

Figure 4 shows an excerpt the manuscript of *Eclat de larme*. On the bottom stave there are chords that make a harmonic development in diachronic way. All the notes above that chord are belonging to the same chord, except glissandos or quartertones.

In the following piece *The White Angel* (Fig. 5) I used a more sophisticated approach to generate the chord progressions. There is an immense similarity to *Eclat*... in the way I use chords as a basis for creation, but here are the chords connected in a system, which I call the *DiffTone* system.

Formalisation and Intuition

To speak about the role of formalisation and intuition in my work I would like to explain beforehand some essential steps that take place during any of my compositional work. My way of composing is usually the following:

Firstly, I never have any idea of music that develops just as a "pure" music, but there is always a relationship to something external to the music: poetry, philosophy,

Fig. 3 *Serenade* for string orchestra (2002), bars 113–116

nature or faith. I don't play piano in an attempt to "find" harmony and I don't listen to other's music either, except very old, church or folk music. However I make an effort to refill my "spiritual batteries"[5] that the music by itself starts to flow from me.

Secondly, at this point, when a sound "through the spirit" is found, the music is still not written, but it is improvised or imagined. However I do manipulate with my

[5] By the spiritual knowledge (γνῶσις) is referred to the knowledge of the intellect as distinct from that of reason. The intellect acts here as the highest faculty of man, through which he knows the inner essences or principles of created things by means of direct apprehension or spiritual

Fig. 4 *Eclat de larme* for chamber ensemble (2005), bars 37–40

improvisations at this stage. I can just record an improvisation, listen to it carefully, and eventually new ideas would come to the mind. I could combine totally different improvisations created without any connection since they were improvised on different occasions.

For the creation of many ensemble parts in my piece *On the Guarding of the Heart* I used the recording of a piano improvisation made one year earlier alongside with other improvisations recorded many years earlier, or also during the composition process. I just juxtaposed them in an audio sequencer and played them together and the result was pleasing for me.

At the third stage I encounter the problem of how to notate what I call "Whatever-Music".[6] Sometimes I feel that I "fight" against my imagination in order to find a proper decision about what to write down, since the "corridors" of the "Whatever-Music" are, for me, enormously complex and endless. When I come to this stage, I definitely come face to face with more rational problems, which is completely natural. The rational problems about the "Whatever-Music" are similar to the problems in science: you have to define and apprehend parts of the universe, to rationalise it so far as it is possible. Only with the rationalisation I would be able to tame and re-create the "Whatever-Music" as a scientist could do with natural observations into mathematical formulae.

(Footnote 5 continued)
perception. Unlike the reason, the intellect does not function by formulating abstract concepts and then arguing on this basis to a conclusion reached through reasoning, but it understands the divine truth by means of immediate experience, intuition or "simple cognition"—term used by St. Isaac the Syrian [4, Glossary].

[6] I could compare the "Whatever-Music" with the freedom of choice. Whatever we choose, we are still "obligated" to respect architecture of the universe, and still we have an infinite freedom of choice. We can choose "whatever" but still in the frame of the natural law. The "Whatever-Music" is also a free choice inside of the frame of the spiritual and esthetical law.

Fig. 5 Excerpt from *The White Angel* for chamber orchestra (2006)

In the fourth stage, because the "Whatever-Music" as such contains infinite possibilities, it is important to now use a logical way of thinking, in order to make the ideas realisable. This is a state of what I view as "cold" decisions amongst the stillness. The "Whatever-Music" needs to be incarnated and is not just a sense of "being" in the Spirit. It is required for decisions to be taken, to start from somewhere, to think logically and use numbers.

My starting point is rather a vision that needs a structure to "unfold", because I see the flashes of ideas, both musical and non-musical, are lead by fantasy. In all cases making the structure from these visions is only on the low-level numeric organisation. For me, the low-level organisation can be total and unlimited[7] if it enlightens the high-level radiant energies. There is not, therefore, any limitation in musical formalised structures. The part of the high-level structures, which are close to the spirit and which are above any "physical" explanation, are not possible to rationalise.[8] The high-level and the low-level structures are of equal importance to me. The word "low" and "high" doesn't allude to any importance of the structure, since the low-level organization by numbers and high-level structures are the part of the same compositional process. Furthermore, in my teaching I always include logical, emotional and spiritual contexts of music and I also teach students to develop the mental power of musical discrimination by analysing music in general, even their own music.

At the last stage I face the rational part of the problem and I confess, that contemplating the spiritual matters I can't solve these problems in reality but it is only through the power of the mental thought. In most cases I start to make a structured concept of my music, by measuring lengths, phrasing, combining chords, planning harmonic progressions, etc. The most time I spend in my composing is endless excessive transforming of the music material until I get satisfied. Sometimes I write very nice material for the end of the piece, but then further material evolves and affects the parameters of the previous and then additional changes become necessary. After each piece is finished I have several hundred working sheets because I almost never erase what is written but, if it is needed, I re-write it again with some alterations. I also use notebooks where I describe each composition, or a part of it, by words or pictures. By just reading these texts I can clearly hear my music even if the notes were left many months behind.

Finally, when I have a very clear idea of the music I want to write, it is perfectly reasonable to be on the pure rationalisation level and "tune" the music's parameters, as a watchmaker makes and tunes his/her clocks. At this state the composing is no longer a hard passion, in the sense of "heavy labour", but a work filled with joy.

---

[7] This is the case, for instance, in the total serialism, or any kind of musical structures, which are manipulated only by numbers which represent the musical matter like pitch, rhythm, form, instrumentation, etc.

[8] It is not possible to rationalise it even through a metaphysical understanding. Here, the spiritual journey has a very strong apophatic character.

Evaluation and Self-reflection

There are numerous layers of evaluation and reflection in my work. These layers are not put into any order, but rather they depend on the working flow from a particular moment. First, there is a physical process of daily composing, when the intellect wanders and alters from minute to minute. This is a short-time reflection. It is very hard to be critical during this phase, but some "flashy" ideas can be crystallized. Another layer is the spiritual contemplation, an extremely important state of self-deduction and stillness. At this moment my intellect asks for peace, silence and calmness to be able to develop from the "inside", to be able to descend into a zero-ego state, to communicate with the pure thought. This stage is a moment when within self-denial the crucial decisions are executed.

The third layer is a critical returning to the finished parts of music, after some hours, days or weeks. In some cases performing composed fragments for musicians or non-musicians and getting their feedback can be very helpful. Finally, the most important point is to stimulate someone's soul in the similar way as strings tuned at the same pitch stimulate each other.

The evaluation can be also possible through another art form. I believe that two, or more kinds of art do not obstruct or distract each other when performed together, since they can express an integrated wholeness. This is found particularly clearly in opera, ballet, church with painting and music, when applicable, exhibitions offering the space for concerts or poetry reading, and so on. Thus, if a composer reads poetry and gets inspired by it to compose music, we can finally try to compare these two: is there any slight feeling of non-comfort when we listen to the composed music and read the poetry that inspired that music? If yes, then something went wrong.

The self-reflection can finally be on a yearly basis: "How do you like your music written years ago? How did it move other people? Is there any need to play a piece again?" are such questions.

There are advantages in using the computer as a tool for finding possibilities to reflect in a new way upon the own structures by providing interesting, or not so interesting solutions. One of these advantages I find right now in the harmonic calculus, where the computer shows some result that can lead me to a further research, which I find also very inspiring. If there are some interesting outcomes it can lead me toward a more flexible understanding and re-thinking of my own system.

Project Approach: The DiffTone System

In the last eight years I became more interested in harmonic progressions, building the chords and searching for their generic evolution. I called it *the harmonic system of difference tones*, sometimes abbreviated to the *DiffTone* system. I started using it partly in my piece *Eclat de larme* but a definitive use was in *The White Angel*. I became more obsessed about the harmonic organization and powers that lay in each chord and the energetic flow between them. All other pieces are, so far, based

Fig. 6 A part from the harmonic progressions in *The White Angel*

on that system or its further development. During the time I have developed a strong desire for the harmonic development in my music and now I cannot compose if I don't have a very clear harmonic system or background. Figure 6 shows the *DiffTone* system in action. It represents the harmonic progressions in *The White Angel*.

Project Expectations

My exploration of the *DiffTone* system continued and by chance, during my development of the details of this system with many music designers, programmers, composers and mathematicians around the globe, I received an invitation to join this project. I see this as a kind of musical "fate". My expectations from the project are high: will it be possible to enter into the deep exploration of something that I have been thinking about for a long time—the harmonic system of difference tones?

Some improvements of the system, I am sure, can definitely be found. The consequences of using a computer could be the development of a coherent system that would bring a possibility for other people to get involved by using and developing that technique, expanding the system that could bring me in another, not yet explored, development of the harmonic progressions.

As my pieces in the last 8 years relate very strongly to harmonic development it would be absolutely fantastic to explore this approach in the course of an on-going project. My aspiration is to see how this project could assist the development of the *DiffTone* system and to get new insights into my own compositional practice or even to develop a new approach to composition.

Exploring a Compositional Process

POINT: We investigated the possibility to build chord progressions with Zivkovic's system—mainly within a 9-tone-scale, which is strongly related to Messiaen's second mode, the octatonic scale. We compared an automated generation of chord progressions to his own pen-and-pencil practice of building chords. From our discussions it turned out that a quality of richness concerning tonal associations is one of his main goals in his compositional formalisation. We looked at the potential of scales to find tonally connoted chords and came to the suggestion of developing a measure for the tonal potential of scales. With several alternative ways to measure, Zivkovic's scale gave optimum results in that respect among all scales of nine tones.

The so-called *combination tone* phenomenon has been known for a long time and is mostly related to the Italian Violinist Guiseppe Tartini,[9] who described it [6]. For sufficiently loud pitches at frequencies f_1 and f_2 the difference tones $f_1 - f_2$, more seldom also $2f_1 - f_2$, $3f_1 - 2f_2$ etc. can be heard. Sum tones at frequencies $f_1 + f_2$ and $2f_2 + f_3$ might also be heard, but are more rarely reported than the difference tones. It is commonly stated that the phenomenon is physically caused by the inner ear by so-called *otoacoustic emissions* [2], in contrast to the psycho-acoustic *phantom fundamental* phenomenon [1]. And although the psychoacoustic effect of binaural beats is evident [3], similar reasons for the combination tone phenomenon don't seem to be proven yet.

Zivkovic's system is based on the most frequently heard combination tone, the difference $f_1 - f_2$. For calculating the physically exact difference tones of course it is not the same if you have given intervals in equal temperature or just intonation. Zivkovic is referring to the simplest relations of just intonation intervals. But let's regard difference tones of tempered intervals first. Below one sees the difference tones in the bass clef, deviations from notated pitches are written in cents (Fig. 7).

Regarding just intonation, there are some difference tones that are closer to other equally tempered pitches than those represented above. Some intervals also have multiple interpretations in terms of just intonation. But at least for fifths and fourths

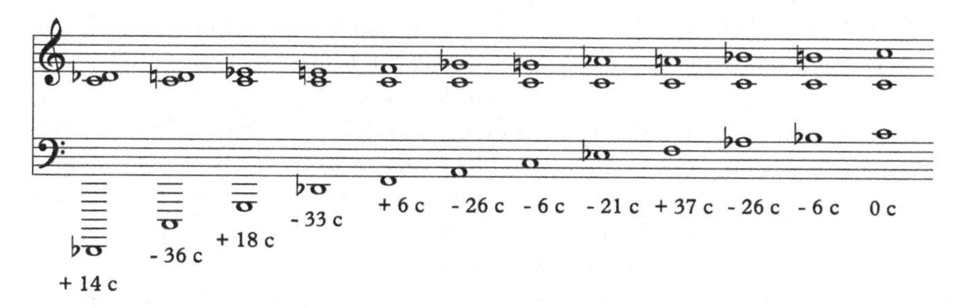

Fig. 7 Difference tones of tempered intervals up to the octave, their deviations from equal temperature are notated in cents

[9] Guiseppe Tartini (1692–1770), Italian violinist, composer and music theorist.

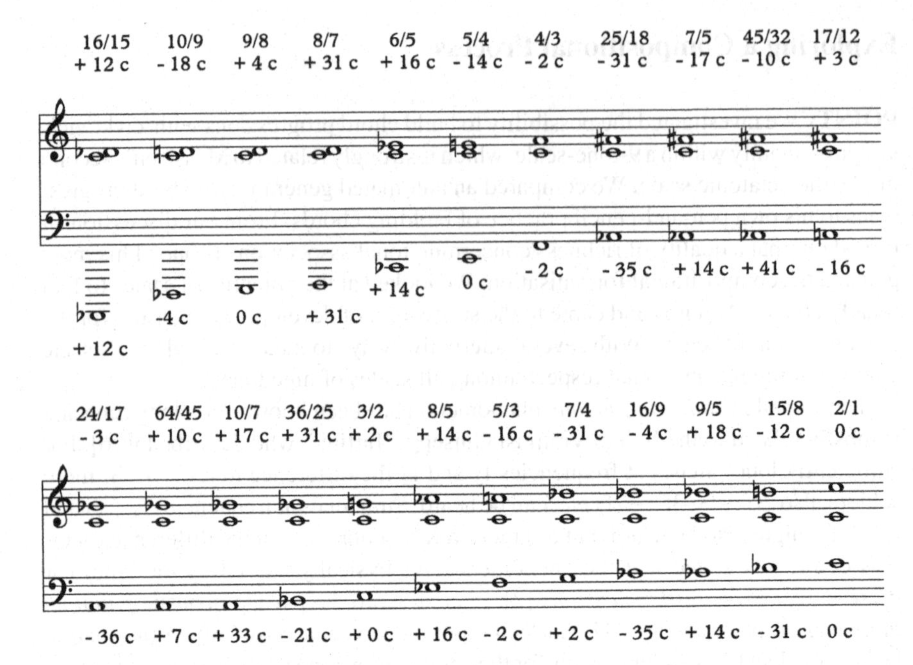

Fig. 8 Difference tones of just intonation intervals up to the octave, rationals and their deviations from equal temperature notated above, difference tones and their deviations from equal temperature notated below staves

it is clear to perceive that 3:2 and 4:3 are the corresponding frequency ratios, also 5:4 and 6:5 for major and minor thirds and 5:3 and 8:5 for major and minor sixths, 16:15 and 15:8 for minor seconds and major sevenths. However things are not so clear for major seconds, minor sevenths and the tritone. Figure 8 shows common alternative interpretations of just intonation intervals and their corresponding difference tones.

If we take 9:8 as the most widely used interpretation of a major second also the difference tone of the corresponding minor seventh is unique. There remains an ambiguity however for the various interpretations of the tritone.

POINT: How did you decide what equally tempered approximations to difference tones you would take into your system, especially for tritones?

Zivkovic: I simply take the difference tone that is naturally found in the overtones, or closest to them, or found in the considered interval or closest to it. For example with a major second, I choose 9/8 rather than 10/9 or 8/7, because the pitch class of the difference tone is itself contained in the considered interval. For the tritone I choose 7/5 since this is the nearby interpretation with small integer numbers, also, starting from a tritone on C, Ab is closer to C–F# than A.

POINT: Figure 9 shows tempered approximations of difference tones of tempered intervals. Within *DiffTone* difference tones, derived from bespoken interval ratios, are used as pitch classes (Fig. 10). At this point Zivkovic builds chord progressions based on the above difference relations. Successor chords are only allowed to consist of pitch classes of difference tones of compressed predecessor chords. E.g. for a

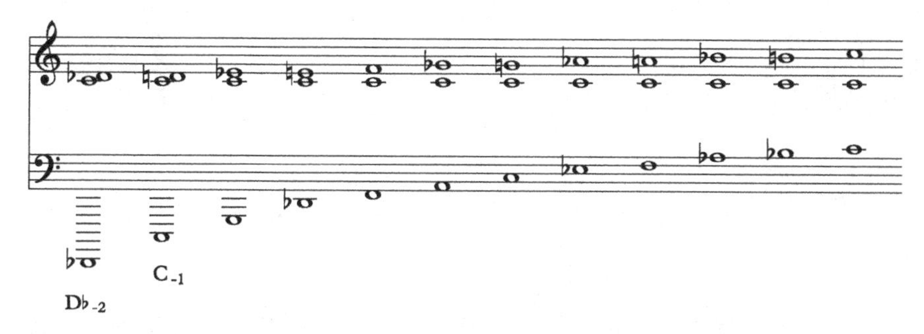

Fig. 9 Tempered approximations of difference tones of tempered intervals

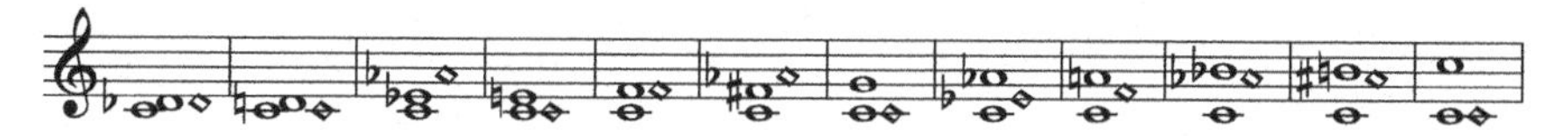

Fig. 10 Difference tone pitch classes used by Djuro Zivkovic in his *DiffTone* system, derived from just intonation intervals

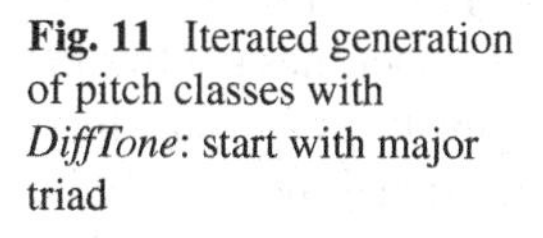

Fig. 11 Iterated generation of pitch classes with *DiffTone*: start with major triad

Fig. 12 Iterated generation of pitch classes with *DiffTone*: start with minor triad

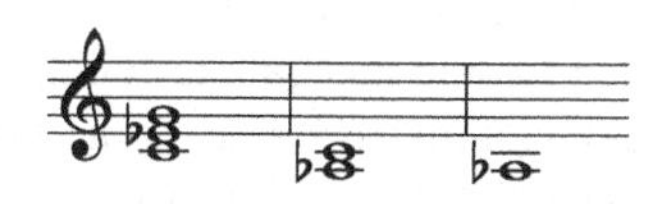

major triad CEG or any of its inversions, the unique compressed form is the root position. There are three possible combinations of its pitches (CG, CE, EG) and all have C as resulting difference tone class (Fig. 11). For a minor triad the procedure might be applied twice before it sticks (Fig. 12). For chords containing more pitches there will be a larger number of difference tones and a larger number of possible successor chords, e.g. for a compressed hexachord F Gb A Bb C Db (Fig. 13).

Following the Gaussian formula for a sum of numbers[10] we get 15 combinations $(n * (n - 1)/2)$ of difference tones with the following occurrences, see Table 1.

Successor chords can be chosen from those pitches. For the sequence they will be taken in compressed form, i.e. an inversion of minimal ambitus. If there is more than one compressed form, the one with the lowest tone nearest to the lowest tone of the predecessor is chosen. If even this selection is not unique, the compression with most pitches in common with the predecessor is selected.[11]

[10] Used by Carl Friedrich Gauß (1777–1855), a German mathematician and physical scientist, at the age of 9 on the occasion of having to sum the numbers from 1 to 100 at school. However the formula has been known since ancient times.

[11] By enumeration we found that there is one single case in which even this doesn't break the tie, it comes from splitting the integer 12 into the tuple (1 3 1 3 3 1). Then for a chord with interval

Fig. 13 Compressed hexachord giving difference tones of Table 1

Table 1 Occurrences of difference tones in compressed hexachord of Fig. 13

Gb	4×
F	3×
Bb	3×
D	2×
Ab	1×
A	1×
Db	1×

POINT: The system produces sequences of compressed chords. Why are you using these sequences in your compositions?

Zivkovic: When I started to work with chord progressions I used only my ear to discover new chords. However, this didn't satisfy me because I sometimes got lost in so many variants. Finally, the chords should in the end serve music, and even when I got very nice progressions I changed my mind and changed some tones. There is nothing wrong with that. But then I realized that there must be some other principles that I could take into account as well. I was aware of the difference tones very long as a violinist, because it is a tool for intonating on the string instruments, particularly the high tones. So I investigated that principle and found that the chords are as good as the free progressions. As I am a person that likes to follow rules, and break the rules if needed, I liked that idea. I just put the chords into a system that worked, and was able to focus on my work.

POINT: A chord in its compressed form in general gives another set of difference tones than any another inversion. Also stretching an interval by an octave in general changes the pitch (and pitch class) of the resulting difference tone. So why are you calculating the difference tones from the idealised compressed form and not from the pitches actually used? Wouldn't the latter be a more straight way to get pitches related to the chord by difference tones and thus to find possible successors?

Zivkovic: Any system needs time to develop. My first use and tests were very simple. I stuck with compressed form since it seemed to me much too diverse to explore so many possibilities. Another reason was more idealistic. I was impressed by the major chord. If we have c–e–g chord in the first octave, we get C as the resulting difference tone. But I would guess that if we put these tones in inversions

(Footnote 11 continued)
vector (1 3 1 3 3), e.g. (C Db E F Ab B) there exist two equidistant maximally compressed chords with the interval vectors (1 3 3 1 1) and (3 1 1 3 1), here (E F Ab B C Db) and (Ab B C Db E F). In that very special case a choice would have to decide with which chord to proceed.

and different octaves we would definitely get different results in the theory but in the practice we would hear again the C-major chord! I respected the audible experience.

POINT: Is it really like this in any case? If we take a sixth chord (E C G) the real difference tone pitch classes are C, Eb and G. And in fact the more far you spread the chord the more it goes away from the C major impression.

Zivkovic: I agree that the three tones of C-maj chord spread away, in different octaves and in other inversions, (can) give a weaker C-maj impression. However, I had never faced a musical texture in my composition that required such a re-consideration of the difference tones' progressions—in the most cases the texture of my music was very compact. I also stuck with the compressed form since in that way it doesn't belong to the voice-leading part of my work: if I agree that all chords are used in the compressed form, the tones can be used more freely in the composition process; but when I think about the different positions, like inversions which give other results, then I have to think how in voice-leading, the free use of the tones are suppressed. In that case I have to respect the different inversions or placement of the tones. I would be forced to use exact pitches—even the correct frequencies in Hertz. When thinking more in that direction I would feel that the whole world of the improvisation is more controlled, and I would be more limited in my creation.

POINT: We suggested a revised procedure to find difference tones (resp. difference pitch classes) of compressed chords. When having two different pitches in a compressed chord and these pitches would, later on, be used in an arbitrary register it could also not be foreseen which of the two pitches of the compressed form would be the high and which the low one in actual usage, it looks reasonable to suppose equal probability. Thus, besides from the fact that different octave registers would lead to different difference tones, it would be nearby to calculate difference tones for all interval relations within the compressed form plus from their inversions.

So the hexachord from Fig. 13 leads to 30 occurrences of difference tone pitch classes. Among them there are 10 different pitch classes, instead of 7 different pitch classes from 15 occurrences in the original form of the procedure, see Table 2.

While generating chord sequences out from one distinct scale, as described below, we were using both variants of the procedure.

POINT: What do you think about this extension of your system?

Table 2 Occurrences of difference tones in compressed hexachord of Fig. 13, regarding complementary intervals also

Gb	6×
F	5×
Bb	4×
Ab	4×
Db	3×
D	3×
E	2×
C	1×
A	1×
B	1×

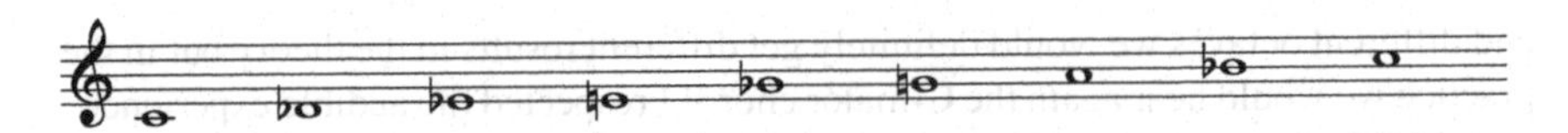

Fig. 14 Octatonic scale

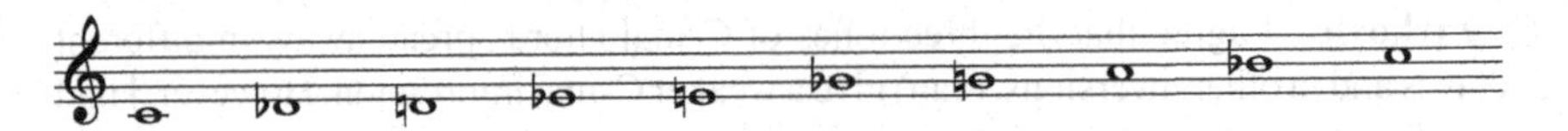

Fig. 15 Extended octatonic scale or *scale antique* (Djuro Zivkovic)

Zivkovic: The extension of the system described above is interesting and it should be tested. What I can recognize as an addition to my system is to pay attention to those tones that occur most in the outcome. I would be interesting to see how they "resonate" in the chord progressions, and if there is any reason to use one of them as a base tone (the lowest tone in the chord).

POINT: Zivkovic has been using scales for quite a while. He now prefers a scale based on Olivier Messiaen's second mode, the octatonic scale, which consists of a sequence of semitones and whole tones (Fig. 14).

Obviously this scale is related to the diminished seventh chord. As with the latter there exist three transpositions and two modes: one can start with a semitone or whole tone. The nine-tone scale Djuro Zivkovic is using, and which he calls *scale antique*, comes from adding an arbitrary of one of the missing pitches of the chromatic entirety to Messiaen's second mode. Why is it arbitrary which pitch to add? Due to the symmetric step structure of the octatonic scale (1 2 1 2 1 2 1 2) the step sequence of the resulting nine-tone scale will be (1 1 1 1 2 1 2 1 2) or one of its rotations. (1 1 1 1 2 1 2 1 2) is the unique representation if we take lexical ordering and the minimum regarding this ordering as representative. We might also want to call this scale *extended octatonic*[12] (Fig. 15).

In contrast to the octatonic scale this scale has no limited transposition, i.e. its 12 transpositions are different.

POINT: What is the relevance of scales for your compositional work?

Zivkovic: The relevance is huge. I want to achieve a very resonant sound when I use that scale. The *scale antique* was a random result during the *DiffTone* calculation. I can't recall precisely how I got it, but I guess by setting up two progression chords together. I have found that 9-tone chord very beautiful and I continued to use it in a longer harmonic part of my compositions. I discovered that just that chord can behave as a scale by using it as a limitation in the *DiffTone* process. I name it the *ancient scale*, since it has, at least for me, some very nice vibrant and colorful "toneä". I also like the components of the scale: there are plenty of major, minor, diminished or augmented chords, plenty of fifths and tritones.

POINT: How are you using scales in connection with the your system?

---

[12] Zivkovic also calls it *mirroring scale* since the interval vectors 1211 and 1121 frame a base tone (not to be confused with a tonic). E.g. Ab is the base tone of D# E F# G Ab A Bb C C# (D#).

Zivkovic: I use the scale in three different ways:

1. By using the entire scale as it is. I do eventually change its base tone.
2. By extracting the *DiffTone* chords inside of one scale.
3. I had the idea to make modulations when *DiffTone* generates notes outside of the scale. But it takes time to investigate, since the 9-tone scale is already very close to 12 tones, and if there were too many modulations, the sense for the scale could potentially be lost. This is still in an experimental phase and I made very limited use of it, for instance in *Le Cimetière Marin*.

POINT: So Zivkovic uses his system to make chord progressions within modes, especially the described. We considered a sequence of chords of given density, but how would we calculate such a sequence? We accomplished this by enumerating all chords of the given density and its possible successors according to the system. As therein only compressed chords are regarded the number of chords within the scale is well treatable. By finding all possible successors for all chords we get the description of a graph and the task of finding chord sequences without repetition reduces to the task of finding long paths in a graph without doublings. As a concrete example within *scale antique* there are 126 possible chords of density 4, and an enumeration within the successor graph results in a maximal length of 7 chords, accomplished by two sequences as shown in Fig. 16, chords in compressed form, thus voicing neglected.

It should be mentioned that finding such sequences just by pencil and paper is hardly possible which might also be a reason for the modifications of the system by Zivkovic in a next step, resulting also in a relaxation of the constraints for the search algorithm. Figure 17 shows two sequences of length 10 and density 5, the chords are represented also in compressed form.

Fig. 16 Maximum length 4-tone chord progressions within *scale antique*, according to *Difftone*, voicing neglected

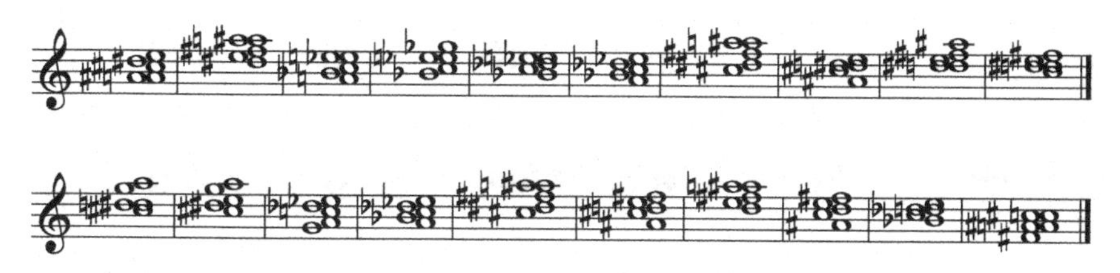

Fig. 17 5-tone chord progressions within *scale antique*, according to *Difftone*, voicing neglected

POINT: Did you try to build such sequences for your compositions and are there specific advantages or problems in building such sequences with pencil and paper?

Zivkovic: As far as I remember I did try to make such sequences (around 2005) but I abandoned it because I couldn't find a solution. There are problems if you want to respect the rule ultimately, breaking it might be an advantage!

POINT: To find such a long chain just by pencil and paper is of course almost impossible, but supposed you could manage to do so, would a chord sequence like in Fig. 17 be an interesting starting point for a composition?

Zivkovic: Yes. I have tested that progression and I have found a very good potential in it. Ten chords can be considered as a short progression, but it can be avoided by slower "modulation" or using these chords in a smaller part of a composition.

POINT: Zivkovic suggested to change the mode or to change the density, he deliberately uses both relaxations in his compositional practice. It is also possible to allow chords to appear more than once in a sequence, but to demand different continuations. So we produced a sequence using *DiffTone* and *scale antique*, but now varying the density between four and six tones. With this easing of constraints it is indeed also far easier to get longer chord sequences without repetition—or equivalently: longer paths without repetition in a larger successor graph. It was one of several occasions during our project where certain aspects of procedures chosen by composers in their pen-and-paper work turned out to be fruitful in computational implementation, when working with larger object sizes and searching for optimal solutions. Figure 18 shows an example of such a sequence of length 30 with chords in compressed form.

As mentioned earlier, richness in tonal associations is a phenomenological aspect which is important for Zivkovic when choosing his harmonic material. But how

Fig. 18 Chord progression of variable density (4–6) within *scale antique*, according to *DiffTone*, voicing neglected

could that be described, maybe even measured, neglecting at first the way a scale is actually used in a composition? One could look at intervals and chords determining a tonal context. Some intervals already achieve that, e.g. a fifth, others less, e.g. a minor third. Much more clearly some three-tone chords establish tonal associations, most striking the major triad. An arbitrarily added tone doesn't completely abolish the tonal effect of a major triad, its gravity is very strong. There are other triads that establish also quite strong tonal relations, e.g. the diminished triad, the dominant seventh chord without fifth and the dominant seventh chord without major third.[13]

These four chords can be seen as approximations to all three-tone groups within the first seven partials. So "richness of tonal association" could be seen founded in closeness to western musical culture and its employment of the dominant seventh chord and at the same time on occurrence of basic segments of the sequence of overtones.

We are now looking at these tonal triads and, in consequence, their occurrence in scales. Here are the first seven partials, based on C, rounded to equal temperature as pitch classes:

C C G C E G Bb

Reduced to pitch classes and sorted (dominant seventh chord):

0 4 7 10

As possible three-tone groups we get (0 4 7), (0 4 10), (0 7 10), (4 7 10), representing major triad, seventh chord without fifth, seventh chord without third and diminished triad. In compressed form they are equivalent to the interval vectors (4 3), (2 4), (3 2) and (3 3).

We counted all occurrences of above triads in scales of same size and ordered them accordingly. Rotations of the scale can be neglected, as they don't change the number of occurrences. Thus scales on the top of the following lists and their rotationally related ones are the "most tonal" ones in that sense. The roughness of this approach is obvious, as the use of scales in real world music (choice of pitch classes, registering, instrumentation etc.) can change a lot. Also the sum of counts of tonal triads of above kinds doesn't say how many fundamentals are related to them, this can be read from a separate counting and gives interesting additional information.

It might be said that the minor triad has an important tonal relevance as well. We omitted this chord for its ambivalence concerning a fundamental, as which we can regard the chord's base tone or—a major third below—the non-sounding base tone of a major seventh chord. However including the minor triad in below statistics doesn't significantly change the order, for regarded scale sizes the outstanding scales keep their place.

---

[13] It should be mentioned that for a long historical period triads without thirds were not considered valid in 4-part voice leading. However, as we are not developing a measure for music in a certain tradition itself, but rather for the association to a certain tradition, we include the third-less seventh chord for its strong association to a fundamental.

```
(1 2 1 2 1 3 2) (3 3 3 5) (4 0 0 4 0 0 4 0 1 1 0 0) 14 5
(1 1 2 2 1 3 2) (2 3 2 4) (4 0 1 1 0 0 4 0 0 1 0 0) 11 5
(1 1 2 3 2 1 2) (2 2 3 4) (4 0 1 1 0 0 1 0 0 4 0 0) 11 5
(1 2 1 2 1 2 3) (2 2 2 5) (1 0 1 1 0 1 1 0 1 4 0 1) 11 8
(1 2 1 3 1 2 2) (3 2 2 4) (4 0 0 4 0 0 1 0 1 1 0 0) 11 5
(1 1 1 1 3 3 2) (2 2 2 4) (4 0 0 4 0 0 1 0 0 1 0 0) 10 4
(1 1 1 2 1 3 3) (2 2 2 4) (0 0 4 0 0 4 0 0 1 0 0 1) 10 4
(1 1 1 3 2 1 3) (2 2 2 4) (0 0 4 0 0 1 0 0 4 0 0 1) 10 4
(1 1 2 2 1 2 3) (3 2 3 2) (1 0 4 0 0 0 1 0 0 4 0 0) 10 4
(1 1 2 2 3 1 2) (3 3 2 2) (1 0 4 0 0 0 4 0 0 1 0 0) 10 4
(1 2 1 2 2 2 2) (2 3 3 2) (1 0 0 1 0 0 4 0 4 0 0 0) 10 4
(1 1 1 1 2 3 3) (2 1 2 4) (0 0 4 0 0 1 1 0 1 1 0 1) 9 6
(1 1 1 3 1 2 3) (1 3 1 4) (0 0 4 1 0 1 0 0 1 1 0 1) 9 6
(1 1 1 3 3 1 2) (2 1 2 4) (0 0 4 1 0 1 1 0 1 0 0 1) 9 6
(1 1 2 1 2 3 2) (2 1 2 4) (4 0 0 1 0 0 1 1 0 1 1 0) 9 6
(1 1 2 2 2 2 2) (1 6 1 1) (1 0 1 0 1 0 4 0 1 0 1 0) 9 6
(1 1 2 3 1 2 2) (1 3 1 4) (4 0 0 1 1 0 1 0 0 1 1 0) 9 6
(1 2 1 2 2 1 3) (2 1 2 4) (0 0 1 0 0 1 1 0 4 1 0 1) 9 6
(1 2 2 1 2 2 2) (3 1 4 1) (0 1 0 1 0 1 1 0 4 0 1 0) 9 6
(1 1 1 1 2 4 2) (1 3 2 2) (1 0 1 1 0 0 4 0 1 0 0 0) 8 5
(1 1 1 1 3 2 3) (2 2 2 2) (1 0 1 1 0 1 0 0 0 4 0 0) 8 5
(1 1 1 2 1 2 4) (2 2 2 2) (0 1 1 0 0 1 0 0 4 0 1 0) 8 5
(1 1 1 2 2 2 3) (1 3 3 1) (0 0 1 1 0 4 0 1 0 1 0 0) 8 5
(1 1 1 2 2 3 2) (2 1 4 1) (1 0 0 4 0 1 0 1 0 0 1 0) 8 5
(1 1 1 2 3 1 3) (3 1 2 2) (0 1 1 0 0 4 0 0 1 0 1 0) 8 5
(1 1 1 2 3 2 2) (3 1 3 1) (0 1 0 1 0 1 0 0 1 0 4 0) 8 5
(1 1 1 3 1 3 2) (2 2 2 2) (1 0 1 4 0 0 1 0 1 0 0 0) 8 5
(1 1 1 3 2 2 2) (2 3 2 1) (0 0 1 1 0 0 1 0 4 0 1 0) 8 5
(1 1 1 4 2 1 2) (1 2 3 2) (1 0 1 4 0 1 0 0 0 1 0 0) 8 5
(1 1 2 1 2 2 3) (3 1 3 1) (1 0 1 0 0 1 0 1 0 4 0 0) 8 5
(1 1 2 1 3 2 2) (2 3 1 2) (1 1 0 0 1 0 1 0 0 0 4 0) 8 5
(1 1 2 2 2 1 3) (2 3 2 1) (0 0 4 0 1 0 1 0 1 1 0 0) 8 5
(1 1 3 1 2 1 3) (3 2 1 2) (0 1 4 0 0 1 0 0 1 0 1 0) 8 5
(1 1 3 1 2 2 2) (3 3 1 1) (0 1 1 0 0 0 1 0 1 0 4 0) 8 5
(1 1 3 2 1 2 2) (2 1 3 2) (1 1 0 1 0 0 0 1 0 0 4 0) 8 5
(1 1 4 1 2 1 2) (2 2 2 2) (1 0 4 1 0 0 1 0 0 1 0 0) 8 5
(1 1 1 1 2 2 4) (1 3 2 1) (0 0 1 0 1 0 1 0 4 0 0 0) 7 4
(1 1 1 2 4 1 2) (2 1 3 1) (0 0 1 1 0 4 0 0 0 0 1 0) 7 4
(1 1 1 4 1 2 2) (2 2 2 1) (1 0 0 4 0 0 0 0 1 0 1 0) 7 4
(1 1 2 1 1 3 3) (3 1 2 1) (0 0 4 0 0 1 1 0 0 1 0 0) 7 4
(1 1 2 1 1 4 2) (2 3 1 1) (1 0 1 0 0 0 4 0 0 0 1 0) 7 4
(1 1 2 3 1 1 3) (2 2 2 1) (1 0 1 0 1 0 0 0 0 4 0 0) 7 4
(1 1 3 1 3 1 2) (4 1 1 1) (0 0 4 0 0 1 1 0 0 0 1 0) 7 4
(1 1 3 2 2 1 2) (2 1 3 1) (1 0 1 1 0 1 0 1 0 1 1 0) 7 7
(1 1 1 1 1 4 3) (2 1 2 1) (0 0 1 0 0 4 0 0 0 1 0 0) 6 3
(1 1 1 1 2 1 5) (1 2 1 2) (1 0 1 1 0 0 1 0 1 1 0 0) 6 6
(1 1 1 1 4 2 2) (1 3 1 1) (1 0 0 1 1 0 1 0 1 0 1 0) 6 6
(1 1 1 1 5 1 2) (1 1 2 2) (1 0 1 1 0 1 1 0 0 1 0 0) 6 6
(1 1 1 2 1 4 2) (2 1 2 1) (0 0 1 1 0 1 1 0 1 0 1 0) 6 6
(1 1 1 2 2 1 4) (2 1 2 1) (0 1 0 1 0 1 0 1 1 0 1 0) 6 6
(1 1 1 3 1 1 4) (1 3 1 1) (0 0 1 1 0 0 0 0 4 0 0 0) 6 3
(1 1 2 1 1 2 4) (1 3 1 1) (0 1 1 0 1 0 1 0 1 0 1 0) 6 6
(1 1 2 1 2 1 4) (2 1 1 2) (1 1 0 0 1 0 0 1 0 1 1 0) 6 6
(1 1 2 1 3 1 3) (3 1 1 1) (0 1 1 0 1 1 0 0 0 1 1 0) 6 6
(1 1 2 1 4 1 2) (3 1 1 1) (1 0 1 0 0 1 1 0 0 1 1 0) 6 6
(1 1 2 2 1 1 4) (1 3 1 1) (1 0 1 0 1 0 1 0 1 1 0 0) 6 6
(1 1 3 1 1 3 2) (2 1 2 1) (1 0 1 1 0 0 1 1 0 0 1 0) 6 6
(1 1 1 1 1 2 5) (1 1 2 1) (1 0 0 1 0 1 0 1 0 1 0 0) 5 5
(1 1 1 1 1 3 4) (2 1 1 1) (0 1 0 0 1 1 0 0 1 0 1 0) 5 5
(1 1 1 1 1 5 2) (1 1 2 1) (1 0 0 1 0 1 1 0 0 0 1 0) 5 5
(1 1 1 1 3 1 4) (2 2 0 1) (1 0 0 1 1 0 0 0 1 1 0 0) 5 5
(1 1 1 1 4 1 3) (2 1 1 1) (0 0 1 0 1 1 0 0 1 1 0 0) 5 5
(1 1 1 2 1 1 5) (0 2 2 1) (0 0 1 1 0 1 0 1 1 0 0 0) 5 5
(1 1 1 4 1 1 3) (1 2 1 1) (0 0 1 1 0 1 0 0 1 1 0 0) 5 5
(1 1 1 5 1 1 2) (1 1 2 1) (0 0 1 1 0 1 0 0 1 0 1 0) 5 5
(1 1 1 1 1 1 6) (0 1 2 1) (0 0 1 0 0 1 1 0 1 0 0 0) 4 4
```

Fig. 19 Occurrences of tonal triads (4 3), (2 4), (3 2) and (3 3) within 7-tone scales

```
(1 2 1 2 1 2 1 2) (4 4 4 8) (4 0 1 4 0 1 4 0 1 4 0 1) 20 8
(1 1 2 2 1 2 1 2) (4 4 4 5) (4 0 4 1 0 0 4 0 0 4 0 0) 17 5
(1 1 1 1 2 1 3 2) (3 4 3 5) (4 0 1 4 0 0 4 0 1 1 0 0) 15 6
(1 1 1 1 3 2 1 2) (3 3 4 5) (4 0 1 4 0 1 1 0 0 4 0 0) 15 6
(1 1 1 2 1 2 1 3) (4 3 3 5) (0 1 4 0 0 4 0 0 4 0 1 1) 15 6
(1 1 1 1 2 1 2 3) (3 3 3 5) (1 0 4 1 0 1 1 0 1 4 0 1) 14 8
(1 1 1 1 2 3 1 2) (3 3 3 5) (1 0 4 1 0 1 4 0 1 1 0 1) 14 8
(1 1 1 3 1 2 1 2) (3 3 3 5) (1 0 4 4 0 1 1 0 1 1 0 1) 14 8
(1 1 2 1 2 1 2 2) (3 3 3 5) (4 1 0 1 1 0 1 1 0 1 4 0) 14 8
(1 1 2 1 2 2 1 2) (4 2 4 4) (4 0 1 1 0 1 1 1 0 4 1 0) 14 8
(1 1 2 2 1 1 2 2) (2 6 2 4) (4 0 1 1 1 0 4 0 1 1 1 0) 14 8
(1 1 1 1 1 2 3 2) (3 2 4 4) (4 0 0 4 0 1 1 1 0 1 1 0) 13 7
(1 1 1 1 2 2 1 3) (3 3 3 4) (0 0 4 0 1 1 1 0 4 1 0 1) 13 7
(1 1 1 1 2 2 2 2) (2 6 3 2) (1 0 1 1 1 0 4 0 4 0 1 0) 13 7
(1 1 1 1 3 1 2 2) (3 4 2 4) (4 0 0 4 1 0 1 0 1 1 1 0) 13 7
(1 1 1 2 1 1 2 3) (2 4 3 4) (0 0 4 1 0 4 0 1 1 1 0 1) 13 7
(1 1 1 2 1 2 2 2) (4 3 4 2) (0 1 1 1 0 1 1 0 4 0 4 0) 13 7
(1 1 1 2 1 3 1 2) (4 2 3 4) (0 0 4 1 0 4 1 0 1 0 1 1) 13 7
(1 1 1 2 2 1 2 2) (4 2 5 2) (1 1 0 4 0 1 0 1 1 0 4 0) 13 7
(1 1 1 2 2 2 1 2) (3 3 5 2) (1 0 1 4 0 4 0 1 0 1 1 0) 13 7
(1 1 1 3 2 1 1 2) (3 3 3 4) (0 0 4 1 0 1 1 0 4 0 1 1) 13 7
(1 1 2 1 1 2 2 2) (3 6 2 2) (1 1 1 0 1 0 4 0 1 0 4 0) 13 7
(1 1 1 1 1 1 3 3) (3 2 3 4) (0 0 4 0 0 4 1 0 1 1 0 1) 12 6
(1 1 1 1 1 2 2 3) (3 3 4 2) (1 0 1 1 0 4 0 1 0 4 0 0) 12 6
(1 1 1 2 3 1 1 2) (4 2 4 2) (0 1 1 1 0 4 0 0 1 0 4 0) 12 6
(1 1 1 3 1 1 1 3) (2 4 2 4) (0 0 4 1 0 1 0 0 4 1 0 1) 12 6
(1 1 1 3 1 1 2 2) (3 4 3 2) (1 0 1 4 0 0 1 0 4 0 1 0) 12 6
(1 1 2 1 1 3 1 2) (5 3 2 2) (1 0 4 0 0 1 4 0 0 1 1 0) 12 6
(1 1 1 1 1 3 2 2) (3 3 3 2) (1 1 0 1 1 1 1 0 1 0 4 0) 11 8
(1 1 1 2 1 1 3 2) (3 2 4 2) (1 0 1 4 0 1 1 1 1 0 1 0) 11 8
(1 1 1 2 2 1 1 3) (3 3 3 2) (0 1 1 1 0 4 0 1 1 1 1 0) 11 8
(1 1 2 1 1 2 1 3) (4 3 2 2) (0 1 4 0 1 1 1 0 1 1 1 0) 11 8
(1 1 2 1 2 1 1 3) (4 2 3 2) (1 1 1 0 1 1 0 1 0 4 1 0) 11 8
(1 1 1 1 1 1 2 4) (2 3 3 2) (0 1 1 0 1 1 1 0 4 0 1 0) 10 7
(1 1 1 1 1 1 4 2) (2 3 3 2) (1 0 1 1 0 1 4 0 1 0 1 0) 10 7
(1 1 1 1 1 3 1 3) (4 2 2 2) (0 1 1 0 1 4 0 0 1 1 1 0) 10 7
(1 1 1 1 1 4 1 2) (3 2 3 2) (1 0 1 1 0 4 1 0 0 1 1 0) 10 7
(1 1 1 1 2 1 1 4) (2 4 2 2) (1 0 1 1 1 0 1 0 4 1 0 0) 10 7
(1 1 1 1 3 1 1 3) (3 3 2 2) (1 0 1 1 1 1 0 0 1 4 0 0) 10 7
(1 1 1 2 1 1 1 4) (2 3 3 2) (0 1 1 1 0 1 0 1 4 0 1 0) 10 7
(1 1 1 1 1 2 1 4) (3 2 2 2) (1 1 0 1 1 1 0 1 1 1 1 0) 9 9
(1 1 1 1 4 1 1 2) (2 3 2 2) (1 0 1 1 1 1 1 0 1 1 1 0) 9 9
(1 1 1 1 1 1 1 5) (1 2 3 2) (1 0 1 1 0 1 1 1 1 1 0 0) 8 8
```

Fig. 20 Occurrences of tonal triads (4 3), (2 4), (3 2) and (3 3) within 8-tone scales

```
(1 1 1 1 2 1 2 1 2) (5 5 5 8) (4 0 4 4 0 1 4 0 1 4 0 1) 23 8
(1 1 1 1 1 2 2 1 2) (5 4 6 5) (4 0 1 4 0 4 1 1 0 4 1 0) 20 8
(1 1 1 1 2 1 1 2 2) (4 7 4 5) (4 0 1 4 1 0 4 0 4 1 1 0) 20 8
(1 1 1 2 1 2 1 1 2) (6 4 5 5) (0 1 4 1 0 4 1 0 4 0 4 1) 20 8
(1 1 1 1 1 2 1 2 2) (5 4 5 5) (4 1 0 4 1 1 1 1 1 1 4 0) 19 10
(1 1 1 1 2 2 1 1 2) (4 6 4 5) (1 0 4 1 1 1 4 0 4 1 1 1) 19 10
(1 1 1 2 1 1 2 1 2) (5 4 5 5) (1 0 4 4 0 4 1 1 1 1 1 1) 19 10
(1 1 1 1 1 1 1 2 3) (4 4 5 5) (1 0 4 1 0 4 1 1 1 4 0 1) 18 9
(1 1 1 1 1 1 1 3 2) (4 4 5 5) (4 0 1 4 0 1 4 1 1 1 1 0) 18 9
(1 1 1 1 1 1 2 1 3) (5 4 4 5) (0 1 4 0 1 4 1 0 4 1 1 1) 18 9
(1 1 1 1 1 1 2 2 2) (4 6 5 3) (1 1 1 1 1 1 4 0 4 0 4 0) 18 9
(1 1 1 1 1 1 3 1 2) (5 4 4 5) (1 0 4 1 0 4 4 0 1 1 1 1) 18 9
(1 1 1 1 2 1 1 1 3) (4 5 4 5) (1 0 4 1 1 1 1 0 4 4 0 1) 18 9
(1 1 1 1 3 1 1 1 2) (4 5 4 5) (4 0 1 4 1 1 1 0 1 4 1 0) 18 9
(1 1 1 2 1 1 1 2 2) (5 4 6 3) (1 1 1 4 0 1 1 1 4 0 4 0) 18 9
(1 1 2 1 1 2 1 1 2) (6 6 3 3) (1 1 4 0 1 1 4 0 1 1 4 0) 18 9
(1 1 1 1 1 2 1 1 3) (5 4 4 3) (1 1 1 1 1 4 0 1 1 4 1 0) 16 10
(1 1 1 1 1 3 1 1 2) (5 4 4 3) (1 1 1 1 1 4 1 0 1 1 4 0) 16 10
(1 1 1 1 1 1 1 1 4) (3 4 4 3) (1 1 1 1 1 1 1 1 4 1 1 0) 14 11
```

Fig. 21 Occurrences of tonal triads (4 3), (2 4), (3 2) and (3 3) within 9-tone scales

Figures 19, 20 and 21 show the results for all scales with 7, 8 and 9 tones. This is a complete enumeration, basically the task is equivalent to finding all ordered integer partitions of the number 12 and identifying them when they are related by rotation. E.g. $12 = 2+2+1+2+2+2+1$, but also $12 = 1+2+2+1+2+2+2$, major is represented as locrian (1 2 2 1 2 2 2), being its lexically minimal rotation. In each row the first vector represents the scales's sequence of semitone steps. The following quadruple lists the number of occurrences of tonal triads (4 3), (2 4), (3 2) and (3 3) within the scale. The next array of 12 numbers indicates the counted occurring of pitch classes as fundamentals of triads in the scale, transposed to the base of the represented scale. Note that those fundamentals don't have to be contained in the scale as well (the fundamental of a diminished is not member of the triad). Only numbers 4 and 1 are indicators in this pitch class array, this comes from the fact that every two overlapping of the regarded triads sum up to a complete dominant seventh chord, containing all four. The following number equals the sum of triad occurrences, for each size 7, 8 or 9 scales are ordered according to that measure. Scales with equal sums of triad occurrences are ordered lexically. The last number in each row shows the number of different fundamental tones associated with the counted triads.

E.g. for locrian (1 2 2 1 2 2 2), vector (3 1 4 1) indicates its number of occurrences of major triad, seventh chord without fifth, seventh chord without third and diminished triad. Vector (0 1 0 1 0 1 1 0 4 0 1 0) gives positions of fundamentals in relation to the base of the scale in its specific rotation, e.g. the number 4 at position 9 indicates the only dominant seventh chord in the locrian scale, the fundamental is located four semitone steps below its base. Containing 9 tonal triads on 6 different fundamentals it's not one of the top-listed, a scale rotating lydian flat 7 flat 9 ((1 3 2 1 2 1 2) or lexically ordered: (1 2 1 2 1 3 2)) is outstanding in this regard, having 14 tonal triads on 5 fundamentals. This scale can also be seen as an octatonic or diminished scale with one pitch omitted. Following a suggestion of Alexander Stankovski it might be called a *reduced octatonic scale*. Note that in contrast to extending the octatonic scale, where, neglecting rotation, only one possibility exists, there are two possibilities for reducing: either (2 1) or (1 2) might be embraced to a minor third, the second version is (1 2 1 2 1 2 3). The latter is also remarkable: although it contains only 11 tonal triads they are based on 8 different fundamentals—by far more than with all other scales of 7 tones. It might be seen as lydian flat 7 sharp 9 (or a minor-plus-major variant of lydian flat 7). Among 8-tone scales (Fig. 20) the octatonic scale contains most tonal triads, the high number of diminished chords is one reason for that. Among 9-tone-scales (Fig. 21), finally, most tonal triads are contained in Zivkovic's derivative of the octatonic scale, which he calls *scale antique*. It is also the one which contains most dominant seventh chords (5).

As an experiment we generated random sequences of pitches and chords out of these scales. Zivkovic was asked to judge these sequences in regard to their potential for tonal associations. It confirmed our assumptions about the evidence of the chosen measures, that Zivkovic's rankings mostly mirrored those generated by the computer (Figs. 19, 20 and 21).

Project Review by Djuro Zivkovic

As a composer, I work with many different tasks, including harmonic organisation, but it is only a small part of the work. Nevertheless the exploration of this aspect in the course of the project was extremely impressive to me, and I will definitely try to work with the various results and approaches in my work to come. For me any system is like a living being, it should always be alive, able to develop, to continue in its "crystallisation process". Personally I would like if this research could reach other composers, who could develop the system in their own ways or to include it in their existing workflow. The analytic approach of the team has also structured my own view on the system. The team has forced me to think about the system, and now I can give more clear and complete answers when, for instance, giving a lecture on my system for the wider theoretical audience.

Projects that involve many participants across borders and diverse research levels that span almost three years are difficult to accomplish without compromises. However, it is not true about this project. I haven't experienced any lack of time, stress moments or absence of care from the all people involved in my topic. The project team has had always enough time to realise my questions, showing a great interest in the subject I work with. The team was attracted to my system, in trying to develop it on their own way with curiosity, dedication and enthusiasm. With some team members I shared other thoughts outside the research in our free time.

References

1. Cariani PA, Delgutte B (1996) Neural correlates of the pitch of complex tones. I. Pitch and pitch salience. J Neurophysiol 76(3):1698–1716
2. Kemp DT (1978) Stimulated acoustic emissions from within the human auditory system. J Acoust Soc Am 64(5):1386–1391
3. Oster G (1973) Auditory beats in the brain. Sci Am 229(4):94–102
4. Palmer G, Sherrard P, Ware K (1995) The Philokalia, vol 4. Faber and Faber, London
5. Plato (1997) Philebus. In: Cooper JM, Hutchinson DS et al (eds) Plato: complete works (trans: Frede D). Hackett Publishing, Indianapolis, pp 398–456
6. Tartini G (1754) Trattato di musica secondo la vera scienza dell'armonia. G. Manfré, Padua

Bart Vanhecke/Straightening the Tower of Pisa

Bart Vanhecke, Daniel Mayer and Gerhard Nierhaus

Bart Vanhecke has always been a composer. His first infantile compositions date from the time when he first learned to read and write music, at the age of eight. Although, as a child he was not aware of it, it was clear that music was not a mere hobby for him, but a way of living, a way of being in the world. However, it was only around the age of 16, when he took his first harmony lessons, that he started to compose more intensively.

In his country of birth, Belgium, music and the arts in general is not culturally accepted as a proper job by a substantial amount of the population. Therefore he kept his compositions to himself, and after his secondary education he studied civil engineering for a couple of years in order to train for a "real job" before he realised that being an engineer was not who he really was. He had discovered Alban Berg's Violin concerto by that time, which opened the door for him to twentieth century music and to dodecaphony in particular. Schoenberg soon became his musical grandfather. Schoenberg's music led him to the music of his successors: Anton Webern, Pierre Boulez and Luigi Nono. As soon as he came into contact with the work of Luigi Nono, he felt a strong connection with his aesthetic universe and with the way he managed to balance strictly technical aspects of his music with aesthetic and extra-musical expression. In spite of this strong feeling of connectedness, Vanhecke's music sounds quite different from Nono's. It is the aesthetic idea behind the pieces, not necessarily the extra-musical—in Nono's case this is often a political idea—that

B. Vanhecke
LUCA (Leuven University College of Arts), Leuven University, Leuven, Belgium
e-mail: bart_vanhecke@msn.com

D. Mayer · G. Nierhaus (✉)
Institute of Electronic Music and Acoustics, University of Music and Performing Arts Graz, Graz, Austria
e-mail: nierhaus@iem.at

D. Mayer
e-mail: mayer@iem.at

G. Nierhaus (ed.), *Patterns of Intuition*,
DOI 10.1007/978-94-017-9561-6_12

he perceives as being very similar. If Arnold Schoenberg is his musical grandfather, he considers Luigi Nono to be his musical father.

Since 2009, Vanhecke has been doing a doctoral research in the arts at the Orpheus Institute in Ghent and University of Leuven on "the systematisation of atonality and dissonance in amotivic serial composition", a further development of the personal composition technique he has been using for all his compositions since 1997. In 2010 he received a doctoral research grant from the Leuven University, which enables him to do full time research for four years. Vanhecke is also a doctoral researcher at the Orpheus Research Centre in Music (ORCiM) in Ghent.

Artistic Approach

Statement

Composers are children of their culture and time; even when they work in relative isolation, their work is the result of all the influences of the culture/s and era they live in which shapes their knowledge and ideas. The combination of aesthetic cultural knowledge and personal ideas forms the composers' aesthetic universe, which is expressed through the act of composition.

Musical expression requires a musical code, a musical technique that enables the composers to transform the ideas of their aesthetic universe into structured sound and silence. When the composers' ideas deviate to a considerable extent from culturally accepted ideas, existing compositional techniques may be inadequate, and artistic expression may require the development of new techniques. There lies the origin of my personal compositional technique: chromatic interval serialism. This technique, which I developed in 1997, is a further step in the evolution of serial techniques, influenced by the line of developments that precede it from Arnold Schoenberg's Dodecaphony over the serial techniques of Luigi Nono or Pierre Boulez. It is an amotivic technique that shifts the focus from pitch class to interval class and aims at the systematisation of atonality and dissonance. It is my firm conviction that serialism still has the potential to evolve further in order to serve as a technique for the expression of novel aesthetic ideas. In that sense, Schoenberg is not dead.

Although my serial technique of composition is strictly formalised on a substructural level, I want the music that results from it to sound as if it were completely freely and intuitively composed. The series of my pieces are like the DNA of my music; the pieces themselves are the "organisms" that result from the compositional procedures. I want the sounding phenotype of my music to be "more" than its serial underlying genotype, just like living organisms represent more than their genetic material. I am convinced that the strict structure behind my music doesn't necessarily have to restrain its expressive power, but that, on the contrary, it can add to it.

Personal Aesthetics

The following are three aspects that form the cornerstones of my aesthetic practice as a composer. The first is the search for technical, idiomatic and stylistic purity. The second is the search for organic unity, for music that forms an organic sounding whole. Striving for organic unity in serial composition assumes that the resulting musical works transcend their strictly serial substructure. In this respect, I like to compare my approach, once again, with the principles of genetics: just as living organisms are highly, but not solely, determined by their genetic material, my compositions are highly, but not solely, determined by their series. The series not only provides for the pitch material but also directs the course of the entire structuring and transforming process leading to a piece of music, comparable to the biochemical processes that transform genetic material—the organism's genotype—into the ultimate living organism—its phenotype. Just as living beings transcend their genetic material, my compositions—the musical phenotype—are more than the series—the structural genotype—on which they are based. In this creative process, which to a large extent is based on intuitive aesthetic sensitivity or taste, the serial technique is not more determining than the way it is implemented. Technique and artistic taste cannot be considered separately from each other but should complement each other in a constant interaction. Strictly adhering to rules does not guarantee artistically valuable results; an aesthetic transcendence is indispensable. In this respect again, serial techniques are in no way different from the techniques used in tonal composition.

My serial technique may be cerebral, but that does not mean the music that results from it is not more than a product of the brain, lacking all expressive power and emotion. Serial music is not necessarily less expressive, or no less "coming from the heart" than tonal music. Each composition is a product of cerebral effort. The thought processes of composing, irrespective of the style or idiom or the technique used, are partly conscious but also escape to some extent conscious control, as was mentioned before. It is these uncontrolled processes that are said to come "from the heart". Both conscious and unconscious cerebral processes provide organic structure, coherence, and consistency of a composition.

Structure is a third indispensable cornerstone of composing. Without structure, there can be no question of a composition. Igor Stravinsky noted in this context: "Music's exclusive function is to structure the flow of time and keep order in it."[1] Strictly designed structure is no impediment to expressive power however. Musical expression is a controversial concept. Stravinsky claims that music is not able to express anything at all.

"I consider that music is, by its very nature, essentially powerless to *express* anything at all, whether a feeling, an attitude of mind, a psychological mood, a phenomenon of nature, etc. [...] *Expression* has never been an inherent property of music. That is by no means the purpose of its existence. If, as is nearly always the

[1] Quoted in [5, p. 232], this puts Stravinsky in line with Eduard Hanslick who wrote: "The content of music is tonally moving forms." [5, p. 29].

case, music appears to express something, this is only an illusion and not a reality. It is simply an additional attribute which, by tacit and inveterate agreement, we have lent it, thrust upon it, as a label, a convention—in short, an aspect unconsciously or by force of habit, we have come to confuse with its essential being." [4, pp. 53–54].

On the other hand, if the term "expression" is restricted to "emotional expression",[2] one could argue that there is no music that is *not* expressive; all music has the potential to be expressive. Just like every object of communication, whether it is a poem, a statement or a facial expression, music is a potential source of expression of emotional ideas and the arousal of emotions. This process of expression and arousal is subjective, relative and culture-bound. Musical expression is subjective because each listener reacts in a different way to the musical stimuli, and the response to these stimuli depends on the context in which the music is heard. The expression is relative and cultural-specific because it depends on the relationship between the listener and the music. This relationship is personal and is partly due to the familiarity of the listener with the culture the music belongs to.

Composition has always been searching and researching for me. My task as a composer is to discover and develop the ideas of what I call my personal aesthetic universe, the structured world of knowledge and thought related to aesthetics, to beauty and to art in general. Artistic creation is the expression of the complete meaning of the ideas belonging to my aesthetic universe. Artistic research is the exploration of that aesthetic universe. Only the artists can explore their own aesthetic universe directly, since they are the only ones who have immediate access to, or knowledge of their own aesthetic universe. Everyone else can only get glimpses from the artists' aesthetic universe through the artworks they create. In this sense, an aesthetic universe is like the far away regions of the physical universe we live in; regions of which we can only get an idea through pictures captured by telescopes or space probes. My compositions are like the "space probe pictures" of my personal aesthetic universe.

Every artist's aesthetic universe consists of a cultural and an idiosyncratic part. The cultural part of an aesthetic universe consists of all our knowledge and ideas the artist shares with other people. The idiosyncratic part contains all the artist's aesthetic ideas that deviate from the culturally accepted.

Very soon after I started composing music, I became aware of the fact that many of my aesthetic ideas were rather idiosyncratic, that what I think and what I want to express is at times very different from the ideas of other musicians around me. This awareness resulted in the development of a personal technique necessary to express my idiosyncratic aesthetic ideas. The urge to explore the idiosyncratic part of my aesthetic universe was the incentive for my doctoral research at the University of Leuven and led me to the ORCiM in Ghent in 2010, where my ideas are confronted with the ideas of other artist-researchers. This confrontation not only enhances the development of my own aesthetic universe, it also enables me to situate my aesthetic ideas within a broader culture.

---

[2] More precisely: "expression of emotional ideas or concepts". Strictly speaking, emotions cannot be expressed but only responded to. This may be what Stravinsky refers to.

Formalisation and Intuition

Formalisation is only one possible way of composing, of creating musical structures and musical forms. The strict formalisation of my work through the use of highly atonal CIG-series (see below) and the subsequent formal construction of rhythmic-harmonic substructures is my way of justifying my choice of pitch classes, harmony and rhythm *for myself*. I don't think there is any need to justify my choices for other people and I don't think other composers should do the same. I have noticed, however, that I generally consider my strictly structured pieces my best, years after their completion.

Strict formalisation is for me a way of limiting my choices. In my opinion, limitation is a central element within the concept of composition. Only by making a choice to use certain elements, and by doing so exclude all others, the composer creates a musical structure called a composition. Without this choice, without the determination of limitations, there is no composition, regardless of the strictness of these limitations.

As a composer I grant myself the freedom to diverge from my own rules of formalisation at all times, and even if I respect the strict structure of a piece's rhythmic-harmonic substructure, there is always a gap between the substructure and the resulting score, a gap that leaves space for intuitive artistic choices.

I define intuition as the ability or the capacity to skip or by-pass referential or connective steps in the acquisition of conceptual or procedural knowledge. It results in knowledge or ideas that are not logically inferred from other knowledge or ideas. The thought-processes of intuition play an important role in artistic expression. Intuition, free imagination, and inspiration turn the meaning of an artwork into a continuously changing rhizomatic structure, in a constantly evolving web of free connections between concepts, wherein "any point [...] can be connected to anything other" [1, p. 7].

In my creative practice, intuition is crucial in the transformation of the rhythmic harmonic substructure of the pieces into the final score. In this process of transformation there is much I cannot explain. There is a gap between the strict substructure and the surface structure that is bridge mainly by artistic intuition. The strictly formal substructure of my music too results from intuitive processes to a certain extent. A lot of the choices made in structuring my work can only be explained this way.

Evaluation and Self-reflection

Artists are individuals who possess an outspoken, highly developed and structured world of thought and ideas about art and aesthetics in general, a world I call the artists' "aesthetic universe". I define an artistic practice as the expression of ideas belonging to the artists' personal aesthetic universe. Artists can also do more than express their aesthetic universe; they can explore it in order to better understand and further develop their ideas. This is how I define artistic research. For me personally, artistic practice and research have always gone hand in hand. Every new piece is the result as well as the reflection of the exploration of my aesthetic universe.

Project Expectations

My artistic practice contains many aspects that are intuitive, aspects that I cannot explain, things I do without being aware of the underlying procedures, maybe even things I am not aware of doing at all. I expect that the confrontation of my way of composing with the more digitalised approach of the project might shed some light on these aspects. It might help me understand my own aesthetic universe better. It is, in other words, artistic research for me that might give me some insight into who I am as a composer, in my personal aesthetic universe, even if this insight would itself be intuitive.

Exploring a Compositional Process

Description of CIG-Serialism

My compositions are based on the serial technique I call chromatic interval group serialism, or CIG-serialism. It is a technique I developed in 1997. The series of CIG-serialism are constructed exclusively with the pitch classes of what I call "chromatic interval groups" of order 3 (CIG-3s). A CIG-3 is an ordered pc set containing 3 pitch classes. When such sets are written in prime form[3] in ascending order, at least two of the consecutive pitch classes are a semi-tone (ic1) apart. There are only 9 of these chromatic pc sets in prime form, as shown in Fig. 1.

Each of the nine chromatic pc-sets can be turned into an ordered set in six ways (six permutations), to form six CIG-3s. Figure 3 shows the 6 CIG-3s that can be

Fig. 1 Prime form of all chromatic pc sets

Fig. 2 Chromatic pc-set 3-1

[3] The prime form of a pc-set is its standard representation. It is the most compact form of the set with pc 0 (the pitch C) as its base. For more details: see [2, pp. 3–5] and [3, pp. 57–59].

Fig. 3 The 6 CIG-3s related to pc-set 3-1

Fig. 4 54 CIG-3s

constructed with the pitch classes of the chromatic pc-set with Forte-number 3-1 (Fig. 2). A complete list of all 54 possible CIG-3s is shown in Fig. 4.[4]

[4] In Fig. 4, the CIG-3s are represented in their prime form (with pc C as their base). Of course as intervals are what count in CIGs and not pitch classes, all CIGs may appear in any transposition.

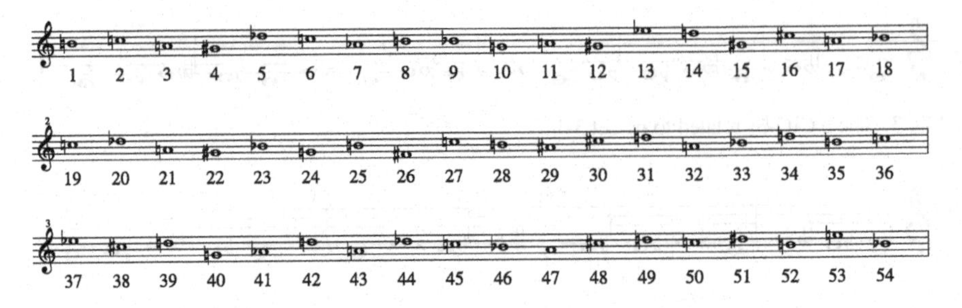

Fig. 5 Series of *A l'image du monde ... originel*

The CIG-series consists exclusively of CIG-3s. This means that every three consecutive notes in a CIG-series form a CIG-3. In addition, every CIG-3 (regardless of transposition) appears exactly once in the series. This is done in order to make the series amotivic. No CIG should be used more than once to avoid CIGs becoming a motive within the series (a structural motive, as opposed to a motive appearing in the score). Notes 53-54-1 and 54-1-2 are also a unique CIG-3. This way, the series has a ring structure, it bites its own tail; it is a "closed series".

An example of a 54-CIG series is shown in Fig. 5. It is the series of the piece *A l'image du monde ... originel* for piano solo (2012).

Notes 1–3 are a CIG-3. Notes 2–4 are different CIG-3. Notes 3–5 another one that did not appear before, and so on until the end of the series (notes 52–54). Notes 53, 54, 1 and 54, 1, 2 are the two remaining CIG-3s that had not been used before.

A l'image du monde ... originel is the first piece I composed with a further restricted CIG-series. Not only do every three consecutive notes in the series form CIG-3s, but every four consecutive notes form a CIG-4. A CIG-4 is an ordered pc set containing 4 pitch classes. When such sets are written in prime form in ascending order, at most one of the interval classes between consecutive pitch classes is not a semi-tone (ic1). In other words, the prime form of a CIG-4 is a cluster with not more than one "gap". The first 4 notes of the series of *A l'image du monde ... originel* (B, C, A, G#) put in ascending order, for instance, form a CIG-4 with a gap (ic2) between A and B only, as Fig. 6 shows. The other pitch classes in the CIG-4 are ic1 apart. I called series consisting entirely of CIG-3s between every three consecutive notes as well as CIG-4s between every four consecutive notes CIG-3/4 series. This adaptation to my compositional technique resulted from the outcome of my doctoral research and yields music with a higher degree of atonality and dissonance in a systematic, structural way. *A l'image du monde ... originel* is the first piece I wrote using a CIG-3/4 series.

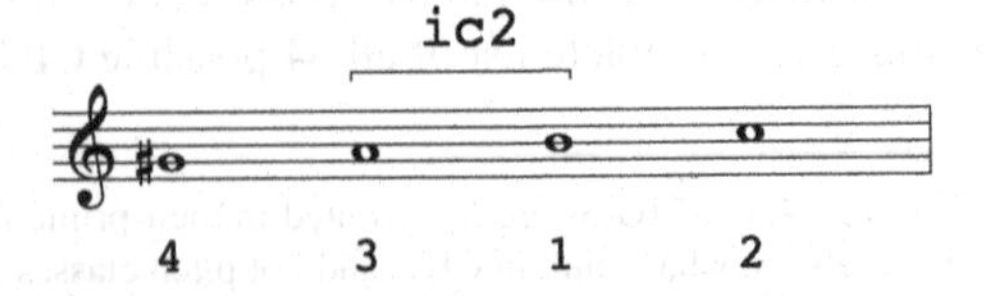

Fig. 6 CIG-4 containing series notes 1–4 of *A l'image du monde ... originel* in ascending order

The Rhythmic Harmonic Substructure

To compose a piece of music from the CIG-series, I first construct a rhythmic-harmonic substructure (RHS), as an intermediate step. To do this I attach a rhythmic cell to every note of the series. The rhythmic cells are then put together in a metric frame. The formulas used for this procedure of transformation from series to RHS are specific for every piece, but always based on the interval content of the series. In this way, the series functions as the genetic material for my pieces. As in biological genetics, the series, or their transformations in the process of composition, may contain voluntary or involuntary mistakes. I call these mistakes "mutations". Note, for instance that note 18 in the form of the series, which is used in the piece, is a B, whereas it had to be a Bb. This is an example of a mutation.

The starting information for the construction of the RHS for *A l'image du monde ... originel*, hereafter named *AIMO* is based on the interval classes "surrounding" each series note, i.e. the interval class of the series note with the preceding note ($-i$) and with the following note ($+i$). For series note 1 (B), the value for $-i$ and $+i$ is $+1$, because the intervals between note 54 and 1 and between notes 1 and 2 are both ascending minor seconds ($ic + 1$) (see Fig. 7).

In the RHS of *AIMO*, the number of note lengths per rhythmic cell (N) for series note x is determined as:

$$N = \text{integer}[S/4 + W/2]$$

where

$$S = |-i| + |+i|$$

and

$$W = |+i + -i| + \text{number of times the pitch class x appears in the series}$$

(e.g. how many times Bb appears in the whole series).

N varies between 3 and 8.

For the first note in the series (B, appearing 6 times in the whole series), for instance, $S = 2$, $W = 2 + 6 = 8$, and therefore $N = \text{integer}\ (2/4 + 8/2) = 4$.

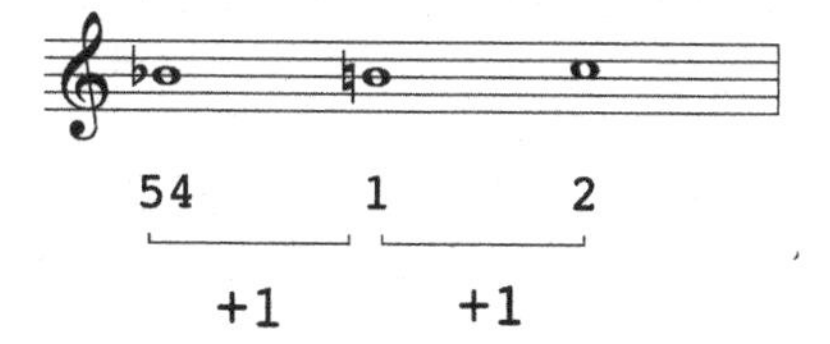

Fig. 7 Interval classes surrounding series note 1 in *AIMO*

With these formulas, the number of notes of the rhythmic cell for every note in the series can be determined.

The next step in the construction of a RHS is the determination of the note lengths of every note within a rhythmic cell. In the case of *AIMO* the series of absolute values of interval classes between series notes (the ic-series) is run through forward for the prime form (P) and inversion (I) of the series, and backward for the retrograde (R) and retrograde inversion (RI) of the series. Since the second half of the series of *AIMO* is the inversion of the first half, the ic-series contains 27 values before it repeats itself. The values are:

1 3 1 5 1 4 3 1 3 2 1 5 1 6 5 4 1 2 1 4 1 2 3 4 5 6 1

The first 4 values of the ic-series are attached to the rhythmic cell of the first series note (since for that note $N = 4$), the rhythmic cell of the next note starts with the fifth value in the ic-series (5), and so on. When the end of the ic-series is reached, it starts over from the beginning:

Note 1 ($N = 4$): 1 3 1 5
Note 2 ($N = 5$): 1 4 3 1 3

A complete list of rhythmic cells is given in Tables 1 and 2. To attach actual note lengths to the values thus obtained, first a "length unit" or "augmentation" is determined for every series note. This augmentation is the note length corresponding to

Table 1 Rhythmic cells Prime and Inversion

Note	N	Augmentation	P	I
1	4	1 3 1 5	3	4
2	5	1 4 3 1 3	1	6
3	6	2 1 5 1 6 5	5	2
4	6	4 1 2 1 4 1	6	1
5	7	2 3 4 5 6 1 1	3	4
6	7	3 1 5 1 4 3 1	1	6
7	5	3 2 1 5 1	1	6
8	5	6 5 4 1 2	1	6
9	6	1 4 1 2 3 4	4	3
10	3	5 6 1	1	6
11	4	1 3 1 5	3	4
12	7	1 4 3 1 3 2 1	4	3
13	6	5 1 6 5 4 1	3	4
14	8	2 1 4 1 2 3 4 5	3	4
15	6	6 1 1 3 1 5	2	5

(continued)

Table 1 (continued)

Note	*N*	Augmentation	P	I
16	6	1 4 3 1 3 2	5	4
17	6	1 5 1 6 5 4	3	4
18	5	1 2 1 4 1	4	3
19	5	2 3 4 5 6	5	2
20	6	1 1 3 1 5 1	1	6
21	7	4 3 1 3 2 1 5	6	1
22	4	1 6 5 4	3	4
23	5	1 2 1 4 1	4	3
24	3	2 3 4	4	3
25	5	5 6 1 1 3	3	4
26	3	1 5 1	6	1
27	7	4 3 1 3 2 1 5	6	1
28	4	1 6 5 4	3	4
29	5	1 2 1 4 1	4	3
30	6	2 3 4 5 6 1	4	3
31	6	1 3 1 5 1 4	4	3
32	7	3 1 3 2 1 5 1	3	4
33	7	6 5 4 1 2 1 4	2	5
34	5	1 2 3 4 5	4	3
35	5	6 1 1 3 1	1	6
36	6	5 1 4 3 1 3	2	5
37	3	2 1 5	5	2
38	4	1 6 5 4	3	4
39	7	1 2 1 4 1 2 3	5	2
40	6	4 5 6 1 1 3	5	2
41	8	1 5 4 3 1 3 2 1	5	2
42	6	5 1 6 5 4 1	3	4
43	6	2 1 4 1 2 3	6	1
44	6	4 5 6 1 1 3	5	2
45	5	1 5 1 4 3	5	2
46	5	1 3 2 1 5	1	6
47	6	1 6 5 4 1 2	6	1
48	7	1 4 1 2 3 4 5	5	2
49	4	6 1 1 3	2	5
50	5	1 5 1 4 3	5	2
51	3	1 3 2	1	6
52	5	1 5 1 6 5	1	6
53	3	4 1 2	6	1
54	7	1 4 1 2 3 4 5	5	2

Table 2 Rhythmic cells Retrograde and Retr. Inv.

Note	*N*	Augmentation	R	RI
1	4	1 6 5 4	4	3
2	5	3 2 1 4 1	5	2
3	6	2 1 4 5 6 1	1	6
4	6	5 1 2 3 1 3	3	4
5	7	4 1 5 1 3 1 1	4	3
6	7	6 5 4 3 2 1 4	1	6
7	5	1 2 1 4 5	1	6
8	5	6 1 5 1 2	3	4
9	6	3 1 3 4 1 5	5	2
10	3	1 3 1	5	2
11	4	1 6 5 4	4	3
12	7	3 2 1 4 1 2 1	2	5
13	6	4 5 6 1 5 1	4	3
14	8	2 3 1 3 4 1 5 1	2	5
15	6	3 1 1 6 5 4	2	5
16	6	3 2 1 4 1 2	1	6
17	6	1 4 5 6 1 5	4	3
18	5	1 2 3 1 3	4	3
19	5	4 1 5 1 3	2	5
20	6	1 1 6 5 4 3	2	5
21	7	2 1 4 1 2 1 4	3	4
22	4	5 6 1 5	5	2
23	5	1 2 3 1 3	4	3
24	3	4 1 5	4	3
25	5	1 3 1 1 6	6	1
26	3	5 4 3	6	1
27	7	2 1 4 1 2 1 4	3	4
28	4	5 6 1 5	5	2
29	5	1 2 3 1 3	4	3
30	6	4 1 5 1 3 1	3	4
31	6	1 6 5 4 3 2	3	4
32	7	1 4 1 2 1 4 5	6	1
33	7	6 1 5 1 2 3 1	1	6
34	5	3 4 1 5 1	2	5
35	5	3 1 1 6 5	4	3
36	6	4 3 2 1 4 1	3	4
37	3	2 1 4	1	6
38	4	5 6 1 5	5	2
39	7	1 2 3 1 3 4 1	3	4
40	6	5 1 3 1 1 6	5	2

(continued)

Table 2 (continued)

Note	N	Augmentation	R	RI
41	8	5 4 3 2 1 4 1 2	4	3
42	6	1 4 5 6 1 5	4	3
43	6	1 2 3 1 3 4	2	5
44	6	1 5 1 3 1 1	6	1
45	5	6 5 4 3 2	2	5
46	5	1 4 1 2 1	3	4
47	6	4 5 6 1 5 1	4	3
48	7	2 3 1 3 4 1 5	1	6
49	4	1 3 1 1	6	1
50	5	6 5 4 3 2	1	6
51	3	1 4 1	6	1
52	5	2 1 4 5 6	6	1
53	3	1 5 1	1	6
54	7	2 3 1 3 4 1 5	1	6

rhythm chart

	note length					
augmentation	1	2	3	4	5	6
1						
2						
3						
4						
5						
6						

Fig. 8 Chart for the determination of note lengths in rhythmic cells

value 1 in the rhythmic cells (varying between demi-semiquaver to dotted quaver) as can be seen in the first column of note length in the rhythmic chart in Fig. 8. In *AIMO*, the augmentation (AUG) is determined as follows:

> For R and I: AUG = sum of note lengths of rhythmic cell modulo 6.
> For P and RI: AUG = 7 − (sum of note lengths of rhythmic cell modulo 6).
> Ex: AUG (note 1) in prime form (P) = 7 − mod6 $(1 + 3 + 1 + 5) = 3$.
> (The result is shown in the last two columns in Tables 1 and 2).

Knowing the rhythmic cell of note 1 has four durations (1, 3, 1 and 5) in augmentation 3 (1 = semiquaver, 2 = quaver, etc.) the rhythmic cell for note 1 in the prime form (P) of the series is shown in Fig. 9.

Fig. 9 Rhythmic cell for note 1 in P

Fig. 10 Bar 1–4 of the rhythmic-harmonic substructure of *AIMO*

After determining rhythmic cells for all series notes, the cells are placed in a metric frame, the RHS of the piece, by determining the distance (time delay) between the beginning of the rhythmic cells. In *AIMO*, the distance between beginnings of rhythmic cells of note *x* and next occurring note (DISx) is a number of semiquavers (1/4 beats) equal to *S* for note *x*.

Ex: DIS (note 1) = 2. This means the rhythmic cell of note 2 starts 2 semiquavers later than that of note 1. The first four bars of the RHS resulting from this procedure are shown in Fig. 10.

Turning the RHS into a Score

Next, the RHS is turned into a surface structure, the final score of the piece. In this process, artistic creativity is more important than structural strictness. Still there are rules and constraints in this final step of the composition: Every note in the RHS should appear in the score. The first note of *AIMO*, for instance, should be a B, since it is the first note in the RHS. A "chord rule" is used to construct chords on any series note at any given moment in a piece. It says that series notes can be accompanied by the series note that precedes or follows it (the neighbouring notes). In some pieces this rule is extended: if a neighbouring note appears in the chord, the next or previous note in the series may also be used. This way, chords are build that consist entirely of notes belonging to the CIG's containing the series note.

Figure 11 illustrates this procedure. It shows the opening bars of *AIMO*. The first notes of the score are A#, B and C. This is a chord constructed around the first series

Fig. 11 Opening bars of *AIMO*

note in prime form (B) surrounded by its neighbouring notes A# (note 54) and C (note 2). Apart from the chord rules mentioned above, the construction of chords is completely done in complete artistic freedom.

Calculation of CIG-Series

POINT: In contrast to other participants in the project, Vanhecke was going to write a second version of a piece by using results of our modelling process: *A l'image du monde ... double, AIMD*, following *A l'image du monde ... originel, AIMO*. We decided to concentrate on the task of finding appropriate 54-CIG rows. First attempts showed that a complete enumeration using a backtracking algorithm would lead to an exorbitant number of solutions, powers of 10 above the literal number of twelve-tone rows (which equals $12! = 479001600$, though this can be reduced by classification). So, the main part of the work consisted in finding strategies to limit the problem to a computationally treatable number, perhaps by a useful classification, and at the same time regarding the composer's constraints. It turned out that some strategies, which Vanhecke already used to simplify his search by hand, also greatly simplified computational search and fit the aesthetic preliminaries at the same time.

CIG-4s

Vanhecke also considered pc-sets of four elements and derived CIGs (called CIG-4). Such pc-sets are allowed to have at most one interval not equalling ic1, there are nine of them:

(0 1 2 3) (0 1 2 4) (0 1 3 4) (0 1 2 5) (0 1 2 6) (0 1 2 7) (0 1 4 5) (0 1 5 6) (0 1 6 7)

Each pitch class set can be arranged in 24 permutations, giving all in all 216 CIGs. It would be possible to analogously build 216-CIG rows, but there are some reasons which make this option for Vanhecke less attractive than building 54-CIG series:

1. Larger numbers make the procedure of finding rows unpleasant, especially when working with pencil and paper.
2. CIG-4s may contain groups of three not being a CIG-3, e.g. (5 0 4 6) contains (0 4 6). So locally a higher degree of tonality might occur, which the composer does not desire.
3. Whilst a 54-CIG series is containing all interval tuples exactly once this is not the case for series built from CIG-4s. 54-CIG series are very balanced in the sense that every tuple of interval successions is occurring exactly once (octave shift neglected), hence ensuring avoidance of motifs:

(1 1) (2 −1) (−1 2) (1 −2) (−2 1) (−1 −1)
(1 2) (3 −2) (−1 3) (2 −3) (−3 1) (−2 −1)
(2 1) (3 −1) (−2 3) (1 −3) (−3 2) (−1 −2)
(1 3) (4 −3) (−1 4) (3 −4) (−4 1) (−3 −1)
(3 1) (4 −1) (−3 4) (1 −4) (−4 3) (−1 −3)
(1 4) (5 −4) (−1 5) (4 −5) (−5 1) (−4 −1)
(4 1) (5 −1) (−4 5) (1 −5) (−5 4) (−1 −4)
(1 5) (6 −5) (−1 6) (5 6) (6 1) (−5 −1)
(5 1) (6 −1) (−5 6) (1 6) (6 5) (−1 −5)

This uniqueness is lost with series built from CIG-4s, the exception comes from the transposition-invariant pc-set (0 1 6 7), where all interval sequences of derived CIG-4s occur twice, e.g. in Table 3.

Though CIG-4s may also be useful while trying to find 54-CIG series and Vanhecke applied this restriction in the piece *A l'image du monde . . . originel*. When looking for (parts of) 54-CIG rows in which sequences of 4 elements are CIG-4s, this additionally lowers the average degree of tonality.

Using Second Half Inversion

This is a restriction already used in Vanhecke's earlier pieces. The set of 54 interval vectors (Fig. 5) is symmetrical in the sense that with every tuple (x y) also its inversion (−x −y) is contained. If interval vectors −6 and +6 are identified, so (−1 6) and

Table 3 Transposition-invariant pc-set (0 1 6 7), example of double-occurrence of interval sequences within CIG-4s

CIG-4 (pc sequence)	→	Interval sequence
(7 6 1 0)		(−1 −5 −1)
(1 0 7 6)		(−1 −5 −1)

(6 −1) are the inversions of (1 6) and (6 1). The symmetry of the 54 CIG-3s can be used to build a 54-CIG series with its second half being the first one inverted, then also the second half's CIG-3s must be the inversions of those of the first half (this property is also preserved under a cyclic shift). The procedure to find such a row significantly reduces backtracking. One only has to search candidates within first halves of a 54-CIG series containing no pair of inverted interval vectors. There is also an aesthetic reason, in doing it this way, the inversions of CIG-3s are placed at opposite positions in the row, hence the amotivic character is emphasised.

One might also think of second half retrogrades and retrograde inversions, but the analogy to twelve-tone rows tone doesn't hold completely: for a 54-CIG with second half retrograde the interval vectors would have to be retrograde and inverted (going backwards changes direction), thus the interval vectors hinging the two halves would have to be of the form (x −x) and (−x x). Such CIG-3s don't exist, hence a 54-CIG row with second half retrograde doesn't exist. A similar argument shows that a 54-CIG with second half retrograde inversion would have to have tuples (1 1) and (−1 −1) at hinge positions, we didn't follow the path of searching for those rows any further.

POINT: What are the characteristics of the 54-CIG series you were looking for in *A l'image du monde ... double*?

Vanhecke: My aim was to find out whether it was possible to compose a piece with a series provided by the algorithmic approach of the POINT project that was essentially identical to the original piece, which was completely "mine". Therefore I decided to start from the same criteria in both pieces:

1. The first criterion for the series of *AIMD* was that it had to be a CIG-3/4 series.
2. The second criterion for the construction of the series of *AIMD* was the presence of the cluster A–C around the beginning and end of the series. *AIMO* was an exploration of the simultaneous extreme low and high registers of the piano. This is why the series of *AIMO* was constructed in such a manner that the cluster A–C was present at the "beginning" and "end"[5] of the series (between notes 54 and 3, see Fig. 5).
3. The last initial criterion was that the second half of the series (notes 28–54) had to be the inversion of the first half, because this was also the case in the series of *AIMO* (Fig. 5).

POINT: We did an enumeration of 54-CIG rows with second half inversions and the additional demand that all groups of four tones should be CIG-4s in order to lower to local degree of tonality, in this way we got a collection of a few thousand rows. We were looking forward to find rows for the specific demands of *AIMD* within that corpus. Searching this corpus for rows didn't give us fully satisfying results concerning of the cluster criterion (2).

POINT: Could you use some of our first generated series, are there further or alternative criteria, that could restrict the search space?

---

[5] "Beginning" and "end" are here written between quotation marks, because—as was explained before—a CIG-series has a cycling structure and has no real end or beginning.

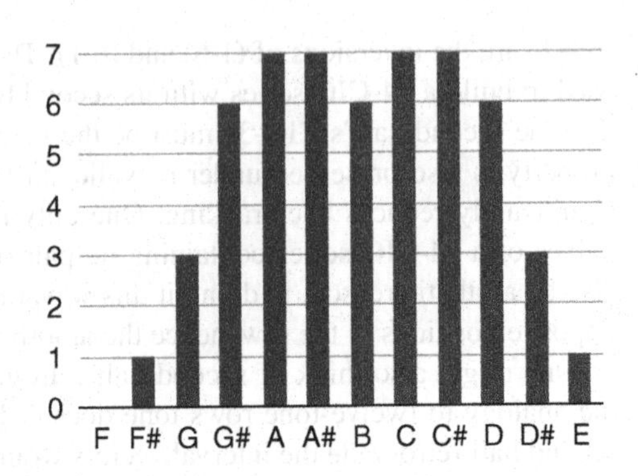

Fig. 12 Pitch class distribution in the series of *AIMO*

Vanhecke: During the process of evaluation of this list, I noticed that the series of *AIMO* had an interesting but unpremeditated feature: it is unbalanced, meaning that the frequency of occurrence of the different pitch classes is not homogeneously distributed. As can be seen in Fig. 12, showing the pitch class distribution of the series of *AIMO*, the pitch classes belonging to cluster A–C occur more frequently (on average) than others. Although this feature was not intended, it serves the aim of exploring the extreme registers of the piano perfectly. A second remarkable feature is the absence of pc F in the series of *AIMO*.

As an additional criterion, I decided that the series of *AIMD* should also be unbalanced, with predominant occurrence of the pitch classes belonging to cluster A–C, just like the series of *AIMO*. The middle of the cluster A–C lies between Bb and B. The pitch class furthest away from this middle would be between E and F, therefore, as a final criterion, I determined that either E or F should be absent from the series of *AIMD*, as is the case in the series of *AIMO*.

POINT: The criterion of an unbalanced, bell-like shape was easy to implement and could restrict the solution space further. We were looking for "real" bell-like shapes, means distributions without dents in the middle as in the row of *AIMO* (Fig. 5), but still the condition of having pitches A–C around beginning and end needed to be more incisive.

POINT: What about the cluster condition, how many of the pitches A–C should appear at beginning and end?

Vanhecke: The series should preferably begin and end with permutations of the complete cluster A–C.

POINT: We then adapted our strategy. As rows with second half inversions allow a cyclic shift, we searched for such rows starting with eight pitch classes being the concatenation of two permutations of A–C in order to shift back by four afterwards, finally having A–C at beginning and end. This little trick dramatically lowered computation time and finally led to a handful of reasonable series, from which Vanhecke chose one (Fig. 13).

POINT: Could you describe the reasons to choose that row from the last series of some dozen rows which were best suited according to your criteria (most distinct bell-shape)?

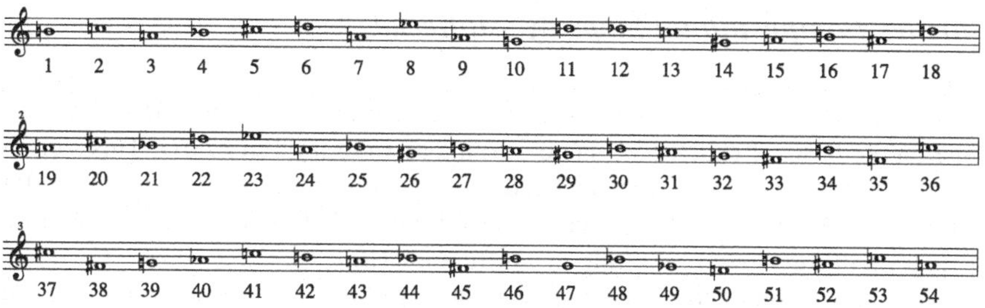

Fig. 13 Series of *AIMD*

Vanhecke: I chose this series for its "perfect" A–C cluster between notes 51 and 4. The cluster is extended over 8 series notes and is distributed in a symmetrical way (for notes at the beginning and four notes at the end). Comparison of the pitch class distributions of both series show that the series of *AIMD* (Fig. 14) is unbalanced in a much more outspoken way than that of *AIMO* (Fig. 12). Again, the former series can be called more "perfect"; it meets the criteria better.

POINT: It was curious that these results were not within the large corpus of the first enumeration—a tiny detail of all rows with cluster conditions, not fulfilling the CIG-4 restriction at hinge positions, excluded them. The finally selected series (Fig. 13) from constraint searching begins with (B C A Bb) and ends with (B Bb C A). In regarding the series as a cycle, the two groups (C A B C) and (A B C A), which contain overlapping hinge positions—and the corresponding inverted two in the middle—are not CIG-4s, but all other 50 groups of four are. So what seemed to be the most restricting criteria, here turned out to be the best strategy—which is not very surprising! However, this exact description of piece-specific criteria also helped evolve our discussions during the process of searching—and finally showed us how to adapt our prior restriction which only seemed general enough at first. In the long run the whole set of restrictions converged and led to a reasonable number of reasonable results.

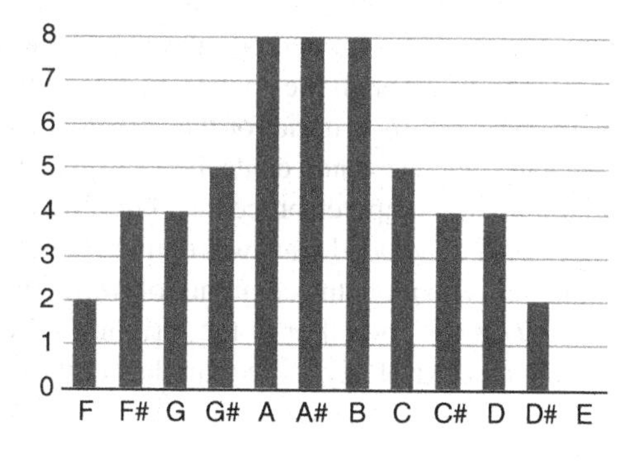

Fig. 14 Pitch class distribution in the series of *AIMD*

Further Compositional Steps by Vanhecke

Vanhecke: The next step in the project was to construct a RHS for *AIMD*. This was done on the basis of exactly the same formula as were used in *AIMO*. This step in the process could therefore be considered entirely "mine". Since the series of AIMD is different from that of *AIMO*, it has a different interval class content, and therefore starts from different data, resulting in a different RHS.

The contribution of the project stopped at this point. The transformation of the RHS into a surface structure, a score, was left up to me. As was mentioned above, I tried to compose *AIMD* in such a manner that *AIMO* and *AIMD* were essentially identical. Since they start from a different RHS, they could not be strictly identical, but I tried to make them so similar to make it impossible to say which piece is the original and which one is the replica. *AIMD* should not be perceived as a variation of *AIMO* any more than the other way around. In order to achieve this, I used not only the series and RHS of *AIMD* for the composition of the Double, but at all times I had the score of *AIMO* in front of me. I tried to "copy" *AIMO* with the material of *AIMD*. This way, the pieces became as similar to each other as, for instance, the different *Marilyn Monroe portraits* by Andy Warhol, or Arnold Böcklin's five versions of the painting *Die Toteninsel* (*The Isle of the Dead*).

Project Review by Bart Vanhecke

POINT: Is *AIMD* your piece (as is the case with *AIMO*) or is it the computers'?

Vanhecke: To answer this question it is important to distinguish between voluntary and involuntary elements in the process of composition, between controlled and automatic cognitive processes on the series construction level.[6]

The construction of the series for *AIMO* involved controlled processes, such as the deliberate choice for a CIG-3/4 series, the central A–C cluster and the inversive symmetry of the series, but also processes that escaped my control completely, such as the fact that the series is unbalanced and does not contain pitch class F, or to a large extent, such as the limited possibilities of successive CIGs. Comparing *AIMO* with *AIMD* showed that all the controlled essential elements of *AIMO* are also controlled by me through my criteria for the construction of the series of *AIMD*. The essential elements that were not controlled by me in the series of *AIMD* were the result of automatic computer processes, but those elements escaped my control anyway. Therefore I concluded that I was entitled to be called the sole composer of *AIMD*. The computer is no more the co-author of *AIMD* than the pencil I used in the composition of *AIMO* is the co-author of the 'original' piece. The composition of *AIMD* did not escape my control any more than the composition of *AIMO*.

---

[6] The processes of construction of the RHS and of transformation of the RHS into the score were identical on the level of control in both cases, as was discussed above.

POINT: "The computer is no more the co-author of *AIMD* than the pencil I used in the composition of *AIMO* is the co-author of the "original" piece". Is it really like this? The pencil per se does not help in finding the optimal solution, it does not act independently in comparison to the computer program.

Vanhecke: What I mean is that, from my point of view, the computer is not more than a tool, just like the pencil I use. By the way, my pencil does help me in finding the optimal solution. Without a writing tool of some sort, I would not be able to structure my pieces the way I do. It helps me visualise my composition. What the computer did for me is help me construct a series. I normally use a kind of domino card system (with all CIGs) as a tool to construct a series. As far as I was concerned, he computer played just that role: an electronic domino card system. Of course, for your purpose, the fact that it was a computer and not a set of cards was essential.

POINT: It was your decision to compose two similar pieces out of two different series, which is completely understandable in regard to your claim that the result considerably depends on your "post-algorithmic" decisions. Nevertheless for us it would be interesting to know, if the "double" would have become a completely different composition, taking not only one of the finally found CIG-54 rows, but the "perfect" CIG-54 as a single starting point.

Vanhecke: The series of *AIMD* can be called "more perfect" than the series of *AIMO*, in the sense that it meets the construction criteria better in the sense that it is more outspokenly unbalanced and contains a more extended A–C cluster. Does this entail that the "double" is aesthetically more perfect than the "original" version? Certainly not. Perfection is not a criterion for aesthetic value. It is often the voluntary or involuntary imperfections that add aesthetic value to the artworks. The campanile of Pisa would probably not have been as attractive and fascinating, or as famous, if it would not have the imperfection that makes it lean over dangerously. The greater perfection of the series of *AIMD* doesn't make the double piece aesthetically any more valuable than *AIMO* in my eyes.

POINT: But as the algorithms just calculated the optimal result in regard to your criteria, it was not the quest for a leaning tower of Pisa, but for a straight one. Wouldn't you have been much happier finding the "perfect" CIG-54 by yourself?

Vanhecke: I would not have been happier. As long as a series works, I cannot subjectively call one series "more perfect" than another. The same goes for my daughters: their DNA is different, but as long as their DNA is healthy, as long as "it works", any DNA, with all its defects, is perfect to me.

An extremely important issue for me is the fact that the project refutes the claim that strictly constructed serial music is the result of mere mathematical or technical puzzle solving, leaving no room for artistic invention. By composing two "essentially identical" pieces with different calculated material I showed the relative impact of that material on the end result.[7] The strictly calculated starting conditions only have

[7] In another project I composed two completely different pieces (*Un souffle de l'air que respirait le passé*, for piano quartet (2011), and *Danse du feu*, for large orchestra (2012)) with exactly the same series and RHS, to prove the same point where there is restricted importance of the serial material on the end result.

a very limited influence on the end result that is still completely determined by the aesthetic judgement of the composer. Indeed, a piece is determined by the material it is composed with, just like human beings are determined by their purely biochemical genetic material. Still, the same genetic material can result in completely different personalities. External influences, education and experience play a major role in formation of a personality. Similarly, the composer's artistic personality plays the most important role in all artistic creation, even the most rigorously strict one.

POINT: An advocatus diaboli could ask: If characteristic specifica don't have much impact, at least in this case, was the formulation of criteria erroneous? Could it be that the knowledge of using a material found by the computer inhibits to judge that it is better?

Vanhecke: The reader should clearly distinguish between the objective criterion of searching for music with an extremely high degree of dissonance and atonality (which the CIG-3/4 technique yields) or the structural criteria at the basis of the project (the unbalanced A–C cluster) on the one hand, and the subjective criterion of aesthetic beauty or perfection on the other. The project has made it clear to me that greater perfection in the first criterion does not result in greater perfection in the second.

References

1. Deleuze G, Guattari F (2004) A thousand plateaus, 5th edn. Continuum, London
2. Forte A (1973) The structure of atonal music. Yale University Press, New Haven
3. Straus JN (2005) Introduction to post-tonal theory, 3rd edn. Pearson Prentice Hall, Upper Saddle River
4. Stravinsky I (1962) Stravinsky—an autobiography. Norton, New York
5. Szamosi G (1986) The twin dimensions: inventing time and space. McGraw-Hill, New York

Peter Lackner/Tropical Investigations

Peter Lackner, Harald Fripertinger and Gerhard Nierhaus

Lackner first had contact with music in his early youth via autodidactic attempts on the accordion and the guitar.[1] The guitar became an electric guitar and after some detours into the world of rock music and the intense experience of the music of Bach, Beethoven, Bruckner and Mahler, Lackner experienced an initial musical spark through the contact with the music of Josef Matthias Hauer and became empowered for a musical approach, which is mostly determined algorithmically. At the same time he started a phase of intense piano playing and his first compositions emerged. Although Hauer served as a trigger for the development of his own musical systems, it was soon the cyclical-serial colour systems of the Austrian Painter and composer Hans Florey,[2] which was firstly an inspiration for Lackner's own work. Behind this fascination there was initially certainly also the wish to write music, which would get some validity through a system by its clear and determined structure. The related partial abstraction of his own "creative will" however is for Lackner not a restriction in composition but opens up new possibilities to him, to organically let the material grow which is at his disposal—to develop it in a sort of "balance" as he will frequently formulate it later on. Besides piano Lackner returned to the guitar and also to other

[1] Biographical introduction and texts from the composer translated from the German by Tamara Friebel.

[2] Austrian painter, flautist and composer (1931–2013).

P. Lackner
Institute for Composition, Music Theory, Music History and Conduction,
University of Music and Performing Arts Graz, Graz, Austria
e-mail: peter.lackner@kug.ac.at

H. Fripertinger
Institute for Mathematics and Scientific Computing, University of Graz, Graz, Austria
e-mail: fripert@uni-graz.at

G. Nierhaus (✉)
Institute of Electronic Music and Acoustics, University of Music and Performing Arts Graz,
Graz, Austria
e-mail: nierhaus@iem.at

G. Nierhaus (ed.), *Patterns of Intuition*,
DOI 10.1007/978-94-017-9561-6_13

instruments like viola or Schwegelpfeife.[3] In this time there was also the encounter with Swedish folk music amongst others, which fascinated Lackner from the first moment on and whose study he deepened during four stays in Scandinavia. Yet new stimulus did not only arise from music, directors like Ingmar Bergmann, Jean Luc Godard, Lars von Trier, writers like Ivan Goncharov, Friedrich Hölderlin, Gustave Flaubert, and Thomas Bernhard were strong inspirations for his work at this time.

A serious illness forced Lackner to prioritise, what followed was a reduction to playing piano and composing. The subsequent study of composition at the University of Music and Performing Arts Graz with Herman Markus Preßl[4] allowed him to reflect and develop his compositional approach in the context of the manifold directions of new music. Presently Lackner teaches music theory at the University of Music and Performing Arts Graz, composes music almost exclusively entitled "canon" and develops a new approach of a classification of tone series and tropes—an examination already begun several years before beginning his study at the music university.

Artistic Approach

Statement

The starting point for me in the practice of music is locating my awareness, affected by emotions and actions, as a sense of balance within a form. All processes of composition, from the search for material, structural considerations, to the performance instructions, are determined by this sense and by possibly no other creative will. The largest possible freedom doesn't mean the use and the exploration of remarkable versatile and "interesting" material, but rather the confinement onto a set of basic principles that seems apt to represent variety in a condensed form. It is either the sensing of a particular constellation, which I aim to determine with a specific search, or else I try to find an existential form for a discovered musical phenomenon, if this is at all possible.

The "ideal" for a composer, or musician in general, may be similar to one of a botanist. One tries to find a seed and according to experience you recognise its attributes, plant it into the correct ground, into the right depth and water it if necessary to let it grow in its own manner. Out of this weeds can grow, more or less interesting to most people, but for others, it is a tree of life.

"Moreover, it is proper to a substantial form to give matter its act of existing pure and simple, because it is through its form that a thing is the very thing that it is. [...] Therefore, if there is a form which does not give to matter its act of existing pure and simple, but comes to matter already possessing an act of existing through some form, such a form will not be a substantial one."[5]

---

[3] A fife with six tone holes, in our days used mostly in folk music.

[4] Hermann Markus Preßl (1939–1994). Austrian composer, professor at the University of Music and Performing Arts Graz.

[5] "[...] Est autem hoc proprium formae substantialis quod det materiae esse simpliciter. Ipsa enim est per quam res hoc ipsum quod est; [...] Si qua igitur forma est qua non det materiae esse simpliciter,

Personal Aesthetics

In my compositional study I was, for instance, confronted with famous serial pieces like *Modes de Valeurs et d'intensités* from Messiaen, *Structures* from Boulez and *Kreuzspiel* from Stockhausen. The basic conception of these pieces, as far as it was discernible for me back then, I was only able to perceive as inadequate in my youthful self-assurance, because I judged the approach in relation to my own musical goals, which were based on quite different premises. After some experiments these composers changed their minds to having again a clear focus on the desired sound image. My main interest was at first concerned with the realisation, "making audible" a certain structure—the sound image was for me at best a confirmation of my approach rather than the primary task. Even today with all the mistrust of a certainty concerning others or my own composition premises I still retain some of this motivation.

In my work there are a considerable number of singularities in details, to which I do not want to react with the same patterns each time because our position towards the same or similar situation is always changing. Nevertheless I believe that one can find here regularities in order to gain an overview. On the other hand the old saying still holds true: "The more restricted the horizon the larger the overview".

I am frequently asked why I name most of my pieces "canon". Apart from the colloquial musical definition of the term, there is a series of meanings for "canon", which can be connoted with "order", "rule" or "scale". If one considers the term in a musical context, for example where "canon" was used by Josquin, Scheidt, Bach and Webern, and hence tries with mathematical diligence to extract an integral canon terminology, it yields that finally everything, which exists self-identically, can be canon-like in principle. This of course may result, for me, in a questioning of the notion of an individual work.

Formalisation and Intuition

I attempt to counteract certain "desires of the composer's power" by accordingly reflective means. Algorithms, aleatoric music served and serve me frequently to realise compositional concepts, in which the will of expression and design alone would often not have lead to these results, which I suppose deserving of coming into the "light" through my work—and this doesn't imply that my respect for intuitively developed music is reduced. I do note here that it is my wish not to polarise too much, because intuition of course always plays a certain role in my music. To sense a result out of the dark, where it may often later also be approximately formalised, at the moment of "finding", you have something, but it is also certainly the case when

(Footnote 5 continued)
sed adveniat materiae jam existenti in actu per aliquam formam, non erit forma substantialis." ("De Anima"; Thomas of Aquinas, Paris 1269). English translation from [25].

you identify a possible underlying structure. On the whole it is thus also about the differentiation between "invention" and "discovery".

Another example of inadequate polarisations: for one person a number could be "extra-musical", but for another, it is the embodiment of music. I personally believe rather towards inclusive positions because I do not want easily to denote something as definitely "extra-musical". For that purpose I would first need to know what music is in order to be able to define an exact line of separation, so I think it's better not to do this! It might sound a little bit esoteric, but I realise often at a close examination of a subject—here it is music—that linguistic means can be insufficient.

This problem of definition is certainly also due to the fact that I assimilate the categories of a quasi-separable-way-of-thinking into terms such as "form" and "content", but also "space" and "time" come into question, as before in order to classify with respect to a "musical" or an "extra-musical" context.

Evaluation and Self-reflection

I see myself as part of the tradition of almost always "too-late-finished pieces" in very good company and also quite at home with this situation. It is often an external factor that forces me to think, stop, terminate! This is for now, in this very moment, the most viable technical realisation. Composing in the "last-possible moment" is not envisioned out of laziness; rather it plays with the idea that I can become smarter with the passing time, on the basis of the work and its processes. I can accordingly never be entirely satisfied, although I consider, as already indicated earlier, an individual piece is rather more seen as "journaling, recording" than "work". It is possible that each piece works or fails for various reasons, but for my work there are in this regard certainly important and quite long-lasting criteria. Much would be gained if an analytic as well as an intuitive approach would come up with at least similar content.

Musical Results from the Development of the Systematisation

Initially I had a focus on the search for sound material, which satisfied—in the farthest sense—canon-like or "balanced" forms. This led to the search of tone-series or tone-cycles, respectively, which I manually determine through means of geometry. Thus, I could systematically detect symmetries, asymmetries, and zones of densities, systematising "traces" which I have been able to do since the 80s and 90s, in order to use the most appropriate material.

In recent work, with the continued development of a systematic approach, the spotlight has been more on the search for sound material as well as sound progressions, space-time equivalences, degrees of density, tone series and their analysis, where I can now react more intuitively in form finding and the subsequent composition work.

Fig. 1 Magic *square* used for *Kanon für Violine und Klavier* (1984)

My piece *Kanon für Violine und Klavier* (1984)[6] (Fig. 2) used two twelve-tone series, which are applied for different purposes. The first constructs a harmonic field based on a magic square[7] (Fig. 1).

The second series sounds melodically within the associated field of the first series. Additional parameters like octave registers, rhythm and playing techniques are deduced from the two series. The priority is not on a particular sound result, but is rather about a musical reflection of a principle: the clash of the individual within the harmonic landscape.

Since 1986 I have been concerned with the juxtaposition and multi-layered linking of cyclical time-processes. Figure 3 shows a graphical illustration of these principles in the case of "*Hexentanz*" *für Violine und Klavier* (*September* 1986).[8]

---

[6] *Kanon für Violine und Klavier* translates as "canon for violin and piano".

[7] A certain array of elements, which are balanced in a square grid.

[8] *Hexentanz für Violine und Klavier* translates as "witch-dance for violin and piano".

Fig. 2 Excerpt from *Kanon für Violine und Klavier* (1984)

Fig. 3 Graphical illustration of "*Hexentanz*" *für Violine und Klavier* (*September* 1986)

The specific search for fundamental material of certain properties based on *tropes*[9] was central for the piece *Kanon für Violine solo*[10] (1988). The grouping of twelve-tone series or arrays into two complementary pairs of hexachords (tropes) goes back to Josef Mattias Hauer, according to whom tropes can be represented in 44 fundamentally different pictures. Further essential stimuli for my work on this topic came from George Perle[11] and Hans Florey. Perle and Florey reduce, independently from each other, the number of tropes by 9 mirroring tropes to 35, see Table 1.

My continuation of these systems consists of researching the possibilities of a cyclical linking of "neighbouring sounds" under various aspects. In order to find possible links, a matrix (Fig. 2) serves to represent the results of a series of previous investigations.[12]

Fundamental in the composition *Kanon für Violine solo* (1988) was the search for cyclical twelve-tone series, which are at once identical under inversion, retrograde and inverse-retrograde. Based on one of these 96 possibilities a one-voice piece is created, which also forms a four-part canon. These properties are also reflected in the rhythmic and formal construction (Table 2).

In *Kanon für drei Bratschen*[13] (1991) several of the previously used principles were combined. The inversion-invariant structure of the piece was made visible by a score in form of a Moebius strip, see Fig. 4.

Since 1994 I have worked on several musical interpretations of the I Ching, which fascinates me particularly in the way change is represented, and the representative plurality that is based on simple constellations.

Amongst others, I approach the idea of the "binary" through an exploration of sound/non-sound, tone/counter-tone and completeness/cancellation also as partial representation of a sound continuum. Figure 5 shows an excerpt from the *Kanon für zwei Gitarren.*[14] *9. September* 2001 which applies these principles.

Since 2004, as a result of targeted searches for material an alternative approach of a systematic of tropes and tone-series was developed amongst others, which was in the following extended in collaboration with Harald Fripertinger. Since this time the constellations found have induced a "reacting", rather than a "deliberate" transformation.

[9] See the definition from Fripertinger in section "Tropes" (see Sect. Tropes).

[10] *Kanon für Violine solo* translates as "canon for solo violin".

[11] American composer and music theorist (1915–2009).

[12] Two tropes are called connectable (see Sect. Tropes) if for each of the two hexachords of the first trope there exists a (uniquely determined) hexachord of the second trope so that the two hexachords have exactly five pitch classes in common.

[13] *Kanon für drei Bratschen* translates as "canon for three violas".

[14] *Kanon für zwei Gitarren* translates as "canon for two guitars".

Table 1 Tropes in lexicographic order. Wilfried Skreiners exhibition catalogue [27] lists as number 36 a picture of Florey [5] from 1965/67 which is probably the first instance that the 35 tropes were listed in lexicographical order

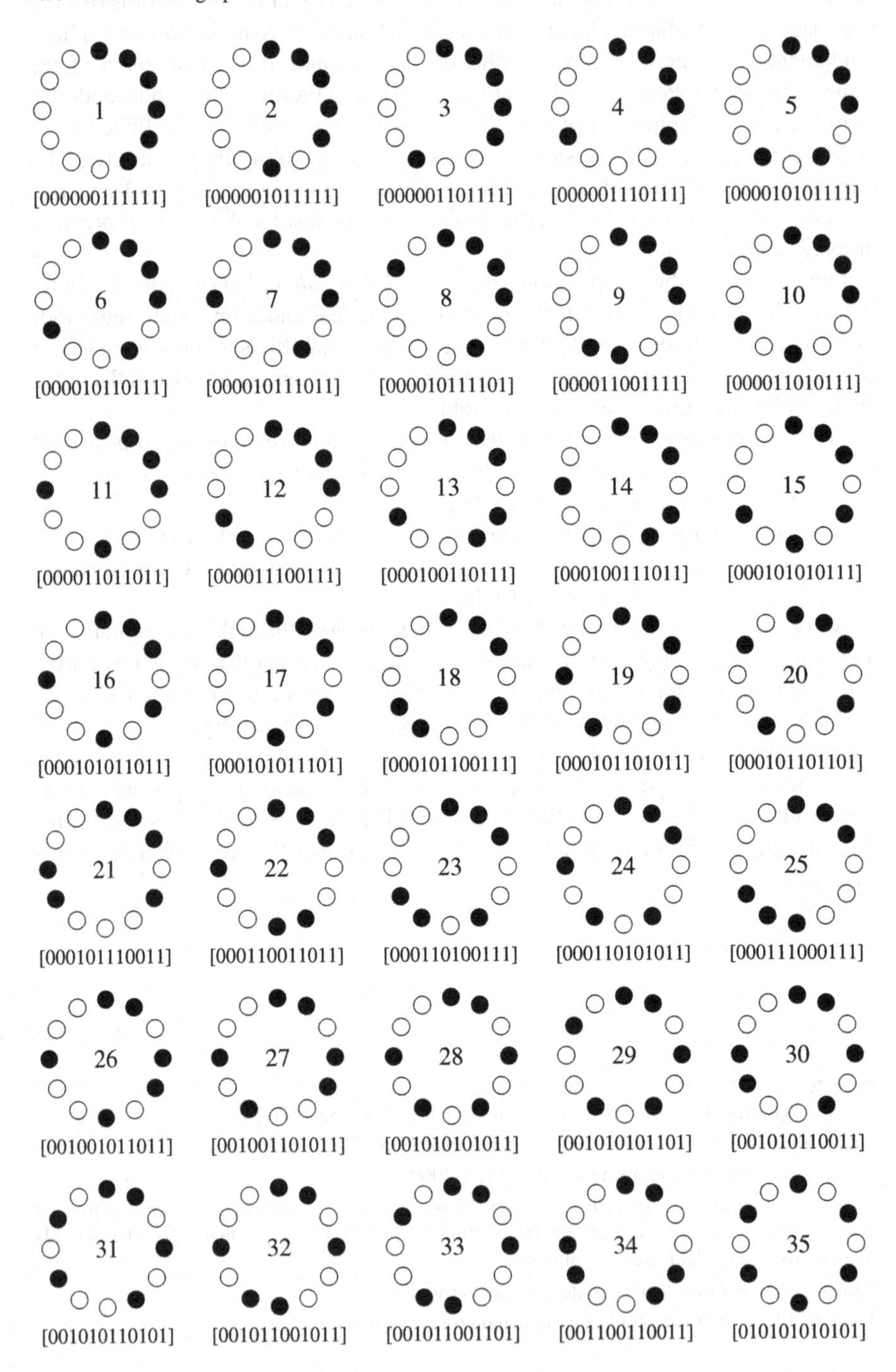

Table 2 Matrix $a_{i,j}$ indicating the number of possible connections from the trope i to the trope j, for $i, j \in \{1, \ldots, 35\}$ (Computed by Peter Lackner in 1991)

	1	**2**	**3**	**4**	**5**	**6**	**7**	**8**	**9**	**10**	**11**	**12**	**13**	**14**	**15**	**16**	**17**	**18**	**19**	**20**	**21**	**22**	**23**	**24**	**25**	**26**	**27**	**28**	**29**	**30**	**31**	**32**	**33**	**34**	**35**	
1	2	6	8	4	2	4	4	2					2	2																						**1**
2	3	2	6	2	2	2	4	1	1	2	2				1	2	3			2						1										**2**
3	2	3	2	2	1	3	3	1	2	3	2	1	1	1		1		1	2	1	1	1				1	1									**3**
4	2	2	4		2	2	2		2	2		2	2	2	2	2	2	2			2	2														**4**
5	1	2	2	2		4	2	2	1	4		2			2	2	1		2				1	2				1	2		1					**5**
6	1	1	3	1	2	1	2	1	1	2	1	1	1	1		2		2	3	2		1	1	1		1			1	1	1		1			**6**
7	1	2	3	1	1	2	1	1	1	1	1		2	2	1	1	2	1	1	1	1	2		1		1	1	1	1	1			1			**7**
8	2	2	4		4	4	4							2		4			4		2								2		2					**8**
9		1	4	2	1	2	2			4	2	4						4			1	2	2		1					1		1	2			**9**
10		1	3	1	2	2	1		2	1	1	2			2	1	1	3	3	1			2	1	1		1	1	1			1	1			**10**
11		2	4			2	2		2	2			2			2		2		2		2	2			4	2					2	2			**11**
12			2	2	2	2			4	4		2						4	2		2	2	4	2	2											**12**
13	1		2	2		2	4				2			2	1	2	2	2			2	4	1			1	2	1		2				1		**13**
14	1		2	2		2	4	1					2	2		4			2		2	6		2						2	1			1		**14**
15		1		2	2		2			4			1		2	2	4	4			2			2	1			4	1	1					1	**15**
16		1	1	1	1	2	1	1		1	1		1	2	1	1	1	1	3	1	1	2	1	1		1	1	2	2	2	1		1			**16**
17		3		2	1		4			2			2		4	2	2	2		2	1						2	4		1		1			1	**17**
18			1	1		2	1		2	3	1	2	1		2	1	1	1	1	1	2	4	2	1	1		1	1		1		1	1			**18**
19			2		1	3	1	1		3		1		1		3		1	2	1	1	1	2	2		1	1		3	1	2		2			**19**
20		2	2			4	2			2	2					2	2	2	2							4	2	2	2			2	2			**20**

(continued)

Table 2 (continued)

	1	2	3	4	5	6	7	8	9	10	11	12	13	14	15	16	17	18	19	20	21	22	23	24	25	26	27	28	29	30	31	32	33	34	35	
21			2	2			2	1	1			2	2	2	2	2	1	4	2			4	1	2				1		2				1		**21**
22			1	1		1	2		1		1	1	2	3		2		4	1		2	5	1	1		1	2			2			1	1		**22**
23					1	2			2	4	2	4	1			2		4	4		1	2		2	1				1			1	2			**23**
24					2	2	2			2		2		2	2	2		2	4		2	2	2					2	2	2	2					**24**
25									4	8		8			4			8					4													**25**
26		1	2			2	2				4		1			2			2	4		2				2	4		1	1		2	4			**26**
27			2				2			2	2		2			2	2	2	2	2		4				4		2		2		2	2			**27**
28					1		2			2			1		4	4	4	2		2	1			2			2	2	3	2		1			1	**28**
29					2	2	2	1		2					1	4			6	2			1	2		1		3	2		3		2			**29**
30						2	2		1				2	2	1	4	1	2	2		2	4		2		1	2	2			1		2	1		**30**
31					2	4		2						2		4			8					4					6	2	2					**31**
32									2	4	4						2	4		4			2			4	4	2					4			**32**
33						2	2		2	2	2					2		2	4	2		2	2			4	2		2	2		2				**33**
34													6	6							6	12								6						**34**
35															12		12											12								**35**
	1	**2**	**3**	**4**	**5**	**6**	**7**	**8**	**9**	**10**	**11**	**12**	**13**	**14**	**15**	**16**	**17**	**18**	**19**	**20**	**21**	**22**	**23**	**24**	**25**	**26**	**27**	**28**	**29**	**30**	**31**	**32**	**33**	**34**	**35**	

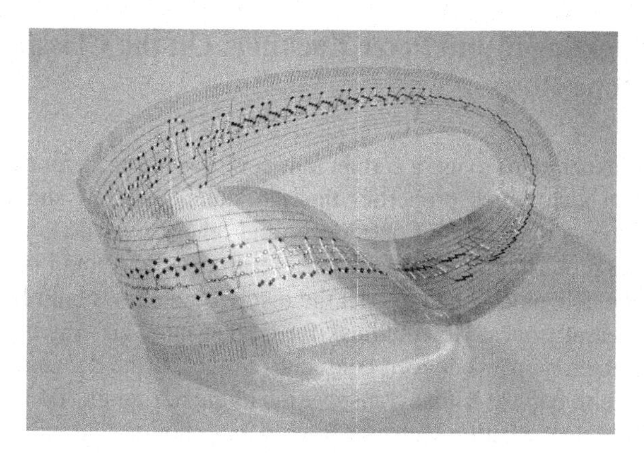

Fig. 4 *Kanon für drei Bratschen* (1991)

Fig. 5 Excerpt from *Kanon für zwei Gitarren. 9. September* 2001

Harald Fripertinger and Peter Lackner: On the Classification of Twelve-Tone Rows

When Peter Lackner plans to use a certain tone row[15] as the basis for a composition, he has a certain idea which properties this particular tone row should have. For instance, he expects that certain tropes occur and other tropes do not occur in this tone row. Or he prescribes a certain sequence of tropes that should be present. For instance, maybe he would like to restrict the result to tone rows built from a certain number of different tropes. He could ask for certain intervals which should occur somewhere in the tone row between consecutive tones, or he demands that certain musical operations, e.g. the tritone transposition of the retrograde, do not change the tone row. It is not clear whether such tone rows exist, or how to find a tone row with all these properties. Therefore, it is desirable to get a complete overview of all tone rows.

Taking into account that the number of different tone rows is so big that a complete database of all tone rows is not feasible, first we describe a similarity relation on the set of all tone rows. Similar tone rows are collected into a single similarity class. This way we reduce the number of tone rows to the number of different similarity classes of tone rows. Then we have to find and define properties of tone rows so that similar tone rows have the same properties. In other words we are looking for meaningful and useful properties of the similarity classes. Or the other way round, if we have certain interesting properties of tone rows, then we must find a suitable definition of similarity so that similar tone rows share all these properties. In the next step from each similarity class of tone rows we select a representative. The representatives of all similarity classes are collected in a database. From each representative we determine all its properties. These properties are also stored in the database so that a user can search for these properties. This way it is possible to decide whether tone rows with prescribed properties exist. If so, the user also obtains a complete list of all similarity classes of tone rows with the prescribed properties.

The remaining part of the present manuscript is mainly an excerpt of [11].

Classification of All Tone Rows

12-Scale and Pitch Classes

A tone in music is described by its fundamental frequency $f > 0$, which we call its pitch. It is usually given in Hertz (Hz). Two tones with frequencies f and $2f$ form the interval of an octave. In equal temperament an octave is divided into 12 equal parts. We speak of a 12-*scale*. Therefore the frequencies f_i, $1 \leq i \leq 11$, of the 11 tones between f and $2f$ would be $f \cdot 2^{i/12}$. Disregarding the fact that human beings can hear tones only in the range from 20 to 20000 Hz, in general the set of all tones

[15] From now on this implies twelve-tone rows.

in a 12-scale (which contains a tone with frequency f) is countably infinite and is given by $\{f \cdot 2^{k/12} \mid k \in \mathbb{Z}\}$. Since we are not interested in the particular frequencies we omit the factor f and each tone is represented by an integer k. Consequently, $\mathbb{Z}$ is a model of a 12-scale.

From the musical perception we deduce that tones which are an integer multiple of an octave apart have a similar quality. We speak of octave equivalence. Tones being a whole number of octaves apart are considered to be equivalent and are collected to a *pitch class*.

Let n be an integer, then by $\overline{n}$ we denote the subset $\{12k + n \mid k \in \mathbb{Z}\}$. It is the residue class of n modulo 12. Of course $\overline{n} = \overline{n + 12k}$ for any $k \in \mathbb{Z}$.

Using $\mathbb{Z}$ as the model of a 12-scale the twelve pitch classes are the subsets $\overline{i}$ for $0 \leq i < 12$. Given an arbitrary tone $n \in \mathbb{Z}$, by integer division there exist uniquely determined integers q and r so that $n = 12 \cdot q + r$ and $0 \leq r < 12$. Then the tone n belongs to the pitch class $\overline{r}$, or in other words $\overline{n} = \overline{r}$. It is now clear that the pitch classes in the 12-scale $\mathbb{Z}$ coincide with the residue classes in $\mathbb{Z}_{12} := \mathbb{Z} \bmod 12\mathbb{Z}$.

This model can easily be generalized to n-tone music, for $n \geq 1$. The set of tones in an n-scale containing a tone of frequency f is then $\{f \cdot 2^{k/n} \mid k \in \mathbb{Z}\}$ and the set of pitch classes in the n-scale is $\mathbb{Z}_n := \mathbb{Z} \bmod n\mathbb{Z}$.

Tone Rows

A *tone row* is a sequence of 12 tones so that tones in different positions belong to different pitch classes. Therefore, we describe a tone row by a mapping

$$f\colon \{1, \ldots, 12\} \to \mathbb{Z}, \qquad \overline{f(i)} \neq \overline{f(j)}, \quad i \neq j,$$

where $\overline{f(i)}$ is the residue class of $f(i)$, $i \in \{1, \ldots, 12\}$. The set $\{1, \ldots, 12\}$ is the set of all order numbers or time positions. The value $f(i)$, $i \in \{1, \ldots, 12\}$ is the tone in ith position of the tone row f. Since different tones in f must belong to different pitch classes and since the actual choice of $f(i)$ in its pitch class is not important, we consider tone rows as functions $f\colon \{1, \ldots, 12\} \to \mathbb{Z}_{12}$. A function $f\colon \{1, \ldots, 12\} \to \mathbb{Z}_{12}$ is a tone row, if and only if f is bijective. Therefore, the set of all tone rows coincides with the set of all bijective functions from $\{1, \ldots, 12\}$ to $\mathbb{Z}_{12}$. It will also be indicated as $\mathcal{R}$.

This leads to a total of $12! = 12 \cdot 11 \cdots 2 \cdot 1 = 479\,001\,600$ tone rows. By introducing certain equivalence relations on the set of all tone rows we collect different tone rows into sets of equivalent tone rows. Equivalence of tone rows will be used as a synonym for similarity of tone rows. The equivalence relation must be carefully determined, because afterwards we are just interested in non-equivalent tone rows, i.e. in the equivalence classes of tone rows. In the present manuscript (see Remark 3) we try to motivate a natural equivalence relation, which generalizes Schönberg's notion of similarity of tone rows. Applying this equivalence relation to the set of all tone rows reduces the number of essentially different tone rows to

836 017. This number is small enough so that it is possible to collect information on all non-equivalent tone rows in a database.

The description of tone rows as bijective functions from $\{1, \ldots, 12\}$ to $\mathbb{Z}_{12}$ given above shows that tone rows are mathematical objects which we call discrete structures.

Classification of Discrete Structures

In general *discrete structures* are objects which can be constructed as subsets, unions, products of finite sets, as mappings between finite sets, as bijections or linear orders on finite sets, as equivalence classes on finite sets, as vector spaces over finite fields, etc. For example we can describe graphs, necklaces, designs, codes, matroids, switching functions, molecules in chemistry, spin-configurations in physics, or objects of local music theory as discrete structures (Cf. [18]).

Sometimes the elements of a discrete structure are not simple objects, but they are themselves classes of objects which are considered to be equivalent. Then each class collects all those elements which are not essentially different. For instance, in order to describe mathematical objects we often need labels, but for the classification of these objects the labelling is not important. Thus all elements which can be derived by relabelling of one labelled object are collected to one class.

Besides relabelling also naturally motivated symmetry operations give rise to collect different objects to one class of essentially not different objects. This will soon be done, when we introduce equivalence relations on the set of all tone rows. Consequently, we rather classify the corresponding equivalence classes of tone rows.

The process of *classification* of a discrete structures provides more detailed information about the objects in a discrete structure. We distinguish different steps in this process:

Step 1: Determine the number of different objects.
Step 2: Determine the number of objects with certain properties.
Step 3: Determine a complete list of the elements of a discrete structure.

In general, step 3 is the most ambitious task, it needs a lot of computing power, computing time and memory.

As it was already mentioned above, it does not make sense to classify all the 12! different tone rows. On the one hand, this number is quite big, on the other hand it is common practice to consider certain tone rows as equivalent, whence, it is more interesting to classify equivalence classes of tone rows. This means, to determine the number of these classes, which is the number of pairwise non-equivalent tone rows, or to determine properties of all the tone rows which are equivalent to a given tone row. Hence, it is important to find a suitable equivalence relation on the set of all tone rows. In this context "suitable" means, both that this notion of equivalence must be properly motivated, and that there should be interesting properties of the equivalence classes of tone rows. The equivalence classes should not be too large, in order not to lose too much information about the individual tone rows. On the other hand they also should not be too small. We will present different notions of equivalence and we will explain which of these notions is the standard situation used in our database.

Already Schönberg and his pupils considered tone rows as equivalent whenever they can be constructed by transposing, inversion and/or retrograde from a single tone row. A formal definition of these operations is given in the next sections.

Transposing, Inversion and Quart-Circle

Let us turn back to the 12-scale. Transposing by a semitone up is the replacement of the tones $f \cdot 2^{k/12}$, $k \in \mathbb{Z}$, by the tones $f \cdot 2^{(k+1)/12}$. Thus, it is described by the mapping

$$T\colon \mathbb{Z} \to \mathbb{Z}, \qquad k \mapsto k+1.$$

Transposing by n semitones up, $n \in \mathbb{N}$, is the same as n-times transposing by a semitone up, whence, $T^n(k) = k + n$, $k \in \mathbb{Z}$. Transposing by a semitone down is the inverse of transposing by a semitone up, thus it is the mapping $T^{-1}(k) = k - 1$, $k \in \mathbb{Z}$. It is possible to define a transposing operator on the set of pitch classes. By abuse of notation we also call it T. It is defined by

$$T(\overline{i}) = \overline{i+1}.$$

Iterating transposing we obtain from the pitch class $\overline{0}$ the pitch classes $T(\overline{0}) = \overline{1}$, $T^2(\overline{0}) = \overline{2}, \ldots, T^{11}(\overline{0}) = \overline{11}$, and $T^{12}(\overline{0}) = \overline{12} = \overline{0}$. This motivates that the pitch classes are cyclically arranged. In conclusion T is a bijective mapping from $\mathbb{Z}_{12}$ into itself, thus it is a permutation. From the iteration process we deduce that T is a cyclic permutation of order 12, actually it is a single cycle of length 12 which is given by $(\overline{0}, \overline{1}, \ldots, \overline{11})$ or by any other cyclic arrangement of the form $(\overline{i}, \overline{i+1}, \ldots, \overline{i-1})$ for $\overline{i} \in \mathbb{Z}_{12}$. Since T has order 12, its inverse T^{-1} is equal to T^{11} and similarly $T^{-k} = T^{12-k}$ for $k \in \{1, \ldots, 11\}$.

In a similar way inversion can be introduced as an operation on the 12-scale $\mathbb{Z}$ and also on the set of pitch classes, where we define

$$I\colon \mathbb{Z}_{12} \to \mathbb{Z}_{12}, \qquad I(\overline{i}) = \overline{-i}.$$

Since $I(\overline{0}) = \overline{0}$ and $I(\overline{6}) = \overline{6}$, the operator has two fixed points. The remaining elements of $\mathbb{Z}_{12}$ are interchanged pairwise. Thus, I is a permutation of $\mathbb{Z}_{12}$ and its cycle decomposition is $(\overline{0})(\overline{1}, \overline{11})(\overline{2}, \overline{10})(\overline{3}, \overline{9})(\overline{4}, \overline{8})(\overline{5}, \overline{7})(\overline{6})$.

Studying the composition of the two permutations I and T we get that $I \circ T^k = T^{-k} \circ I$, $k \in \{0, \ldots, 11\}$. All inversion operators on $\mathbb{Z}_{12}$ can be written as compositions $T^r \circ I$ for $r \in \{0, \ldots, 11\}$. If r is even, then $T^r \circ I$ consists of exactly two fixed points and five cycles of length two, otherwise it consists of six cycles of length two.

Sometimes we also consider the quart-circle Q and quint-circle $I \circ Q$ defined by

$$Q\colon \mathbb{Z}_{12} \to \mathbb{Z}_{12}, \qquad \bar{i} \mapsto \overline{5i}.$$

The quart-circle replaces the pitch classes in the chromatic scale by a sequence of pitch classes of quart intervals, the quint-circle replaces a chromatic scale by a sequence of quints, since $(I \circ Q)(\bar{i}) = \overline{7i}$, $\bar{i} \in \mathbb{Z}_{12}$. The cycle decomposition of Q is given as $(\bar{0})(\bar{1},\bar{5})(\bar{2},\overline{10})(\bar{3})(\bar{4},\bar{8})(\bar{6})(\bar{7},\overline{11})(\bar{9})$.

Permutation Groups on $\mathbb{Z}_{12}$

When we consider a certain set of permutation operators on a set X we are always interested in the set G of all permutations which can be computed as iterations of these operators, all their inverse operators and all their compositions. All these operators together form the group G generated by these operators. It is acting on the given set X. The identity element id belongs to G. Applying it to any element x of the set X does not change x. We write $\mathrm{id}x = x$ for all $x \in X$. The composition $g_2 \circ g_1$ of any two operators $g_1, g_2 \in G$ is also in G. Applying it to any element $x \in X$ we obtain $(g_2 \circ g_1)x = g_2(g_1 x)$. In other words, applying the composition $g_2 \circ g_1$ to x is the same as first applying g_1 to x obtaining and element $x' \in X$ and then applying g_2 to x'.

From the previous section we know that the (permutation) operators T, I and Q are acting on the set $X = \mathbb{Z}_{12}$.

Considering just the transposition operator T, we have deduced that it has order 12, whence, the group G containing T and all its iterates, inverses and compositions is $G = \{T^0 = \mathrm{id}, T, T^2, \ldots T^{11}\}$. It is a cyclic group of order 12, since it is generated by one element, namely T. We write $G = \langle T \rangle$ and usually we abbreviate this group by C_{12}.

If we consider both T and I acting on $\mathbb{Z}_{12}$, the group $\langle T, I \rangle$ generated by T and I consists of exactly 24 elements, namely the elements of the form $T^r I^j$ where $0 \leq r \leq 11$ and $0 \leq j \leq 1$. This group is the dihedral group D_{12} acting on 12 elements.

Often the elements of $\mathbb{Z}_{12}$ are drawn as vertices of a regular 12-gon.

In Fig. 6 we see that the pitch class $\bar{i}$ has the neighbors $\overline{i-1}$ and $\overline{i+1}$. The dihedral group is the biggest group preserving all these neighbor relations.

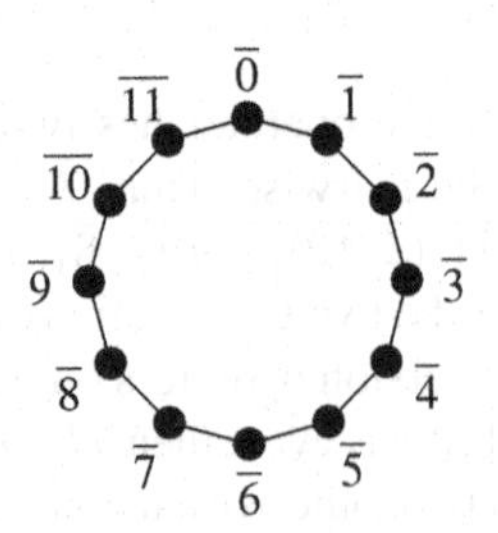

Fig. 6 The 12-scale as a regular 12-gon

Theorem 1 *Let π be a permutation of $\mathbb{Z}_{12}$, then $\pi(i+\bar{1}) = \pi(i)+\bar{1}$ for all $i \in \mathbb{Z}_{12}$, or $\pi(i+\bar{1}) = \pi(i) - \bar{1}$ for all $i \in \mathbb{Z}_{12}$, if and only if π is an element of D_{12}.*

The group $\langle T, I, Q\rangle$ is the group of all affine mappings on $\mathbb{Z}_{12}$ which we abbreviate by $\mathrm{Aff}_1(\mathbb{Z}_{12})$. It is the set of all mappings $f(\bar{i}) = \bar{a}\bar{i} + \bar{b}$, where $a, b \in \mathbb{Z}$, $\gcd(a, 12) = 1$. It is enough to choose a just from $\{1, 5, 7, 11\}$ and b from $\{0, \ldots, 11\}$. It is easy to check that $I \circ Q = Q \circ I$, and $Q \circ T = T^5 \circ Q$. Consequently the 48 elements of $\mathrm{Aff}_1(\mathbb{Z}_{12})$ can be uniquely written as $T^r I^j Q^k$ with $r \in \{0, \ldots, 11\}$ and $j, k \in \{0, 1\}$.

Now we can explain how to construct tone rows being equivalent to a given one.

Equivalence Classes of Tone Rows Described as Discrete Structures

Let $f\colon \{1, \ldots, 12\} \to \mathbb{Z}_{12}$ be a tone row, i.e. a bijective mapping, then

- the *transposed* of f is the tone row of the form $T \circ f$,
- the *inversion* of f is the tone row of the form $I \circ f$,
- the *quart-circle* of f is the tone row of the form $Q \circ f$,

where T, I, Q are the operators on $\mathbb{Z}_{12}$ introduced above. From the definition it is clear that these functions are again bijective mappings from $\{1, \ldots, 12\}$ to $\mathbb{Z}_{12}$, whence, they determine tone rows.

The operators T, I, Q are acting on the range of the functions f, i.e. they are acting on the level of pitch classes. We say that this action on the set of tone rows is induced by the action on the set of pitch classes. In [17] these operations are called *pitch class operations*.

Next we introduce some operators acting on the domain of these f, i.e. they are acting on the time level of tone rows.

We assume cyclicity in the time domain, so that after having played a tone row we repeat it by starting again from the beginning.

- We define the *cyclic shift* as the mapping

$$S\colon \{1, \ldots, 12\} \to \{1, \ldots, 12\}, \qquad S(i) := \begin{cases} i+1, & \text{if } i \leq 11 \\ 1, & \text{if } i = 12. \end{cases}$$

  It is a permutation of $\{1, \ldots, 12\}$. Its cycle decomposition consists only of one cycle given by $(1, 2, \ldots, 12)$, whence, S is of order 12.

- The *retrograde* is the mapping

$$R\colon \{1, \ldots, 12\} \to \{1, \ldots, 12\}, \qquad R(i) := 12 - i + 1.$$

  It is a permutation of $\{1, \ldots, 12\}$. Its cycle decomposition is given by $(1, 12)$ $(2, 11)$ $(3, 10)$ $(4, 9)$ $(5, 8)$ $(6, 7)$, whence, it has the order two.

- The permutation *five-step* $F\colon \{1,\ldots,12\} \to \{1,\ldots,12\}$ is defined by

$$F := \begin{pmatrix} 1 & 2 & 3 & 4 & 5 & 6 & 7 & 8 & 9 & 10 & 11 & 12 \\ 1 & 6 & 11 & 4 & 9 & 2 & 7 & 12 & 5 & 10 & 3 & 8 \end{pmatrix}.$$

It satisfies $F(1) = 1$ and $F(i+1) \equiv F(i) + 5 \bmod 12$, $i \in \{1,\ldots,11\}$.

The group $\langle S \rangle$ is a cyclic group of order 12, whence, it is isomorphic to C_{12}, and the group $\langle S, R \rangle$ is a dihedral group isomorphic to D_{12}. The group $\langle S, R, F \rangle$ is isomorphic to the group of all affine mappings $\mathrm{Aff}_1(\mathbb{Z}_{12})$. It contains also the seven-step $F \circ S \circ R$.

The operators S, T and F determine further operations on the set of all tone rows. Let $f\colon \{1,\ldots,12\} \to \mathbb{Z}_{12}$ be a tone row, i.e. a bijective mapping, then

- the *cyclic shift* of f is the tone row of the form $f \circ S$,
- the *retrograde* of f is the tone row of the form $f \circ R$,
- the *five-step* of f is the tone row of the form $f \circ F$.

These operations on tone rows are induced by a group action on the domain of the tone rows. In [17] they are called *order-number operations*.

From the mathematical point of view it is not important to consider the domain of tone rows as the set $\{1,\ldots,12\}$. The cyclicity of this set would be better modeled by taking $\mathbb{Z}_{12}$ also as the domain of tone rows. Then tone rows are bijective mappings from $\mathbb{Z}_{12}$ to $\mathbb{Z}_{12}$, in other words permutations of $\mathbb{Z}_{12}$. Identifying the integer $i \in \{1,\ldots,12\}$ with the residue class of $i-1$ in $\mathbb{Z}_{12}$, the permutation S corresponds to T, R corresponds to IT, and F corresponds to Q. Since the cardinality of $\mathbb{Z}_{12}$ is 12, the set of all permutations of $\mathbb{Z}_{12}$ is isomorphic to the symmetric group S_{12}, whence, tone rows are just elements of S_{12}.

Coming back to A. Schönberg's notion of equivalence we see that the tone rows of the form $T^r \circ I^s \circ f \circ R^t$ for $r \in \{0,\ldots,11\}$, $s, t \in \{0, 1\}$, are the tone rows being equivalent to the given row f. In this situation, we have 48 different operators which can be applied to f, therefore at most 48 tone rows are collected to the equivalence class of f. Using the notion of group actions, which will be briefly introduced in the next section, the equivalence class of f is the orbit of f under the action of the direct product $\langle T, I \rangle \times \langle R \rangle$ on $\mathcal{R}$, the set of all tone rows.

Group Actions

Now we briefly describe the theory of group actions. For more details on group actions see [18]. A multiplicative group G with neutral element 1 acts (from the left) on a set X if there exists a mapping

$$*\colon G \times X \to X, \qquad *(g, x) \mapsto g * x,$$

such that

$$(g_1 g_2) * x = g_1 * (g_2 * x), \qquad g_1, g_2 \in G, \quad x \in X,$$

and

$$1 * x = x, \qquad x \in X.$$

We usually write gx instead of $g * x$. A group action of G on X will be indicated as ${}_GX$. If G and X are finite sets, then we speak of a *finite group action.*

A group action ${}_GX$ determines a group homomorphism ϕ from G to the symmetric group $S_X := \{\sigma \mid \sigma : X \to X \text{ is bijective}\}$ by

$$\phi : G \to S_X, \qquad g \mapsto \phi(g) := [x \mapsto gx],$$

which is called a *permutation representation* of G on X. Usually we abbreviate $\phi(g)$ by writing $\overline{g}$, which is the permutation of X that maps x to gx. For instance $\overline{1} = \phi(1)$ is always the identity on X. (The reader should realize that $\overline{1}$ is now a permutation of X and not the residue class of 1.) Accordingly, the image $\phi(G)$ is indicated by $\overline{G}$. It is a *permutation group* on X, i. e. a subgroup of S_X.

A group action ${}_GX$ defines the following equivalence relation on X. Two elements x_1, x_2 of X are called equivalent (under G), we indicate it by $x_1 \sim x_2$, if there is some $g \in G$ such that $x_2 = gx_1$. The equivalence class $G(x)$ of $x \in X$ with respect to $\sim$ is the *orbit* of x under G or the *G-orbit* of x. Hence, the G-orbit of $x \in X$ is $G(x) = \{gx \mid g \in G\}$. The set of G-orbits on X is indicated as $G\backslash\backslash X := \{G(x) \mid x \in X\}$. In general, classification of a discrete structure means the same as describing the elements of $G\backslash\backslash X$ for a suitable group action ${}_GX$.

If X is finite, then $\overline{G}$ is a finite group since it is a subgroup of the symmetric group S_X which is of cardinality $|X|!$. For any $x \in X$ we have $G(x) = \overline{G}(x)$, whence, $G\backslash\backslash X = \overline{G}\backslash\backslash X$. Hence, whenever X is finite, each group action ${}_GX$ can be described by a finite group action ${}_{\overline{G}}X$.

Let ${}_GX$ be a group action. For each $x \in X$ the *stabilizer* G_x of x is the set of all group elements which do not change x, thus $G_x := \{g \in G \mid gx = x\}$. It is a subgroup of G. If ${}_GX$ is a group action, then for any $x \in X$ the mapping $\phi : G/G_x \to G(x)$ given by $\phi(gG_x) = gx$ is a bijection. As a consequence we get: If ${}_GX$ is a group action where G is a finite group, then the size of the orbit of $x \in X$ is equal to $|G(x)| = \frac{|G|}{|G_x|}$. Thus, the number of elements equivalent to x can easily be obtained as soon as we have described the equivalence relation by a finite group action.

Finally, as the last notion in connection with group actions, we introduce the *set of all fixed points* of $g \in G$ in X which is denoted by $X_g := \{x \in X \mid gx = x\}$.

Let ${}_GX$ be a finite group action where G is finite. The main tool for determining the number of G-orbits on X is the Cauchy-Frobenius Lemma. Sometimes it is misleadingly called Burnside's Lemma. It can be found in many text books for combinatorics or algebra.

Theorem 2 (Cauchy–Frobenius Lemma) *The number of orbits under a finite group action ${}_GX$, where G is finite, is the average number of fixed points:*

$$|G\backslash\backslash X| = \frac{1}{|G|}\sum_{g\in G}|X_g|.$$

The most important applications of classification under group actions can be described as symmetry types of mappings between two sets. Group actions ${}_GX$ and ${}_HY$ on the domain X and range Y of functions $f: X \to Y$ induce group actions on $Y^X := \{f \mid f: X \to Y \text{ is a function}\}$, the set of all functions from X to Y, in the following way:

- G acts on Y^X by

$$G \times Y^X \to Y^X, \qquad g * f := f \circ \overline{g}^{-1}. \tag{1}$$

- H acts on Y^X by

$$H \times Y^X \to Y^X, \quad h * f := \overline{h} \circ f. \tag{2}$$

- The direct product $H \times G$ acts on Y^X by

$$(H \times G) \times Y^X \to Y^X, \quad (h, g) * f := \overline{h} \circ f \circ \overline{g}^{-1}. \tag{3}$$

From the Lemma of Cauchy–Frobenius it is possible to determine enumeration formulae for these group actions on Y^X. Here, we don't want to go into details. The interested reader is referred to original manuscripts and textbooks describing Pólya's theory of enumeration and its generalizations to group actions of the form (2) or (3) by Nicolaas Govert de Brujin, Frank Harary, Edgar Milan Palmer. See e. g. [2–4, 14, 18–20].

Equivalence of Tone Rows Expressed by Group Actions

The set of tone rows is the set of all bijective functions from $\{1, \ldots, 12\}$ to $\mathbb{Z}_{12}$. The group actions of the form (1)–(3) can be restricted to actions on the set of bijective functions.

By introducing suitable group actions on the set of all tone rows we describe equivalence relations on the set of all tone rows. I. e. different notions of similarity are expressed by different groups acting on the set of all tone rows. The similarity classes or equivalence classes are henceforth called orbits. The bigger the operating group is, the bigger are the orbits of tone rows, and the smaller is the number of different orbits, i. e. the number of pairwise non-equivalent tone rows. (For similar and other applications of group actions to the enumeration of tone rows see [7, 8, 22–24].) The mathematical notions of groups and group actions and in particular group actions on the set of all tone rows are thoroughly described by Tuukka Ilomäki in [17, Sect. 2.2]. For the computation of complete lists of orbit representatives we used standard

Table 3 Number of orbits under various group actions on the set of all tone rows

	Acting group	Double coset	# of orbits
(1)	$\langle T\rangle \times \langle R\rangle$	$C_{12}\backslash S_{12}/\langle R\rangle$	19 960 320
(2)	$\langle T, I\rangle \times \langle R\rangle$	$D_{12}\backslash S_{12}/\langle R\rangle$	9 985 920
(3)	$\langle T\rangle \times \langle S\rangle$	$C_{12}\backslash S_{12}/C_{12}$	3 326 788
(4)	$\langle T, I\rangle \times \langle S\rangle$	$D_{12}\backslash S_{12}/C_{12}$	1 664 354
(5)	$\langle T, I\rangle \times \langle S, R\rangle$	$D_{12}\backslash S_{12}/D_{12}$	836 017
(6)	$\langle T, I, Q\rangle \times \langle S, R\rangle$	$\mathrm{Aff}_1(\mathbb{Z}_{12})\backslash S_{12}/D_{12}$	419 413
(7)	$\langle T, I\rangle \times \langle S, R, F\rangle$	$D_{12}\backslash S_{12}/\mathrm{Aff}_1(\mathbb{Z}_{12})$	419 413
(8)	$\langle T, I, Q\rangle \times \langle S, R, F\rangle$	$\mathrm{Aff}_1(\mathbb{Z}_{12})\backslash S_{12}/\mathrm{Aff}_1(\mathbb{Z}_{12})$	211 012

methods and generalizations for group actions of the form (2) or (3) as orderly generation (cf. [1, 21]) and Sims chains (cf. [26]).

There are several other notions of similarity of tone rows. In the monograph [17] the author presents various kinds of similarity measures, which determine a degree of similarity between two given tone rows. Depending on the similarity measure different measures can yield different degrees of similarity for two given tone rows. Our approach of similarity is motivated by the similarity operations actually used in musical composition. It perfectly reflects symmetries of these objects, and in general it can always be applied for the classification of objects when similarity is described by the action of a group.

In Table 3 we are enumerating tone rows with respect to different symmetry groups. Describing tone rows as mappings from $\{1, \dots, 12\}$ to $\mathbb{Z}_{12}$, the orbits are symmetry types of mappings. Representing tone rows as elements of S_{12}, the orbits correspond to double cosets (second column in Table 3). In the third column we give the number of orbits of tone rows, i.e. the number of essentially different tone rows under the corresponding group actions. These numbers were computed by using the computer algebra system SYMMETRICA [28].

A. Schönberg's model of equivalence corresponds to the settings in (2) of Table 3.

According to Theorem 1 the dihedral group is the biggest group which preserves the neighbor relations in $\mathbb{Z}_{12}$. Therefore, we consider the settings of 5. as the standard settings for our classification. In this situation both the cyclic orders of the pitch classes in $\mathbb{Z}_{12}$ and of the (time) positions, or order numbers, in $\{1, \dots, 12\}$ are preserved. There is also big evidence that Josef Matthias Hauer was considering all elements of $D_{12} \times D_{12}$ as symmetry operators on the set of tone rows. (He used a closed circular representation of a tone row in [15] which is similar to the concept of Fig. 8 in the present manuscript.) Also Read considers in [23, p.546] this notion as the natural equivalence relation on the set of all tone rows. He also determines 836 017 as the number of pairwise non-equivalent tone-rows. Previously, this number was already determined in a geometric problem by Golomb and Welch [13]. In their manuscript [16] Hunter and von Hippel also consider the cyclic shift as a symmetry operation for tone rows. For the enumeration of the $D_{12} \times D_{12}$-orbits they give reference to [13].

Remark 3

1. If not specified in another way, a tone row f' is considered to be *equivalent* to a tone row f if f' can be constructed from f by any combination of transposing, inversion, cyclic shift and retrograde.
2. The equivalence classes of tone rows coincide with the orbits under the group action

$$(\langle T, I\rangle \times \langle S, R\rangle) \times \mathcal{R} \to \mathcal{R}, \quad ((\varphi, \pi), f) \mapsto \varphi \circ f \circ \pi^{-1}.$$

   The acting group is the direct product of two dihedral groups D_{12}, whence, it consists of $24^2 = 576$ elements. Consequently, there are at most 576 tone rows in the orbit of a given tone row. The elements of the orbit $(\langle T, I\rangle \times \langle S, R\rangle)(f)$ of $f \in \mathcal{R}$ are of the form $T^k \circ I^j \circ f \circ S^\ell \circ R^m$ with $k, \ell \in \{0, \ldots, 11\}$ and $j, m \in \{0, 1\}$.
3. There exist 836017 pairwise non-equivalent tone rows. Each of them can be found in the database.

In order to present a tone row as a graph we draw the 12 pitch classes as a regular 12-gon and we connect pitch classes which occur in consecutive position in the tone row. E. g., the tone row $f := (f(1), \ldots, f(12)) = (0, 5, 10, 3, 8, 1, 7, 2, 9, 4, 11, 6)$ is represented by Fig. 7.

Since we allow transposing as an operation on tone rows we delete the labels of the 12 nodes. The inversion of f is the mirror of the given graph which can be visualized by looking at it from the back side of the paper. Since we allow the retrograde we do not show directions and since we allow cyclic shifts of f we insert the

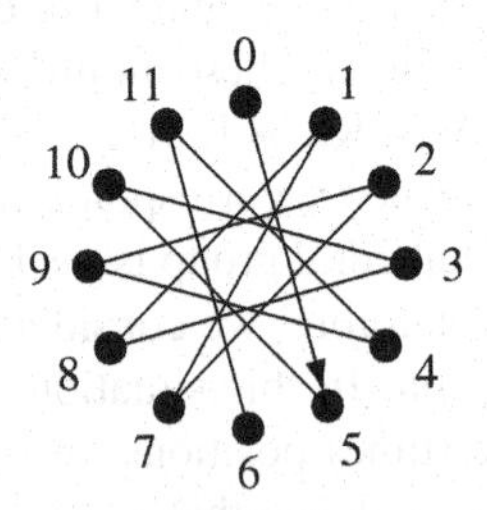

Fig. 7 Representation of a tone row as an oriented open polygon

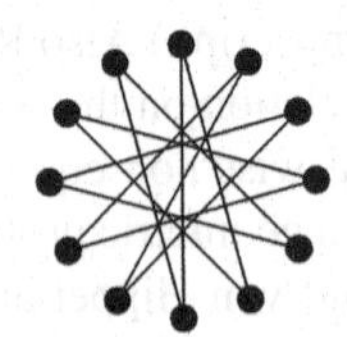

Fig. 8 Representation of a tone row as a closed polygon

missing edge connecting the pitch classes 6 and 0. This way we obtain a graph of the $D_{12} \times D_{12}$-orbit of f as Fig. 8 which is called the chromatic circular representation of a tone row.[16]

The Orbit of a Tone Row

Tone rows are bijective mappings from $\{1, \ldots, 12\}$ to $\mathbb{Z}_{12}$. Now we define a total order on $\mathbb{Z}_{12}$. We assume that $\overline{0} < \overline{1} < \cdots < \overline{11}$. Then it is convenient to represent the tone row f as a vector of length 12 of the form $(f(1), \ldots, f(12))$. E. g., the chromatic scale from pitch class 0 to pitch class 11 is represented as the vector $(0, 1, \ldots, 11)$. Using the total order introduced above, the set of tone rows written as vectors is totally ordered by the lexicographical order. We say the tone row f_1 is smaller than the tone row f_2, and we write $f_1 < f_2$, if there exists an integer $i \in \{1, \ldots, 12\}$ so that $f_1(i) < f_2(i)$ and $f_1(j) = f_2(j)$ for all $1 \leq j < i$. For any two tone rows f_1 and f_2 we have either $f_1 < f_2$, or $f_2 < f_1$, or $f_1 = f_2$. For example, it is easy to prove that the chromatic scale above is the smallest tone row which is possible.

Given a group G which describes the equivalence of tone rows and a tone row f, we compute the orbit $G(f)$ of f by applying all elements of G to f. By doing this we obtain the set $\{gf \mid g \in G\}$ which contains at most $|G|$ tone rows. As the *standard representative* of this orbit, or as the *normal form* of f, we choose the smallest element in $G(f)$ with respect to the lexicographical order.

In connection with the orbit of f we solve the following problems:

- Determine the set of elements of the orbit $G(f)$.
- Determine the standard representative of the orbit $G(f)$.
- Given two tone rows f_1 and f_2 belonging to the same orbit, determine an element $g \in G$ so that $f_2 = gf_1$.

Remark 4 The set of normal forms of tone rows is also totally ordered by the lexicographical order. Hence we can produce a list of all the 836017 representatives of $D_{12} \times D_{12}$-orbits of tone rows. In this list the chromatic scale $(0, 1, 2, 3, 4, 5, 6, 7, 8, 9, 10, 11)$ is in first position and $(0, 5, 10, 3, 8, 1, 7, 2, 9, 4, 11, 6)$ turns out to represent the last orbit, i. e. the orbit with number 836017.

Moreover, each individual tone row is uniquely determined by the number of its orbit (or normal form) and by its position in its orbit. E.g. the main tone row of Peter Lackner's *Kanon für Violine solo* (1988) (see Sect. Musical Results from the Development of the Systematisation) is $(10, 9, 0, 5, 8, 7, 1, 2, 11, 6, 3, 4)$ what stands for (a♯, a, c, f, g♯, g, c♯, d, b, f♯, d♯, e). It is the 129-th tone row in the 683 320-th orbit of tone rows.

[16] Already in 1924 (cf. [15]) Hauer introduced the circular representation of tone rows. Instead of the chromatic order of the pitch classes he used the order according to the quint-circle.

The Stabilizer of a Tone Row

Let $f\colon \{1, \ldots, 12\} \rightarrow \mathbb{Z}_{12}$ be a tone row and let G be a group describing the equivalence classes of tone rows. We have seen that the size of the orbit of f depends on the size of its stabilizer G_f. The stabilizer G_f is a subgroup of G. The stabilizer type of the orbit $G(f)$ is the conjugacy class $\tilde{G}_f$ of G_f.

Assume that $(\varphi, \pi) \in \langle T, I\rangle \times \langle S, R\rangle$ belongs to the stabilizer of f, then $(\varphi, \pi)f = f$, which means that $\varphi \circ f = f \circ \pi$. In other words, applying the permutation φ of pitch classes to the tone row f gives the same row as applying the permutation π^{-1} of order numbers to f. We also say that (φ, π) is a symmetry of f.

It was shown in [9, p. 1.7.3.4] that under the equivalence considered by Schönberg there are 9 972 480 orbits of tone rows with trivial stabilizer, i. e. the identity is the only symmetry of all these tone rows. Moreover, there are 11 520 orbits of tone rows with stabilizer type $\tilde{U}$ for $U = \langle (TI, R)\rangle$ and 1 920 with stabilizer type $\tilde{V}$ for $V = \langle (T^6, R)\rangle$. Consequently, in these two cases either the inversion $TI \circ f$ or the transposition by the tritone $T^6 \circ f$ coincides with the retrograde $f \circ R$ of the tone row f (of stabilizer type $\tilde{U}$ respectively $\tilde{V}$).

In [16] the authors study how rare is symmetry in tone rows under the action of $\langle T, I\rangle \times \langle S, R\rangle$. They compute exactly the same number of orbits having no symmetry or symmetries of type $\tilde{U}$ or $\tilde{V}$ and they conclude that 99.93 % of all $\langle T, I\rangle \times \langle R\rangle$-orbits have no symmetry.

In situation 5 of Table 3 which is our standard situation we have 17 different stabilizer types. They are shown in Table 4. The first column contains the name $\tilde{U}_i$ of the different stabilizer types, the second column generators of the group U_i. The third column shows the order of the group U_i, i. e. the number of elements in U_i, the fourth the size of the conjugacy class $\tilde{U}_i$, i. e. the number of subgroups of $D_{12} \times D_{12}$ which are conjugate to U_i. Finally, the last column presents the number of $D_{12} \times D_{12}$-orbits of tone rows which have stabilizer type $\tilde{U}_i$. These numbers were computed by applying Burnside's Lemma (cf. [18, Chap. 3]). For doing this, we used the computer algebra system GAP [12]. As was already mentioned the tone row of *Kanon für Violine solo* (1988) (see Sect. Musical Results from the Development of the Systematisation) is of stabilizer type $\tilde{U}_9$.

For the situation of $D_{12} \times D_{12}$-orbits of tone rows in [16, p.130] the numbers of orbits of size $576/d$ for $d \in \{1, 2, 3, 4, 6, 8, 12, 24\}$ are computed. For $d = 1$ the authors obtain the number of orbits of stabilizer type $\tilde{U}_1$, for $d = 2$ the number of orbits which are either of type $\tilde{U}_2$ or $\tilde{U}_3$ or . . . or $\tilde{U}_6$ and so on. These numbers coincide with our numbers from Table 4 when we sum up the corresponding cardinalities of the $\tilde{U}_i$-strata for all i so that $|U_i| = d$. In this situation 99.48 % of all $D_{12} \times D_{12}$-orbits have no symmetries.

In the situations (6) and (7) of Table 3 there are exactly 29 different conjugacy classes of subgroups which occur as stabilizer types of these orbits. From our computations it follows that 98.31 % of these orbits have no symmetries.

In situation 8 of $\mathrm{Aff}_1(\mathbb{Z}_{12}) \times \mathrm{Aff}_1(\mathbb{Z}_{12})$-orbits of tone rows there are exactly 90 different conjugacy classes of subgroups which occur as stabilizer types of these orbits. Again the percentage of orbits with no symmetry is decreasing, now to 97.17 %.

Table 4 Stabilizer types for $D_{12} \times D_{12}$ orbits of tone rows

Name	Generators	Size of the group	Size of the class	$\lvert\tilde{U}_i \backslash\!\backslash \mathcal{R}\rvert$
$\tilde{U}_1$	Identity	1	1	827 282
$\tilde{U}_2$	(TI, S^6)	2	6	912
$\tilde{U}_3$	(T^6, R)	2	6	912
$\tilde{U}_4$	(T^6, S^6)	2	1	130
$\tilde{U}_5$	(I, SR)	2	36	942
$\tilde{U}_6$	(TI, R)	2	36	5 649
$\tilde{U}_7$	(T^4, S^4)	3	2	11
$\tilde{U}_8$	(T^3, S^3)	4	2	2
$\tilde{U}_9$	$(TI, S^6), (T^6, R)$	4	36	96
$\tilde{U}_{10}$	$(I, SR), (T^6, S^6)$	4	18	12
$\tilde{U}_{11}$	$(TI, R), (T^6, S^6)$	4	18	42
$\tilde{U}_{12}$	$(I, SR), (T^4, S^4)$	6	24	2
$\tilde{U}_{13}$	$(TI, R), (T^4, S^4)$	6	24	15
$\tilde{U}_{14}$	$(I, SR), (T^3, S^3)$	8	36	6
$\tilde{U}_{15}$	$(TI, R), (T^2, S^2)$	12	12	2
$\tilde{U}_{16}$	$(I, SR), (T, S)$	24	12	1
$\tilde{U}_{17}$	$(I, SR), (T, S^5)$	24	12	1

The Interval Structure of a Tone Row

The *interval* from pitch class a to pitch class b for $a, b \in \mathbb{Z}_{12}$ is defined as the difference $b - a$ as an element of $\mathbb{Z}_{12}$. This is the minimum number of steps in clockwise direction from a to b in the regular 12-gon of Fig. 6. The tone row $f\colon \{1, \ldots, 12\} \to \mathbb{Z}_{12}$ determines the following sequence of eleven intervals

$$(f(2) - f(1), f(3) - f(2), \ldots, f(12) - f(11)). \qquad (*)$$

Since in our main setting (situation 5 of Table 3) we consider a tone-row as a closed polygon we also have to add the closing interval $f(1) - f(12)$. Consequently, the *interval structure* of the tone-row $f\colon \{1, \ldots, 12\} \to \mathbb{Z}_{12}$ is the function $g\colon \{1, \ldots, 12\} \to \mathbb{Z}_{12} \setminus \{0\}$, defined by

$$g(i) := \begin{cases} f(i+1) - f(i) & \text{for } 1 \leq i \leq 11 \\ f(1) - f(12) & \text{for } i = 12. \end{cases}$$

Let g be the interval structure of f. It is easy to check that the interval structures of all tone rows in the $D_{12} \times D_{12}$-orbit of the tone row f correspond to the $\langle I \rangle \times D_{12}$-orbit of the interval structure g of f. Here we have the natural action of $\langle I \rangle \times D_{12}$ on the set of all functions from $\{1, \ldots, 12\}$ to $\mathbb{Z}_{12} \setminus \{0\}$ as defined in (3).

The vector $(a_1, \ldots, a_{11})$ is called the *interval type* of f if for each $j \in \{1, \ldots, 11\}$ the interval j occurs exactly a_j times among $g(1), \ldots, g(12)$. Obviously, we have $\sum_{j=1}^{11} a_j = 12$.

A tone row f is called an *all-interval row* if all elements of $\mathbb{Z}_{12} \setminus \{0\}$ occur in the sequence (*). Then each element of $\mathbb{Z}_{12} \setminus \{0\}$ occurs exactly once in (*). Hence, $\{g(j) \mid 1 \leq j \leq 11\} = \{1, \ldots, 11\}$ and, therefore,

$$g(12) = -\sum_{j=1}^{11} g(j) = -\sum_{i \in \mathbb{Z}_{12} \setminus \{0\}} i = 6.$$

The $D_{12} \times D_{12}$-orbit of f contains all-interval rows if and only if each element of $\mathbb{Z}_{12} \setminus \{0\}$ occurs in the interval structure of f. In this situation the interval 6 occurs exactly twice and all other intervals exactly once in the interval structure of f. Thus the interval type of f looks like $(1, 1, 1, 1, 1, 2, 1, 1, 1, 1, 1)$. In this situation we call the orbit $(D_{12} \times D_{12})(f)$ an all-interval orbit.

Let f be a tone row where the interval structure of f contains all possible values from $\mathbb{Z}_{12} \setminus \{0\}$. An element h of the $D_{12} \times D_{12}$-orbit of f is an all-interval row in the common sense if and only if the closing interval $h(1) - h(12)$ is equal to 6. From our database on tone rows we deduce that there are exactly 519 $D_{12} \times D_{12}$-all-interval-orbits of tone rows. They are either of stabilizer type $\tilde{U}_1$, $\tilde{U}_2$ or $\tilde{U}_3$.

As a generalization of all-interval rows we want to propose the following rows: Consider $i, j \in \mathbb{Z}_{12}$. The interval from i to j was defined as $j - i \in \mathbb{Z}_{12}$, thus it depends on the direction from i to j. The *distance* $d(i, j) \in \mathbb{Z}_{12}$ between i and j is defined as follows. If $i - j \in \{0, \ldots, 6\}$ then $d(i, j) = i - j$, otherwise $d(i, j) = j - i$. Therefore $d(i, j) \in \{0, \ldots, 6\}$ and $d(i, j) = d(j, i)$. This is the minimum number of steps in clockwise or anti-clockwise direction from i to j in the regular 12-gon of Fig. 6.

Consider a tone-row f and determine the list $(d_1, \ldots, d_{12})$ of distances between consecutive tones of f, i.e. determine $d_i := d(f(i), f(i+1))$ for $i \in \{1, \ldots, 11\}$ and $d_{12} := d(f(1), f(12))$. These are all together twelve distances. The vector $(a_1, \ldots, a_6)$ is called the *distance type* of f if for each $j \in \{1, \ldots, 6\}$ the distance j occurs exactly a_j times among $d_1, \ldots, d_{12}$. Obviously, we have $\sum_{j=1}^{6} a_j = 12$.

If f is an all-interval row, then the interval structure of f contains each interval once with exception of 6 which occurs twice. Therefore, the distance type of f looks like $(2, 2, 2, 2, 2, 2)$. We call an arbitrary tone row an *all-distances-twice row* if its distance type is of the form $(2, 2, 2, 2, 2, 2)$. This notion can easily be generalized to tone rows in $\mathbb{Z}_n$ for even n.

Our database shows that there are exactly 4 162 $D_{12} \times D_{12}$-orbits of all-distances-twice rows. They occur in 27 different interval types.

Tropes

A *hexachord* in the 12-scale $\mathbb{Z}_{12}$ is a 6-subset of $\mathbb{Z}_{12}$. There exist $\binom{12}{6} = 924$ different hexachords in the set $\mathcal{H} = \{A \subset \mathbb{Z}_{12} \mid |A| = 6\}$.

Let A be a hexachord, then its complement $A' := \mathbb{Z}_{12} \backslash A$ is also a hexachord. Now we consider "pairs" $\{A, A'\}$ of hexachords which we call *tropes*. (We use quotation marks around the word pair, since $\{A, A'\}$ is actually not a pair, but a 2-set of hexachords!)

The set of all tropes will be indicated by $\mathcal{T} := \{\{A, \mathbb{Z}_{12} \backslash A\} \mid A \in \mathcal{H}\}$. In total there exist $924/2 = 462$ tropes in the 12-scale. If a group G acts on $\mathbb{Z}_{12}$, then the induced action of $g \in G$ on the set of tropes is given by $g * \{A, A'\} := \{g * A, g * A'\}$, $\{A, A'\} \in \mathcal{T}$.

This way we obtain the numbers of G-orbits of tropes presented in Table 5. For more details on the enumeration of tropes see e. g. [7, 8].

A complete list of D_{12}-orbits of tropes was given in Table 1. For each orbit we present the standard representative $f \in \{0, 1\}_{6,6}^{\mathbb{Z}_{12}}$, which is the lexicographically smallest element in the orbit $(S_2 \times D_{12})(f)$. Moreover we provide a graphical representation of f as a coloring of the 12-scale given in Fig. 6 with two colors. In its center we indicate the name of the orbit of the trope, which is a number from the set $\{1, \ldots, 35\}$. These numbers are called *trope numbers* or *number of the D_{12}-orbit* of a trope.

The conjugacy classes of subgroups of D_{12} are displayed in Table 6. Tropes with number 4, 20, 24, 27 have stabilizer type $\tilde{V}_2$, whereas 2, 5, 9, 13, 15, 17, 21, 23, 26, 28, 29, 30, 11, 12, 33 have stabilizer type $\tilde{V}_3$. Moreover 14 is of stabilizer type $\tilde{V}_4$ and 1, 8, 31, 32 of stabilizer type $\tilde{V}_5$. The stabilizer types of 25, 34, and 35 are $\tilde{V}_{12}$, $\tilde{V}_{15}$, respectively $\tilde{V}_{16}$.

Consider a tone row f which is a bijective mapping from $\{1, \ldots, 12\}$ to $\mathbb{Z}_{12}$. We obtain in a natural way a "pair" of hexachords defined by f by taking the sets of the first six and the last six pitch classes of f,

$$\tau_1 := \big\{\{f(1), f(2), f(3), f(4), f(5), f(6)\},\\ \{f(7), f(8), f(9), f(10), f(11), f(12)\}\big\}.$$

Similarly, we deduce the "pairs" of hexachords defined by the cyclic shifts $f \circ S$, $f \circ S^2, \ldots, f \circ S^5$ and obtain the tropes $\tau_2, \ldots, \tau_6$. The further shifts $f \circ S^6, \ldots, f \circ S^{11}$ again yield the tropes $\tau_1, \ldots, \tau_6$, where the two hexachords of each trope are just interchanged. Therefore, a tone row f induces a *trope sequence* $t_f \colon \{1, \ldots, 6\} \to \mathcal{T}$,

Table 5 Number of G-orbits of tropes

G	$\lvert(S_2 \times G)\backslash\backslash\{0, 1\}_{6,6}^{\mathbb{Z}_{12}}\rvert$
C_{12}	44
D_{12}	35
$\mathrm{Aff}_1(\mathbb{Z}_{12})$	26

Table 6 Conjugacy classes of subgroups of D_{12}

Name	Generators	$\lvert V_i \rvert$	$\lvert \tilde{V}_i \rvert$
$\tilde{V}_1$	$\langle 1 \rangle$	1	1
$\tilde{V}_2$	$\langle I \rangle$	2	6
$\tilde{V}_3$	$\langle TI \rangle$	2	6
$\tilde{V}_4$	$\langle T^6 \rangle$	2	1
$\tilde{V}_5$	$\langle T^4 \rangle$	3	1
$\tilde{V}_6$	$\langle T^3 \rangle$	4	1
$\tilde{V}_7$	$\langle T^6, I \rangle$	4	3
$\tilde{V}_8$	$\langle T^6, TI \rangle$	4	3
$\tilde{V}_9$	$\langle T^2 \rangle$	6	1
$\tilde{V}_{10}$	$\langle T^4, I \rangle$	6	2
$\tilde{V}_{11}$	$\langle T^4, TI \rangle$	6	2
$\tilde{V}_{12}$	$\langle T^3, I \rangle$	8	3
$\tilde{V}_{13}$	$\langle T \rangle$	12	1
$\tilde{V}_{14}$	$\langle T^2, I \rangle$	12	1
$\tilde{V}_{15}$	$\langle T^2, TI \rangle$	12	1
$\tilde{V}_{16}$	$\langle T, I \rangle$	24	1

$t_f(i) = \tau_i$, $1 \leq i \leq 6$. If we replace in the trope sequence of f the tropes by the numbers of their D_{12}-orbits, we obtain a function $s_f : \{1, \ldots, 6\} \to \{1, \ldots, 35\}$, the *trope number sequence*, where $s_f(i)$ is the number of the orbit $D_{12}(\tau_i)$, $1 \leq i \leq 6$.

It is easy to observe that the $D_{12} \times D_{12}$-orbit of the tone row f coincides with the D_{12}-orbit of s_f where the dihedral group D_{12} acts on the domain of s_f as introduced in (1). Moreover, the permutation representation of D_{12} on the set $\{1, \ldots, 6\}$ is the dihedral group D_6 generated by the cyclic shift $(1, 2, 3, 4, 5, 6)$ and the retrograde $(1, 6)(2, 5)(3, 4)$. We call this orbit of s_f the *trope structure* of the orbit $(D_{12} \times D_{12})(f)$. The D_{12}-orbit (or D_6-orbit) of s_f is represented by the smallest element with respect to the lexicographical order.

From the database it is possible to deduce that there are 538 139 different trope structures. There are trope structures, e. g. $(1, 1, 1, 1, 1, 1)$, which determine a unique $D_{12} \times D_{12}$-orbit of tone rows. But there exist also two trope structures namely (10, 18, 22, 14, 22, 18) and (10, 18, 22, 14, 22, 27) which belong to 48 different $D_{12} \times D_{12}$-orbits of tone rows.

So far we have explained how a tone row defines a trope structure. Conversely, given a function $h : \{1, \ldots, 6\} \to \{1, \ldots, 35\}$ we investigate whether there exists a tone row f so that $D_{12}(h)$ is the trope structure of the orbit $(D_{12} \times D_{12})(f)$.

First we analyze when two (numbers of) orbits of tropes can occur in consecutive places. In this situation we call the two (numbers of orbits of) tropes *connectable*.

Table 7 Number of $D_{12} \times D_{12}$-orbits of tone rows with given number of distinct trope numbers

Number of distinct trope numbers	1	2	3	4	5	6
Number of orbits of tone rows	4	276	5 251	60 196	290950	479 340

Two tropes $\tau_1 = \{A_1, A'_1\}$ and τ_2 are connectable if there exist $i \in A_1$ and $j \in A'_1$ so that $\tau_2 = \{(A_1 \setminus \{i\}) \cup \{j\}, (A'_1 \setminus \{j\}) \cup \{i\}\}$. The "pair" $\{i, j\}$ is called *pair of moving elements* or shorter *moving pair* between τ_1 and τ_2.

Two (numbers of) orbits of tropes are connectable if there exist representatives τ_1 and τ_2 of these orbits which are connectable.

Given two connectable tropes $\{A, A'\}$ and $\{B, B'\}$ we form the four intersections $A \cap B$, $A \cap B'$, $A' \cap B$, and $A' \cap B'$. Exactly two of them have cardinality 5 and two of them have cardinality 1. The two elements belonging to the 1-sets form the moving pair between the two tropes.

In Table 2 we determine the number of possible connections (i. e. the number of moving pairs) between any two tropes. For computing the ith row of this table we choose one representative $\tau = \{A, A'\}$ of the ith D_{12}-orbit of tropes, determine all 36 possible moving pairs of the form $\{k, \ell\}$, $k \in A$, $\ell \in A'$. Then we construct all tropes of the form $\{(A \setminus \{k\}) \cup \{\ell\}, (A' \setminus \{\ell\}) \cup \{k\}\}$ and count which D_{12}-orbits these tropes belong to. (If there are no moving pairs from a trope belonging to the ith orbit to a trope of the jth orbit, then the corresponding field in the table is left empty.) For example the first line indicates that from a trope of the first D_{12}-orbit there are two moving pairs leading to tropes of the first D_{12}-orbit, 6 moving pairs leading to tropes of the second D_{12}-orbit and so on. Hence, the numbers in each line sum up to 36.

The number of different trope numbers occurring in the trope number sequence is also an interesting property of tone rows. According to Florey [6] it is a measure for the quality of a tone row (Table 7).

Theorem 5 *There exists a tone row f so that $\sigma: \{1, \ldots, 6\} \to \{1, \ldots, 35\}$ is the trope number sequence of f, if and only if for $1 \le r \le 6$ there exists a representative τ_r of the $\sigma(r)$-th D_{12}-orbit of tropes, so that*

- *τ_r and τ_{r+1} are connectable with the moving pair $\{i_r, j_r\}$, $1 \le r \le 5$, and*
- *τ_6 and τ_1 are connectable with the moving pair $\{i_6, j_6\}$, and*
- *each element of $\mathbb{Z}_{12}$ is moving exactly once, i. e.*

$$\bigcup_{r=1}^{6} \{i_r, j_r\} = \mathbb{Z}_{12}.$$

Proof If f is a tone row, then it is clear from the construction that the assertions on the trope number sequence $\sigma = s_f$ are satisfied.

Conversely, assume that σ is a function with the given properties. We have to find a tone row f with trope number sequence σ. For $1 \le r \le 6$ we write τ_r as $\{A_r, A'_r\}$. Without loss of generality we have:

- $i_r \in A_r$ and $j_r \in A'_r$, $1 \le r \le 6$,

- $A_r \setminus \{i_r\} \subseteq A_{r+1}, A'_r \setminus \{j_r\} \subseteq A'_{r+1}, 1 \leq r \leq 5$.

Then $A_6 \setminus \{i_6\} \subseteq A'_1$, $A'_6 \setminus \{j_6\} \subseteq A_1$, and $A_1 = \{i_1, \ldots, i_6\}$, $A'_1 = \{j_1, \ldots, j_6\}$. Finally, the sequence $(i_1, \ldots, i_6, j_1, \ldots, j_6)$ is a tone row which determines the trope sequence $(\tau_1, \ldots, \tau_6)$ and the trope number sequence σ. □

There is a close connection between the stabilizer type of a tone row and its trope structure. Here we present just one result. For more details see [11].

Theorem 6 *Let f be a tone row. The pairs (TI, S^6) and (T^6, R) belong to the stabilizer of f, if and only if the following assertions hold true.*

- *f has exactly four different trope numbers which belong to the set $M = \{1, 2, 5, 8, 9, 13, 15, 17, 21, 23, 25, 26, 28, 29, 30, 31, 32, 34, 35\}$.*
- *The trope number sequence of f is of the form $(t_1, t_2, t_3, t_4, t_3, t_2)$, where t_1 belongs to $\{1, 8, 31, 34\}$, which are the numbers of those tropes $\{A, A'\}$ so that $T^6(A) = A'$ and $TI(A) = A'$, and t_4 is an element of $\{25, 32, 35\}$, which are the numbers of those tropes so that $T^6(A) = A$ and $TI(A) = A'$.*
- *There exists a trope sequence $(\tau_1, \ldots, \tau_6)$ where τ_r belongs to the t_r-th D_{12}-orbit of tropes, $1 \leq r \leq 6$, which satisfies the properties of Theorem 5, $TI(\tau_r) = \tau_r$, $1 \leq r \leq 6$, and $\tau_6 = T^6(\tau_2)$ and $\tau_5 = T^6(\tau_3)$.*

Figure 9 was designed by Peter Lackner already in the year 1988. It shows the trope number sequences of all 96 orbits of tone rows of stabilizer type $\tilde{U}_9$. We should explain how to read this table. The first section describes the six trope number sequences: (2, 1, 2, 9, 25, 9), (2, 1, 2, 15, 25, 15), (5, 1, 5, 23, 25, 23), (5, 1, 5, 9, 25, 9), (13, 1, 13, 23, 25, 23), (13, 1, 13, 15, 25, 15).

Exchanging the Parameters

If we have a look at the tone rows of stabilizer type $\tilde{U}_2$ or $\tilde{U}_3$ we realize that in both situations there exist exactly 912 different $D_{12} \times D_{12}$-orbits of tone rows. As we will soon see, there is a simple way to construct from a tone row invariant under (TI, S^6) a tone row invariant under (T^6, R) and vice versa. Here it is useful to consider tone rows as bijective mappings from $\{1, \ldots, 12\}$ to itself, whence they are permutations of $\{1, \ldots, 12\}$, thus elements of S_{12}. The permutations T and S are now identified with the cyclic permutation $\zeta = (1, 2, 3, 4, 5, 6, 7, 8, 9, 10, 11, 12)$. The inversion TI without fixed points, and the retrograde R are identified with $\rho = (1, 12)(2, 11)(3, 10)(4, 9)(5, 8)(6, 7)$. Finally the quart-circle Q and the five-step F are identified with $\varphi = (1, 5)(2, 10)(3)(4, 8)(6)(7, 11)(9)(12)$. If ζ, ρ or φ are permutations of the domain of tone rows, then they are considered as cyclic shift S, retrograde F, or five-step F. If they are permutations of the range, then they correspond to transposition T, inversion TI, or quart-circle Q. The following theorem holds true:

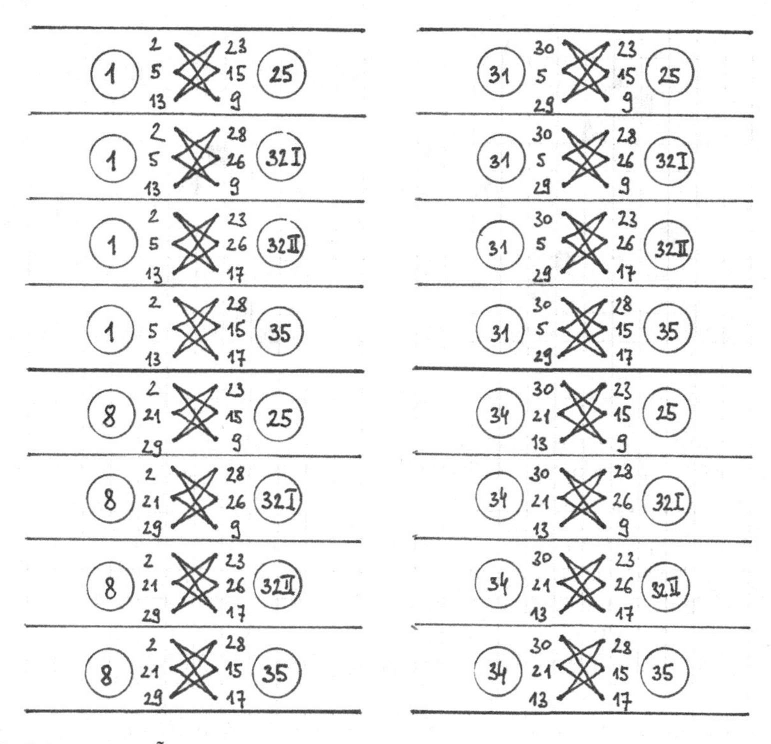

Fig. 9 Orbits of type $\tilde{U}_9$

Theorem 7 *The tone row $f \in S_{12}$ is invariant under (TI, S^6) if and only if f^{-1}, the inverse permutation of f, is invariant under (T^6, R).*

Proof The tone row f is invariant under (TI, S^6), i.e. $\rho \circ f \circ \zeta^6 = f$, if and only if $f^{-1} = (\rho \circ f \circ \zeta^6)^{-1} = \zeta^6 \circ f^{-1} \circ \rho^{-1}$, which means that f^{-1} is invariant under (T^6, R). □

Switching from f to f^{-1} means interchanging the two parameters time and pitch of a tone row. Let us use the convention of music notation that the time parameter is indicated horizontally on the x-axis and the pitch parameter vertically on the y-axis, then a tone row is described by a 12×12-matrix containing in each column and in each row exactly one 1 and eleven 0s. The 1 stands in the i-th row of the j-th column if and only if $f(j) = i$. E.g., the tone row f given in its vector representation $(f(1), \ldots, f(12)) = (1, 2, 5, 10, 11, 9, 4, 3, 12, 7, 6, 8)$ is represented as the matrix[17] M_f where 1 is replaced by a black and 0 by a white square.

[17] Already in 1924 (cf. [15]) Hauer introduced this matrix representation for tone rows.

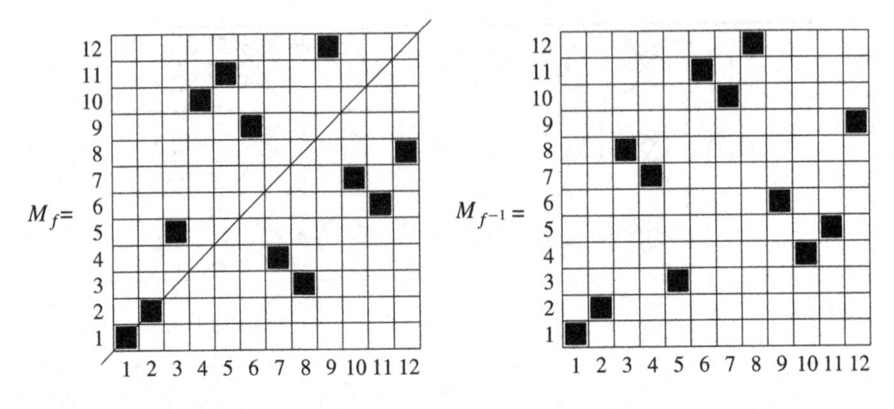

The inverse f^{-1} of f corresponds to the matrix which can be obtained by reflecting the matrix M_f along the diagonal from right up to left down. This gives the matrix $M_{f^{-1}}$ and the tone row $f^{-1} = (1, 2, 8, 7, 3, 11, 10, 12, 6, 4, 5, 9)$. Since f had stabilizer type $\tilde{U}_2$ this tone row has stabilizer type $\tilde{U}_3$. If f has stabilizer type $\tilde{U}_9$, then f^{-1} also has stabilizer type $\tilde{U}_9$. In this case, it is possible that either f and f^{-1} belong to the same or to two different $D_{12} \times D_{12}$-orbits. E.g., if $f = (1, 2, 3, 4, 11, 5, 10, 9, 8, 7, 12, 6)$, then $f^{-1} = (1, 2, 3, 4, 6, 12, 10, 9, 8, 7, 5, 11)$ which represent two different orbits of stabilizer type $\tilde{U}_9$, whereas, if $f = (1, 2, 3, 4, 5, 6, 12, 11, 10, 9, 8, 7)$, then $f^{-1} = f$.

The mapping $P: \mathcal{R} \to \mathcal{R}, f \mapsto f^{-1}$, is called the *exchange of parameters*. Now we study the actions of $\mathfrak{D}_{12}$ which is the smallest permutation group on $\mathcal{R}$ containing the action of $D_{12} \times D_{12}$ and the exchange of parameters P, and the action of $\mathfrak{A}_{12}$ which is the smallest permutation group on $\mathcal{R}$ containing the action of $\mathrm{Aff}_1(\mathbb{Z}_{12}) \times \mathrm{Aff}_1(\mathbb{Z}_{12})$ and the exchange of parameters P. Extending Table 3 we obtain the following numbers of orbits of tone rows given in Table 8. For more details see [11].

The stabilizer types, normal forms and orbits of tone rows under the actions of $\mathfrak{D}_{12}$ or $\mathfrak{A}_{12}$ are thoroughly described in the database. E. g., there exist 31 respectively 93 stabilizer types of $\mathfrak{D}_{12}$- respectively $\mathfrak{A}_{12}$-orbits of tone rows. Only 94.65 % of all $\mathfrak{A}_{12}$-orbits of tone rows have trivial stabilizer.

Table 8 Number of orbits under various group actions on the set of all tone rows

	Acting group	# of orbits
(9)	$\mathfrak{D}_{12}$	420948
(10)	$\mathfrak{A}_{12}$	106986

The Database

The *Database on tone rows and tropes* is publicly available[18] via the address http://www.uni-graz.at/fripert/db/. The main objects in this database are the $D_{12} \times D_{12}$-orbits of tone rows.

With the included software it is possible to compute the orbit, the stabilizer, or the normal form of a tone row. The database contains information on the 836 017 $D_{12} \times D_{12}$-orbits of tone rows. It is possible to search for or to retrieve information on the normal forms of tone rows, the interval structure of tone rows, the trope structure of tone rows, tone rows with prescribed stabilizer type, all-interval rows, orbits invariant under the quart-circle, orbits invariant under the 5-step, and orbits invariant under the parameter exchange.

In addition to the action of $D_{12} \times D_{12}$ we also study the actions of other groups on the set of all tone rows. Among these there are the following five groups which all contain $D_{12} \times D_{12}$ as a proper subgroup:

- $\mathrm{Aff}_1(\mathbb{Z}_{12}) \times D_{12}$ which contains the quart-circle Q.
- $D_{12} \times \mathrm{Aff}_1(\mathbb{Z}_{12})$ which contains the 5-step F.
- $\mathrm{Aff}_1(\mathbb{Z}_{12}) \times \mathrm{Aff}_1(\mathbb{Z}_{12})$ which contains the 5-step F and the quart-circle Q.
- $\mathfrak{A}_{12}$ which contains the 5-step F, the quart-circle Q, and the exchange of parameters P.
- $\mathfrak{D}_{12}$ which contains the exchange of parameters P.

Using the database the decomposition of the $D_{12} \times D_{12}$-orbit of f into the $D_{12} \times \langle R \rangle$-orbits (Schönberg-situation) can be determined.

Moreover we were collecting musical information on tone rows appearing in works of various composers. Hence it is also possible to search for musical information on a given tone row. This opens the door for new research: Since we have normal forms of tone rows, it is easy to check, whether similar tone rows appeared in different compositions. Or knowing certain properties of tone rows it is interesting to study whether we can deduce from the composition that the composer was aware of these properties.

As a matter of fact, at the moment we have more than 1 200 entries of musical information in our database. Of course this is not enough for doing some statistical analysis or to suggest trends in the usage of certain types of tone rows. Therefore, we try to collect further tone rows and data.

There are more than 500 entries with tone rows by Hauer. All tone rows from the Second Viennese School and a selection of compositions until today are input.

A manual describing the interaction with the interface will be published as [10].

[18] Accessed July 5, 2014.

References

1. Colbourn CJ, Read RC (1979) Orderly algorithms for graph generation. Int J Comput Math 7:167–172
2. de Bruijn NG (1964) Pólya's theory of counting. In: Beckenbach EF (ed) Applied combinatorial mathematics, Chap. 5. Wiley, New York, pp 144–184
3. de Bruijn NG (1971) A survey of generalizations of Pólya's enumeration theorem. Nieuw Archief voor Wiskunde 2(XIX):89–112
4. de Bruijn NG (1972) Enumeration of mapping patterns. J Comb Theory (A) 12:14–20
5. Florey H (1965/67) Tropentafel mit drehbaren Farbkreisen (Neuordnung der Tropen in Gestalt von 35 Grundbildern nach dem Verhältnis der Tropenhälften zu Gegenhälfte und Spiegelbild, woraus sich 5 Tropen-Typen ableiten). Picture, Acryl/Papierelemente, Karton; Plexiglas, Novopan, Schraubverbindungen 98 × 67 cm
6. Florey H (1988) Analytische Bemerkungen zu Josef Matthias Hauers letztem Zwölftonspiel. Beilage zu einer Schallplatte herausgegeben von der Hochschule für Musik und darstellende Kunst in Graz
7. Fripertinger H (1992) Enumeration in musical theory. In: Beiträge zur Elektronischen Musik Heft 1. Hochschule für Musik und Darstellende Kunst, Graz
8. Fripertinger H (1992) Enumeration in musical theory. Séminaire Lotharingien de Combinatoire 476(S–26):29–42
9. Fripertinger H (1993) Endliche Gruppenaktionen auf Funktionenmengen. Das Lemma von Burnside–Repräsentantenkonstruktionen–Anwendungen in der Musiktheorie. Bayreuth Math Schr 45:19–135
10. Fripertinger H, Lackner P (2015) Database on tone rows and tropes, a short user's guide. J Math Music (to appear)
11. Fripertinger H, Lackner P (2015) Tone rows and tropes. J Math Music (to appear)
12. GAP—Groups, Algorithms, and Programming, Version 4.5.7 (2012) The GAP Group. http://www.gap-system.org
13. Golomb SW, Welch LR (1960) On the enumeration of polygons. Am Math Mon 87:349–353
14. Harary F, Palmer EM (1966) The power group enumeration theorem. J Comb Theory 1:157–173
15. Hauer JM (1924) Zur Lehre vom atonalen Melos. In: Westheim P (ed) Das Kunstblatt, vol 8(12), pp 353–360
16. Hunter DJ, von Hippel PT (2003) How rare is symmetry in musical 12-tone rows? Am Math Mon 110(2):124–132
17. Ilomäki T (2008) On the similarity of twelve-tone rows. Studia Musica, vol 30. Sibelius Academy, Helsinki
18. Kerber A (1999) Applied finite group actions. Algorithms and combinatorics, vol 19. Springer, Berlin
19. Pólya G (1937) Kombinatorische Anzahlbestimmungen für Gruppen, Graphen und chemische Verbindungen. Acta Mathematica 68:145–254
20. Pólya G, Read RC (1987) Combinatorial enumeration of groups, graphs and chemical compounds. Springer, New York
21. Read RC (1978) Every one a winner or how to avoid isomorphism search when cataloguing combinatorial configurations. Ann Discret Math 2:107–120
22. Read RC (1994) Four new mathematical problems in music theory. An essay written in fulfilment of the requirements for the A.R.C.T. In: Theory of the royal conservatory of music, Toronto
23. Read RC (1997) Combinatorial problems in the theory of music. Discret Math 167–168(1–3):543–551
24. Reiner DL (1985) Enumeration in music theory. Am Math Mon 92:51–54
25. Rowan JP (1949) The soul, a translation of St. Thomas Aquinas' De anima. B. Herder Book Co, St. Louis

26. Sims CC (1970) Computational methods in the study of permutation groups. In: Leech J (ed) Computational problems in abstract algebra. Proc Conf, Oxford, 1967. Pergamon Press, Oxford, pp 169–183
27. Skreiner W (1981) Hans Florey: Farbtotalität in 35 Grundbildern: Neue Galerie am Landesmuseum Joanneum, 9. Juli-27. August 1981, Graz
28. SYMMETRICA—A program system devoted to representation theory, invariant theory and combinatorics of finite symmetric groups and related classes of groups (1987). Lehrstuhl II für Mathematik Universität Bayreuth, Bayreuth. http://www.algorithm.uni-bayreuth.de/en/research/SYMMETRICA/

Interdisciplinary Contributions

Artistic Research in/as Composition: Some Case Notes

Darla Crispin

Introduction

The many case studies presented within the POINT project invite a re-evaluation of aspects of the critical milieu within which the materials are presented. The project is situated in a domain of practical music scholarship that is being shaped by the continuing debates around artistic research and its viability as an organisational concept. More recently, concerns about the nature of judgment in artistic research, and important extrapolations into its relationship with ethics, are being brought into the foreground within arts research and its dissemination platforms.

This chapter outlines some speculations on these developments, first through revisiting aspects of the social context, and then through exploring some new models that pertain to specific aspects of case studies within POINT, as well as to the possible relationship of aesthetics, formalism and ethics in artistic research in music.

Composition and the Academy: The 'Traditional' Model and Current Contexts

Within the evolution of the arguments concerning practice-based research, practice-as-research and artistic research, composition has had a strong base from which to make its claims of legitimacy and, therefore, primacy over performance. The argument goes that creative, as opposed to performative, practice seems self-evidently original; in notated music, at least, it leaves a durable outcome; and, with a small imaginative stretch, it can be seen as contributing to 'knowledge and understanding', provided these terms are understood broadly.

D. Crispin (✉)
Norwegian Academy of Music, Oslo, Norway
e-mail: darcrisp@msn.com

G. Nierhaus (ed.), *Patterns of Intuition*,
DOI 10.1007/978-94-017-9561-6_14

This has meant that composers have been generally more confident than their performing counterparts in leading the campaign for recognition—above all, for recognition within the academic domain as a species of researcher. Furthermore, long before the contemporary arguments about its research status, composition has been assured of its place in the academy because of its historical position as one of the key activities engaged in by distinguished university musicians. The conferring of doctorates of music on the basis of substantial compositional 'test-pieces' has a long tradition in many countries, and it is one that survives in the domain of 'honorary doctorates' for distinguished composers, alongside the newer apparatus of qualifications that treat composition as a practice-based research activity.

Growing Challenges

Even though this is a position that performers might envy—and some honorary doctorates do now go to the most distinguished performers—it has also been a source of criticism, in that 'academic' composition has been, and continues to be, vulnerable to accusations of wider social irrelevance. Nevertheless, within Western art music, the dominance of composition remains, not least because those musician-researchers who do now seek to express their research through performance have not only the 'scientific sceptics' but also the composer hegemony—as manifested in the persistent model of *Werktreue*—to contend with. Composition has historically been able to escape this double-bind, both through focusing upon the constructive principles within the music itself, and, in some cases, through making ethical claims about a specific compositional practice.

The hegemony of the old model is stubborn, as the agents within the composer-performer-audience matrix retain a framework of habits that conditions how most music is made, transmitted and received. However, the hermetically-sealed nature of these claims is increasingly being challenged, not least by the incursion of improvisational and co-creative practices into the arena of composition, some of which are discussed within this volume. In actuality, the model of *Werktreue* is a shorthand that may be illuminating about aspects of Western art music of the past, but becomes less so as we seek to understand the altogether more complex and multi-faceted realities of current compositional practice. In contemporary Western art music, there is no single practice that dominates; yet, a fragile rhetoric of unity persists.

The seemingly easy transition of thought from compositional practice to artistic research is further problematized because much practice has been seen to produce not only questionable research but also questionable composition. At either end of the spectrum of possibilities for composition, an extreme can be identified and described, each with its problems, see Table 1.

Table 1 A sketch of Western art music composition, from conception to reception

Type	Composition as a function of the unique situation surrounding each compositional project. May be systematic but more often contingent. Composer is often, though not always, the performer/practitioner and may use improvisation as part of the creative process (Generative music)	Composition embodying a consistent, although often highly personal, aesthetic universe. Often highly systematized and encoded in complex systems, yet highly referential, using tools of science, but not scientifically
Research problem	Research that resides in the specificity of embodied experience (as in freely improvised composition, for example) is difficult to generalize out to a wider public—it is self-referential, with few or no common references	The system of composing, when developed as a research-related methodology, can act as a 'stand-in' for the creative 'spark' of the compositional process. Means and ends become confused
Artistic problem	Can tend to reinforce notions of poorly-formed, unstructured, dull, real-time events, or 'shallow' compositions (often buttressed by unwieldy superstructures of abstract theorizing)	Can tend to reinforce old problems of the artistic community's perception of an over-privileged/enfranchised, 'ivory-tower' school of scholastic composition, lacking in true communicative merit, but able to 'play the research game'
Reception problem	General public may be indifferent to the existence of this music; it also attracts composer-practitioners who may be ill-at-ease in trying to articulate their embodied knowledge, resorting to silence	General public might be indifferent to the existence of this music; it also attracts composers who may substitute argument for artistic ability and whose superficial ability to 'talk the talk' of academic discourse can lead to 'academic imposture', see [5]

Revisiting the 'Social Situation'

Both of the strands described above have contributed to a sense of crisis in 'high art' composition today. Were this problem simply to do with the concerns of a 'niche' practice for composers, performers and listeners, then the issue would be about how this practice would find its modest place in the larger cultural scheme. But there is still power and prestige attached to 'new music', linked with status positions in educational, social and cultural institutions. As long as these remain predominantly associated with elite, white, male privilege and dominance, there will be pertinent questions to ask concerning the social function of this practice and those institutions that, seemingly unquestioningly, facilitate its continuity.

Of course, not all within the academy have been blind to these matters—far from it; self-critical arguments have certainly been put forward within ethnomusicology and 'New Musicology', and as an adjunct of the practice/artistic turn. But these, themselves, have sometimes become a source of rhetoric that defuses a re-evaluation of aesthetic matters (even within the dominating spheres which they often purport to critique). Artistic research should be addressing these problems—perhaps especially within the conservatoire, where notions of what constitutes 'progress' still often differ widely from those that pertain within the universities. Without this evolving scrutiny from within practice, and cut adrift from wider meaning, music can either seem ubiquitous and mindlessly consumed or be rendered inaccessible. Indeed, there are cases where both dangers apply—where, a fundamentally arcane music is becoming additionally compromised through its zeal to market its products through the mechanisms of pop culture.

In the presentation of music, social agendas are often tied together with those surrounding reception. From this we would have to surmise that the generally negative reception of contemporary music has some kind of social, as well as artistic, signification. This rupture is complex; accounting for it is often confusing. 'Truth' and 'progress' in art surely cannot be pushed forward forever by a tiny, elite group—the age of a thriving, dynamic artistic 'avant-garde' is now as much a historical phenomenon as, say, that of romanticism; nor should defining the artistic 'zeitgeist' be the sole prerogative of those who 'refuse' to encounter the difficult. 'Reception' goes beyond those who actually engage with works; even withdrawal from the sphere of engagement becomes an aspect of reception.

All of this seems to function in a manner that hardens the stratification between musics and exacerbates the notion that different types of music belong to different social groups. Perceiving these in hierarchical terms is the unfortunate, but inevitable, result. The elevation of a set of rules of engagement for these varied strata then follows, and elite music becomes articulated in elite behaviour. But, as already suggested, the abandonment of 'difficulty' within the contemporary idiom is no solution. It remains, and thus becomes part of the 'body of culture/knowledge' that must be understood within a holistic reading. This reading goes beyond the practice and nature of composition itself, becoming a lens through which the inscribed, dialectical processes of history are studied through the composition:

> Associating composers' public utterances with the complex problems they tried to solve any given time would make it even clearer that the principle of autonomy, rather than having suffered an external attack from the joint forces of the cultural industry and postmodernist music, has always been involved in a dialectical interplay with its opposite. This dialectic in turn can be seen as a segment of a historical reality whose investigation requires *a reflection on the process of modernity in music that needs to be more thoroughgoing than it has been hitherto* [2, p. 183], italics mine.

Of Resistance and Consumption

In reality, there is a myriad of choices to be made by the truly attentive listener, of which some will be preferred and some will not. Resistance to some contemporary music does not invariably imply indifference to all its manifestations, or even to its fundamental aspirations; the equation does not work as crudely as this. On the other hand, the social claims of pretty much all the more 'serious' contemporary music can look quite shallow when contrasted with music that is contemporary in a more inclusive sense. This would seem to be one of the many cases that it is possible to make in attacking the exalted status of certain types of music, their listeners and listening practices.

The linkage between complexity in art and the 'ownership' of such art by a dominant cultural intelligentsia remains, and its stubborn hold on art-making should continue to be challenged. However, it is ultimately impossible to boil this down to definite edicts concerning art-making and social value. We should not be aiming for the elimination of high art music for the furtherance of a utilitarian approach to the creation of music in all its manifestations, any more than we would sanction removal of the most complex aspects of physics, mathematics or law, for example. The balancing act is a constant, essential process.

All of this, of course, has informed critical theory around music, perhaps in particular, through the work of Theodor W. Adorno, whose theorising upon mediation (including that of performance) between tradition, creator and musical work remains important to those interested in how art speaks to us, in its 'epistemic character' [*Erkenntnischarakter*], and how, through this, art can be revelatory about social reality (see, for example [1]). For Adorno, the non-conceptual aesthetic content of art is the only thing capable of keeping the hope for a better world alive—of splinting the broken dream of utopia. But implicit in this is a heavy responsibility. Adorno posits a construction that entails autonomy from reified social function, something 'difficult', something to be striven for. In doing so, he emphasises artistic work that exists in a critically-engaged relationship to its own historicity. The quest to understand this complex situation continues within the apparently inexhaustible contemplation of the 'new'. More recent commentators continue these reflections, pointing up specific aspects that retain their resonance:

> The consciousness of the relations between past and present, continuity and discontinuity, has characterized art music in various ways over the last two centuries. In fact, the pursuit of the 'new' is not an abstract principle with ideological nuances but one component of 'time consciousness'; it is expressed in the historical distance that transpires in enquiries into the compositional techniques of earlier periods (undertaken by generations of composers), as well as in the theorizing with which the composer defines the issues he/she is faced with and above all in the creation of sound forms which stimulate new communicative dynamics. These sound forms characterize the *Jetztzeit* not simply for their novelty content but also as the expression of the general subjectivity captured at a given moment.
>
> Opponents of modernism tend to view construction, exemplified by 12-note technique and the serial organization of the sound space, as an end in itself. This assessment fails to take into account the fact that all musical compositions imply construction, and this is defined with respect to a specific realization in sound; thus, the debate should move from the abstract level, where the focus is construction as a principle, to the concrete level involving a discussion of the adequacy of the procedures enacted vis-à-vis the result obtained. In other words, *it should be turned into an aesthetic rather than an ideological judgment* [2, p. 179], italics mine.

The Need for a Fresh, Evaluative Practice, Rethinking Aesthetics

If we are to take up this call for aesthetic re-evaluation, collecting and recording replicable examples of good practice becomes vital, particularly when new music composition also has to contend with attacks on its intrinsic 'quality'. Social progress sometimes sits uncomfortably within this sphere of debate, given the sense that some musically mediocre composers may buttress their hegemonic status by substituting technical sophistication and complex compositional procedures for consideration of the generation of musical meaning. In such cases, the sponsoring of composition by the state becomes questionable.

But this, in turn, means that a debate about the nature of aesthetic judgment and implementation of its outcomes within the wide sphere of the arts is long overdue, especially in light of the shifts that have accompanied the rise of 'artistic research' within the academy. Interestingly, at the time of writing (2014), the Journal for Artistic Research (JAR) has made a call for contributions on 'criticism', while the Platform for Artistic Research Sweden (PARSE) has made 'judgment' the topic of its inaugural conference and online journal. So, the arts disciplines have recognized, to some degree at least, that change is needed. Without these contextualizing debates, the nature of that change, and manner of its process, remains unclear. Given this call for clarity, we may therefore ask: Does the compilation of practices within this volume, and the dialectical process set up through the algorithmic readings, constitute evidence of a reformative disciplinary scrutiny?

The Need for Flexible Listening Practices

> Let us then keep ourselves from forgetting whatever may be able to come to our aid from the arsenal provided us by eye and ear... Moreover, why should we complain about the difficulty of having to make sensibility and organisation, scheme and gesture, or plan and accident coincide? Let us then learn to live out the instability of our condition to the full. As common sense teaches us, we are beings steeped in both instinct and reason.... I see no advantage in getting rid of the one to benefit the other. That is why I resolutely hold that eye and ear, even when they conflict, must each keep their privileges. It is up to the composer to put up with the discomfort of the situation I propose. Sometimes, cordial relations obtained between the two parties. However, let's not delude ourselves; these cases are the exception in this stormy alliance. And the remainder is purgatory! (Boulez trans. Samuels, in [3, p. 222]).

Boulez' pronouncement remains surprisingly apposite in its appeal for a kind of listening that reconciles eye and ear—or, at least, allows the information that they impart to be considered in a holistic fashion. But composition itself has changed since he made this set of statements. The accounts in this POINT volume collectively reveal how the persistence of the tensions he describes has precipitated a spectrum of practices that display open-endedness, non-conclusion and variability. Within the softening boundaries amidst these varied ways of working, attention begins to move away from a text-centred idea, to one in which the untidy situatedness of compositional practice becomes ever more apparent: 'There is simply no all-encompassing agenda to rebel against anymore. Through the internet you can easily find artists and critics who agree or disagree with an aesthetic position regardless of its content.' [Sköld] Nonetheless, one of the most important signs within this volume of the shifting emphasis is increasing attention paid to ideas that relate directly to listening. This is a specific point made by several of the contributors:

1. I believe composing means to work on an understanding of listening, an examination of the manner in how we take in and perceive acoustic information and make it a part of us. This understanding goes far beyond a conceptual realm; it is only achieved through the act of perception, an act that takes place on many simultaneous levels. Embodiment, tactility, space, memory and mimesis are just a few aspects, which are relevant in this process and a starting point of the act of composing as an engagement with the conditions of listening as a process which includes the listening as a necessary precondition [Gadenstätter].
2. What also interests me, is to create with every piece, taking into account the risk of failure, something new, something even for me never fully predictable and to invite others to share this adventure, which is formed from all of that, by listening and reading. Due to this impulse I feel committed to new music, which constitutes itself by the rejection of all outlived musical conventions, expressions and formulas as atonal music in a broad sense [Nachtmann].
3. The possibility to move freely along the time-line when writing, to later exchange what's already written with new findings and insight—to let this influence future sections back in the beginning—leads to a completely different approach compared to the linear time structure of an improvisation. On the contrary the challenge of improvisation lies precisely in the brilliance of the moment since no posteriori correction is possible. Crucial is the aspect of listening, which transfers and takes me into a state of subtle presence. Everything which is heard—the carrier of information and relation—is composed or made up from sudden, imminent direct sensory perceptions and sensations, or of a pensive leaning towards old experiences and intuitive presumptions [Harnik].
4. Determining and capturing the starting point: composing means working with sound material and recognising the quality of the perception that is immanent to the particular material. It comprises an "understanding of listening" as a multi-layered phenomenon and the recognition of one's own preconditions and attitudes. This refers to the insight and the questioning of one's own perspective, which entails the expectation and the desire of the potential "sound-to-emerge" [Reiter].
5. In working with structures, I am constantly guided by intuition. For me, musical intuition is the possibility to perceptually measure, weigh and balance musical material in heard and unheard musical structures. Hearing is the key word here; we tend to rely heavily on our eyes when studying and working with music, but we must remind ourselves that there is no given correlation between the experience of seeing and hearing... At the same time, visualising music is what has made the elaborate structures of western classical music possible. What a composer needs then is not only an intuition for heard musical structures but also an intuition for how a given visual abstraction relates to auditory perception, in other words the capability to imagine sound that is not there [Sköld].

The important point about these examples is that 'listening' is not merely defined as the composers' ability mentally to hear and receive their own work. What is also at issue is the importance of the listener as the 'receiver' of the work, as part of that system within which composition becomes fully realised. The concern with the listener, as part of that larger consciousness changes the nature of what composition is, and what it can do.

It is even arguable the composition may, in some way, enfold the heart of an ethical practice within this listening model:

> . . . Music does not become ethical through ethical text (libretto, song texts, programme notes, its discursive contexts). Neither does it become ethical through a presentation in a context dominated by ethical ideas or moral principles. Nor is ethics an intrinsic quality of (certain) music or sound. Instead, a musical ethics can only come into existence on the basis of a contact with the perceiver—that is, through the act of listening. Thus, ethical moments can only be understood as strategies of engagement, through receptive interpretation, affected and informed by both doubt and astonishment. Unravelling several nodes... does not lead to a musical ethics that explains experience but to an ethics that is grounded on experience.
>
> Listening. That is where it begins, our contact with music. It is the only way to attend to music's call, the only way to experience music.
>
> Only from this process of listening can the articulation of the musical ethics emerge [4, p. 166].

Listening contributes to the potential for inclusive, shared acts of creation. Within this kind of act, those within these pages discuss how 'composing is . . . a way to express wholeness' [Zivkovic]. Furthermore, there is a link between the move toward ethical models of listening and the need to revisit discussions of aesthetics in contemporary music. Some of the evaluative processes that the composers undertake point the way forward for a revivification of the debate on aesthetics:

> Composing is for me . . . a way of aesthetic research about "sounding" material, features thus as an artistic practice at the same time as a so to say "scientific" character [Nachtmann].

For some contributors, the ramifications of the process being 'sounded out' are even more profound:

> Composing is a transformation. Thoughts are converted into sound or vice versa; sonic material causes thoughtful considerations [Klement].

Klement also has some interesting compositional questions showing the dialectical processes between formalism and intuition:

1. How can I find systems which are in accord with my ideas?
2. How can I find rules and processes, which edit sound material in such a way that it unfolds?
3. What do I define as musical material and how do I organise it?

This potential for this kind of systematic evaluation is underlined by the questions concerning 'Formalism' and 'Intuition' within the book. The algorithmic work encodes music that arises from specific practices, which range from highly organised abstract ways of generating material to accounting for the 'real-time' processes of improvisation, as practiced by Harnik, for example. Unsurprisingly, the 'repertoire of gestures' that she has evolved as an improviser leaves a compositional imprint:

> I consider composing and improvising as a kind of interplay between the calculated and the inconceivable [Harnik].

The photographs in this volume that show Harnik's hands as she improvises and composes are reminiscent of the pictures that one sees in historical piano method books. In this instance, a very contemporary creative practice actually fuses with a historical phenomenon that harks back to a time when composers were often not isolated in their creative practice but equally prominent as practising performers. Thus, in this example, the composer is not insulating herself from the past, but experiencing its residues through her embodied work—an inevitable part of her own history and musical training, but revivified through what her contemporary practice entails. Past experience and present composing merge within a new practice.

Other Examples of Permeable Boundaries

1. I like to spend time moving back and forth with the musical data sets (certain pitches, durations, rhythmical fragments, etc.), to repeat them varying, to concatenate, to permutate, to organise them randomly, etc. In doing so I mostly do not use a computer but my head, which constructs such ruling systems with a good amount of individuality [Klement].
2. In the composition process I follow the tendencies that seem to me intrinsic to the material itself, and I search for musical spaces that invite the listener to take on for some moments an entirely new perspective within an already familiar, yet nevertheless current musical language [Reiter].

These examples point to the notion that the algorithmic models in the book, rather than merely being potentially mystifying glosses *on art*, have a dialogical potential *with art*; they propose analytical insight as a locus, or an agent, for transformation. The creative interview exchanges underline this potential, one in which composition may be revivified by a strengthened, yet open set of listening practices, in dialogue with multivalent, open readings. What a contextualisation of the POINT project demonstrates is this prospect of looking more widely at the topics of sound, structure, and meaning, their expression in concepts of aesthetics, formalism and ethics, and how these may be modelled in such a way that conception and reception may be accounted for.

Figure 1 presents one possibility. Here, the 'standard' notion of composer-audience communication is retained, and the performer's role in this minimized. This is not merely an aspect of modelling; it is often the experience of performers of new compositions, who may become invisible agents in the bringing to life of a musical score in a 'live' performance, where the focus remains on the composer.

Nonetheless, the model does make various propositions concerning various kinds of receptive potentialities of the audience, through audience members' capacity to listen, to make some kind of analysis about what has been apprehended through that listening, and therefore to make some kind of judgement about the experience. As a shorthand scheme, this may have some value for the purposes of clarification, but it is not entirely satisfactory in modelling some of the subtleties of this process; moreover, the absence of performer agency here is problematic. Above all, the manner in which receptive apprehension develops is over-simplified, and made too uniform.

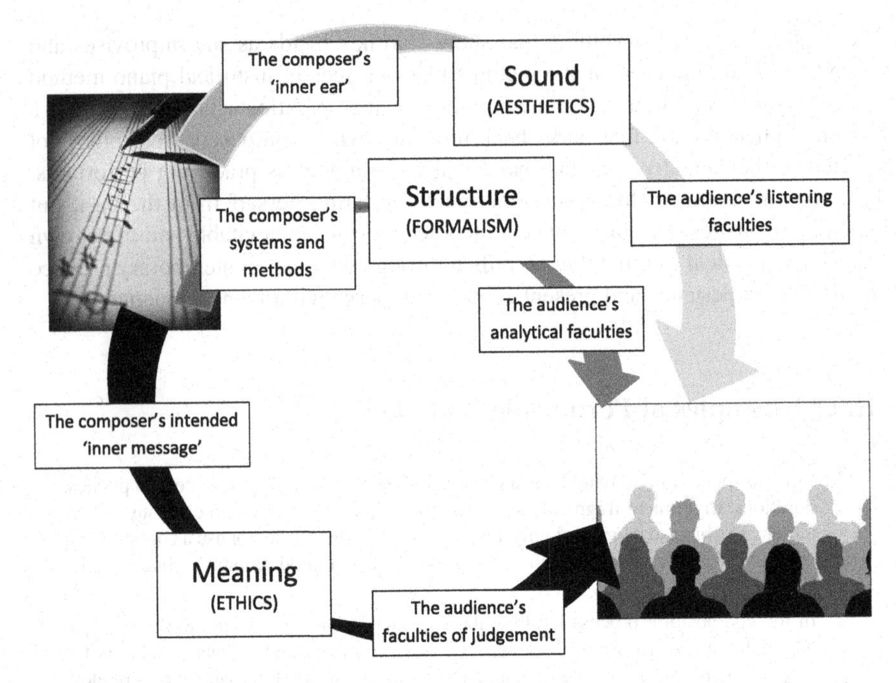

Fig. 1 A uni-directional model for the communication of musical sound, structure and meaning

Figure 2 presents a modification that begins to address these gaps. What is included is a mediating field in which sound, structure and meaning exist in a non-hierarchical setup, in the performative matrix between the conception of the work and its reception. Glossing this further, one may propose that this open area is not merely a performer's domain, but one in which composers and audiences also participate to a greater or lesser extent, in terms of their 'performing' the acts of composition and listening. In this kind of model, the literal nature of 'interpretation' is made more malleable, removed from being a 'performance' phenomenon to being a more broadly creative act in which the 'reading' of the work belongs to no one, but is the domain of all.

Through its processes of mapping exchanges concerning formalism and intuition, the POINT projects begins to indicate (to 'point up') those areas of porousness which break down the unidirectional aspects of Figs. 1 and 2, facilitating a possibility within which the composer-performer-audience directive becomes more of a renewable cycle inclusive of all, and eliminating definitive notions of authorship. Formalism and intuition become phenomena that co-exist—and even overlap—within a larger field of conception in which the sense of contradiction between these two elements begins to dissolve. This blurring of boundaries also allows notions of 'difficulty' to be reframed, since the comprehension of the formal aspects of a musical work becomes interwoven with apprehending how it sounds, and what 'understanding'

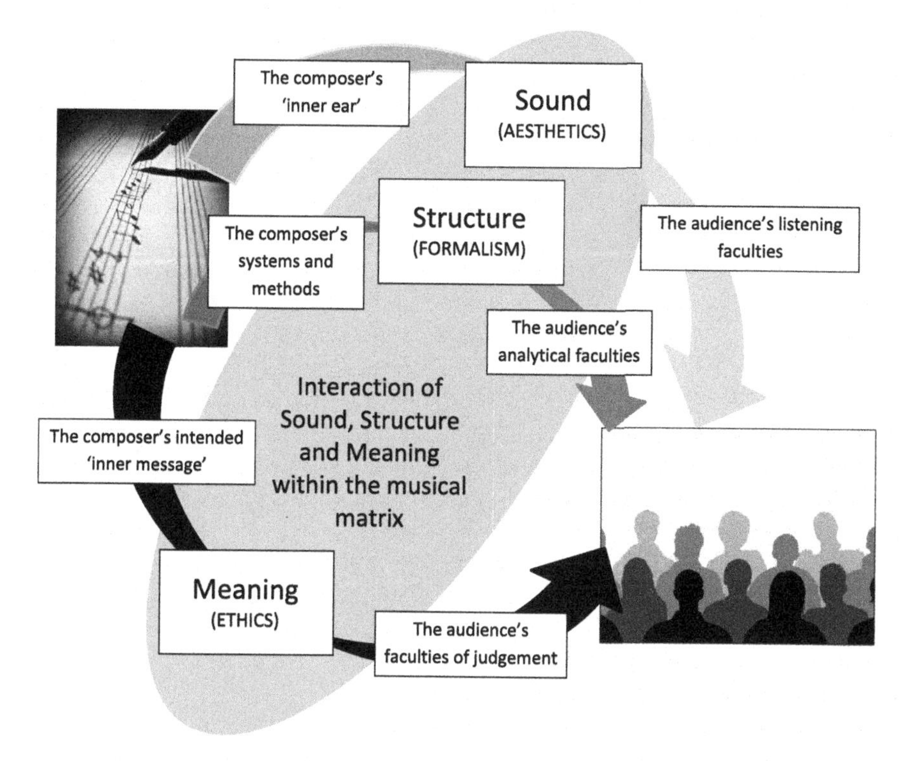

Fig. 2 An interactive field model for a dialogue around musical sound, structure and meaning

those sounds may mean. Meanwhile the judgements concerning this become a shared responsibility, rather than merely an action of a receiver in a predetermined critical role. Criticism, instead of being reactive and creatively impotent, becomes a collaborative—and a culturally responsible—action.

References

1. Adorno TW (1962) Einleitung in die Musiksoziologie. Suhrkamp, Frankfurt am Main
2. Borio G (2014) Musical communication and the process of modernity In: Round table: modernism and its others. J R Music Assoc 139(1):178–183
3. Boulez P (2004) The musicians writes: for the eyes of the deaf? In: Ashbu A (ed) The pleasure of modernist music: listening, meaning, intention, ideology (trans: Samuels R). University of Rochester Press, Rochester, pp 197–222
4. Cobussen M, Nielsen N (2012) Music and ethics. Ashgate Publishing Ltd, Farnham
5. Sokal AD, Bricmont J (1998) Intellectual impostures. Profile Books Ltd, London

In Re: Experimental Education

William Brooks

Arrival

This chapter is borne to the harbour of my being on three currents. The first, a series of gentle swells on the phrase "in re", is manifested thus far in one published article, a second in press, and a third in preparation [5–7]. The second is a more enduring stream of thought deriving from the writings of William James and John Dewey. And the third, with which I shall begin, is a pool that formed and flowed from California in 2012.

The Stanford Symposium

In February of 2012 Stanford University hosted a symposium devoted to the teaching of composition at the doctoral level in the United States. Nine distinguished teachers from major institutions critiqued student works, joined in round-table discussions, and presented papers setting forth their own perspectives and philosophies. The latter, together with an overview of the entire event, were published in *Contemporary Music Review* [11, pp. 249–329].

Talent and Skill

Erik Ulman [34] prefaced the papers themselves with a summary that stressed the diversity represented in pedagogy and curriculum. But in fact all the participants appeared to agree on at least one key point: the practice and therefore the teaching of

W. Brooks (✉)
University of York, Heslington, York, UK
e-mail: w.f.brooks@york.ac.uk

G. Nierhaus (ed.), *Patterns of Intuition*,
DOI 10.1007/978-94-017-9561-6_15

composition includes (or has been thought to include) two quite distinct components. Mark Applebaum [1], the instigator of the symposium, put it thus: "Can talent be taught? I suspect not. . . . We can, however, teach skills" (p. 262). Fred Lerdahl [24] offered a similar assessment: "Instruction in music composition divides roughly into two parts: teaching craftsmanship and guiding a student toward his or her own path. The first can be taught, but the second is a mysterious undertaking" (p. 291). Shulamit Ran [27] argued that "an ideal composition teacher, at any level, is (1) able to help the student listen, critically, and in a deep way, to his/her own music, and (2) considers it a priority to develop and refine the kind of technical tools that will help the student implement his/her personal artistic vision" (p. 307). And Roger Reynolds' [29] more general description rested on a similar dualism: "I feel it is essential for every creative musician to develop what I will call—in a general sense—a 'way'. This is not simply a matter of compositional techniques, characteristic ways of handling pitch, temporal, and timbric resource, for example. It also implies a path from initial impulse to final product; a path that allows a composer to be confident that s/he can proceed with moderate certainty under any circumstances, whatever the resources available, whatever the nature of the desired musical end" (p. 319).

Others approached this duality from a more personal perspective that called clean distinctions into question. Martin Bresnick [2] discussed the difference as embodied in his two principal teachers: John Chowning, who directed him to the technical dimensions of sound itself, and György Ligeti, who demanded originality to the extent of asking for the impossible. The implication was that Bresnick's own compositional self was born only from the integration of these two components. Scott Lindroth [25] also drew on examples from his own studies; he recalled an "inscrutable" teacher and opposed this with one who offered "advice on details of orchestration, counterpoint, texture, proportions, and the like" (p. 298). He argued that advanced students benefit most from a third approach, in which a teacher seeks to cause them to, in effect, lose their ways, their habits. Chaya Czernowin [13] went much further, arguing that all such pedagogical manoeuvres are beside the point, because "every separation between means or technique on one hand and expression or concept or idea on the other is totally false" (p. 285).

History and Diversity

A second recurring issue concerned the place of historical precedent: to what extent, if any, ought students be knowledgeable about existing musics, and what should these be? On such questions there was a clear division of opinion, which Ullman [34] summarised thus: "While some in various ways maintained that the Western musical tradition could or should retain its privileged curricular status, others wanted to identify a core less with this or any other body of work and practices and more with abstract principles assumed to be more fundamental or desirable" (p. 263). Ran [27] was the most uncompromisingly and unrepentantly conservative: "In a deep sense, Bach and Beethoven, Schubert and Brahms, and so many other great composers whose music has withstood the test of time, were my greatest teachers. (As were Stravinsky, Bartok, Schoenberg, Berg, Varèse, the list only begins.) I believe this is so not because this music forms an important tradition, though of course it does, but because it is GREAT MUSIC. Call me old-fashioned, but I continue to consider this kind of training—engaging with great music in any way possible—invaluable" (p. 306). Applebaum [1] was perhaps the most sceptical: "Maslow's 'law of the instrument' warns us that poverty of imagination is the consequence of a narrow palette: *If all you have is a hammer, every problem looks like a nail.* . . . Still hanging about like a miasma is the perpetually re-inherited idea that literacy in Western art music and the ability to do tonal harmony are necessary. . . . I remain queasy about the notion that just because I had to learn how to use a hammer, so should you" (pp. 262–263). Edmund J. Campion [9] cast the historical net more widely with a question that approached belligerence: "[Are you] going to compose inside today's accepted modes of musical expression or are you going to create by exploring and researching emerging music practices?" (p. 280).

Applebaum's and Campion's critiques rested on the observation that today's environment is characterised by a diversity of styles and experiences; students bring other tools than "hammers" to their studies. Lerdahl [24] concurred, but he drew the opposite conclusion: "I firmly believe in traditional instruction in ear training, tonal harmony and counterpoint, tonal analysis and standard instrumentation and orchestration. Such instruction is important for two reasons. First, one learns from the central tradition. . . . Second, contemporary music has not coalesced into a common practice in which basic principles of musical organization are agreed upon. Instead, there are many styles and compositional methods, and each composer must find his or her way through the maze" (p. 292). Lindroth [25], too, advocated conventional training, but primarily "for the inexperienced composer" (p. 299). And Czernowin [13] again pushed the argument much further, looking beyond "practices" and "styles" to focus on "authenticity": "We have endless numbers of common practices, to such a degree that we have actually no common practice at all. . . . Thus, the main problem today is not how to become very good at one or more of these styles, working within a style. It is also not the act of choosing a style or a combination of styles. Today . . . the main problem has to do with the question of authenticity" (p. 284).

Inner and Outer

These two issues would seem to arise from a single source, a sense that composition partakes of two worlds: the *internal* (talent, an authentic "voice") and the *external* (technique, historical understanding). The relationship between the internal and the external, in several composers' minds, would seem to be a circular one—an imprecise and fluctuating loop, a kind of feedback, in which internal impulses produce external actions and external stimuli produce internal responses. Czernowin and Bresnick both implied such a process, but Reynolds [29] presented it explicitly: "Composition is an undertaking of cyclical character: from an initial decision or urging toward a creative act; through the mulling over of resources, scale, form, media, and so on; through the writing out of a score of some sort; then interaction with performers and performance spaces in rehearsal; the performance itself, and finally the reconsideration over time of how the performance 'felt' in relation to what one intended. ... Every stage in the cycle matters, but what matters most is that one engages with the process in a committed and individual manner. ... And the more such cycles a student goes through attentively, and, ideally, interactively with a thoughtful observer—a mentor—at hand, the faster one's craft, one's 'voice', emerges and refines itself" (p. 316).

One thinks of Charles Ives [19], ranting about inner "music" going out and outer "sound" coming in: "My God! What has sound got to do with music! ... Why can't music go out in the same way it comes in to a man, without having to crawl over a fence of sounds, thoraxes, catguts, wood, wire, and brass?" (p. 84). Or, in a more moderate moment, attempting to steer a middle course between conformity and rebellion: "Why tonality as such should be thrown out for good I can't see. Why it should always be present I can't see. It depends... on what one is trying to do, and on the state of the mind, the time of day or other accidents of life" [20, p. 117].

Martin Bresnick [2], discussing the place of history in composition, attempted a similar middle course: "The repertoire of Western music [is] taught by music theorists and musicologists...It seems to me crucial that this valuable historical repository be reclaimed by composers for their own creative purposes" (p. 272). And from this one gets not to Ives but to John Cage, speaking to Richard Kostelanetz in 1968: "We must get ourselves into a situation where we can use our experience no matter what it is. We must take intentional material, like Beethoven, and turn it to non-intention" [23, p. 58].

Cage and Ives; the exteriority of sound; the interiority of music; the cycle that flows first outward, then inward; the reconstruction of history by and into an authentic self; the very concept of music as useful experience—all these suggest that these currents of thought draw on the wellspring of American pragmatism, especially on the writings of John Dewey and William James. Bresnick [2] links himself explicitly to such forbears: "This empirical, pragmatic approach is, I think, largely an American attitude, whose origins may be found in Emerson, Whitman, William James, John Dewey, and many others. ... John Chowning discreetly and transparently taught me his way of working—by the vivid example of his excitement in discovering new sounds, his new ways of sounding them, and by his openness to thoughtful experiment. The value of these efforts was not always immediately apparent: things failed, or proved, in practice, to be ineffectual or unnecessary.... I remain convinced by this sort of 'materialist,' empirical approach to composition" (p. 271).

Bresnick's leap is not surprising. Most of the composers at Stanford were Americans, after all, and many situated their work in relation to (if not as part of) so-called "experimental" music. But they spoke of this only indirectly; the focus of the symposium was their educational approach and philosophy. "Experiment", "empirical", "experience", "education"—these are key terms in writings by the pragmatists, and they played an important part in the Stanford discussions. And pragmatists—Dewey, especially—were deeply concerned with how experience is made into knowledge, or art, and how knowledge or art is embedded in experience. That is a prime educational preoccupation, and Dewey exercised a profound influence in America on what became known as "progressive" education, in which attention to individuality, "authenticity", plays a central role. It seems almost inevitable, really, that a gathering of eminent American teachers of creative practice (composition) would invoke, often indirectly, the progressive thinking that flowed through much of American culture during the previous century; and it thus seems quite reasonable that probing their discourse for underlying issues leads to a terrain where Dewey, James, and the pragmatic tradition of thought are plausible guides.

But what, exactly, are the points of contact? Where are the resonances, and what the struck sound? And what can this have to do with POINT—temporally and geographically distant and in no way conceived as an educational project?

Pragmatism

There is certainly historical evidence for a link between pragmatism and experimental music. The link between Cage and Dewey was explored by Joan Retallack, writing in 1992. "Cage's work... unfolds within the American pragmatist tradition characterized by the aesthetic theory of the philosopher John Dewey", she writes [28, p. 243], and Dewey reappears regularly in the analysis of Cage's "poethics" that follows. In a later essay, Austin Clarkson [10] explored the relationship between Cage's "oriental" aesthetic and William James, mediated by Jung and Suzuki. I myself have touched on this topic [3, 4], and Edward Crooks's superb dissertation [12] explores some of the hidden Western sources in Cage's "oriental" influences.

For Ives there is a surprising gap—possibly because he himself is so resolutely focused on Emerson that other philosophers have gotten short shrift. But four decades ago Rosalie Sandra Perry [26] tied Ives to James and pragmatism in a seminal chapter; and more recently Christopher Bruhn's fine dissertation [8] has picked up and extended the threads. Suffice it to say that there is ample reason to stipulate kinship, if only because both Ives and James trace their lineage to Emerson.

In what follows I will leave such historical matters untouched. I propose rather to discuss the four key terms noted above—experiment, empiricism, experience, and education—with particular reference to James and Dewey, secondary links to the Stanford conversations and, eventually, a return to the POINT project.

Experiment

I first encountered Gerhard Nierhaus and the POINT project at the Orpheus Institute in Ghent, Belgium. The Orpheus Research Centre, at which I have been a senior research fellow since 2009, has fostered an extensive examination of the term "experiment" in recent years, particularly as applied to artistic research in music. The scientist-philosopher Hans-Jörg Rheinberger provided one model for exploration [31]; I have approached the problem with reference to experimental music [5]; and recently Katharine Coessens brought Friedrich Steinle to initiate a discussion of "exploratory experimentation", based in part on Goethe's experimental work with colour [30, 32]. It is no wonder, then, that in drafting the present chapter my thought has drifted to these three ongoing Orpheus conversations and the implications they might have for POINT. All three approaches can be mapped crudely but convincingly onto oppositions between *inner* and *outer* that somewhat resemble the dualism that underlay the Stanford symposium. Rheinberger distinguishes, crucially, between "experiment" and "experimental system"; the latter—the constellation within which the former is manifested—embraces laboratory, social relations, economic factors, and uncounted other forces. The "experiment"—historically a self-contained idea that arises from the mind, from "pure" thought, and is then tested in a "material" world—thereby becomes a much messier thing. The "system" in which it is embedded feeds into the construction of the "idea"; thought ceases to be "pure", and the "material" is constantly reconfigured in a series of cycles inwards, outwards, and around the putative "scientist". The attractiveness of this model to researchers in artistic practice is evident.

Similarly, in my analysis of "experimental music", I found it useful to distinguish between "test" and "observation". The former, I argued, is associated with judgment and value; a test determines the validity of a concept or hypothesis that is formed, in effect, in the mind *prior* to the experiment. "Observation", on the other hand, entails activities undertaken with relatively few preconceptions, and hypotheses (if any) are formed only *after* the experiment has occurred. Tests are responses to questions of the form "Is it the case that . . .?"; observations arise from "What happens if . . .?" Test and observation are two halves of a cycle that (anticipating Dewey) we might call *experience*: the test points outward, from self to world, while observation feeds the self *from* the world.

This is very like the distinction that Steinle and others have drawn between "theory-driven" and "exploratory" experimentation. An experiment driven by theory is determined entirely by a set of predicted outcomes; anything that does not bear on the prediction is ignored. Newton, Steinle argues [30], worked in this manner in evolving his theory of optics: each confirmed prediction gave rise to a new hypothesis to be subjected to experiment (and thus confirmed); if a prediction was disproved, it was the theory (not the experimental experience) that was subjected to examination. "Exploratory" experiment, on the other hand, as practiced by Goethe and others, freely accumulates data within a bounded universe, a "topic"; only after (or in the course of) this open, observational praxis are relationships and consequences inferred.

In all three models, then, there is a rough and slippery distinction between *inner* and *outer* actions and the orders in which they are undertaken. Without plunging into the abyss of metaphysics, it is useful to speak of *mind* and *world*, *idea* and *reality*. To the extent that pragmatism was a response to the non-negotiable conflict between "idealism" and "realism", it effectively sidestepped the import of this distinction without denying it: mind and world, idea and reality form a cycle, a closed system of mutual effects that are apprehended through actions and their consequences—practical actions: *pragmatism*.

Empiricism

Today three philosophers are accepted as the patriarchs of pragmatism: Charles Saunders Peirce, William James, and John Dewey. All three embraced a form of empiricism in which our inner world of judgment, thought, and intelligence is contingent upon an outer world—an apprehended reality. But the contingent process was not simple observation; the mind does not stand above, apart from the outer world, determining properties and rules by applying an innate (inner) logic. The mind is part of the world, and part of the body, and part of the act of sensation; the world is apprehended by action and by consequence: an entity is what it is because someone *does things* to it (or in it) and observes the result. This is a kind of exploratory experimentation, with principles arising from a variety of acts undertaken not quite at hazard. Other than temporal continuity, spatial proximity, and the biological laws of sensation (the "principles of psychology", one might say [21]) there is no need to assume a priori properties.

Peirce, James, and Dewey differed in important respects. Peirce came to call his thinking "pragmaticist"; James embraced "pragmatism", having taken the term from Peirce originally; Dewey preferred "instrumentalism". For Peirce the essential problem was *meaning*; for James, arguably, it was *truth*; and for Dewey it was *learning* (or, in the present context, *creativity*). Peirce was led inevitably to the construction of a semiology; James was required to confront belief; and Dewey devoted himself to society, education, and art. But the thought of all three was *radical*, as James applied the term: all three argued that the subject of the inquiry could not be separated from the inquiry itself. The meaning of "meaning" must be explicated using the properties that "meaning" has come to acquire. An explanation of creativity has to account for the creation of the explication. And even empiricism, James argues, must be *radical*: "the parts of experience hold together from next to next by relations that are themselves parts of experience" [22, p. xiii].

"Radical", too, are the constructions of "experiment" presented above. Rheinberger's "experimental system" must include the prevailing models for experimentation that underpin the laboratory work. But the "experimental system" is itself one such model; thus it enters the world it is describing and hence must account for its own being. Similarly, what Steinle calls "exploratory experiment" (and what I have called "observation") arises only out of its own practice. *Tests*, which are "theory-driven", arise from principles that are assumed to exist in a different realm from the objects being manipulated: the law of gravity has no effect on the motions of the planets. Exploratory experimentation, in contrast, is intrinsically vague and amorphous, itself being constantly reshaped in response to what is *observed*; the construction of an exploratory experiment—of the term itself—is itself an exploratory experiment. Experimental systems and exploratory experiments are inextricably entangled in the shifting relations of the self with the world around it, and those relations are themselves part of the system constructed, the domain explored—in a word, part of *experience*.

Experience

And so we come to "experience", and—in this context—especially to John Dewey's summative work, *Art as Experience* [16]. In this Dewey applies the principles of radical empiricism to artistic expression and esthetic (Dewey's spelling) understanding. Consistent with the pragmatic method, he begins by refusing a priori distinctions—in particular, the distinction between "fine" art and daily life. As an illustration, he observes that "the intelligent mechanic, engaged in his job, interested in doing well and finding satisfaction in his handiwork, caring for his materials and tools with genuine affection, is artistically engaged" (p. 5). It follows that art resides not in objects—paintings, buildings, sounds—but in *experiences*: "Even a crude experience, if authentically an experience, is more fit to give a clue to the intrinsic nature of esthetic experience than is an object already set apart from any other mode of experience" (p. 11).

The first problem, then, is to examine experience. Again Dewey starts from a *tabula rasa*: the only assumption is that experience is necessary to life, and "life goes on in an environment; not merely *in* it but because of it, through interaction with it" (p. 13). In that sense experience is continuous and indivisible. But we speak of having *an* experience—something that is bounded and extracted from the continuity. What creates the boundaries? The pragmatic answer is that *we* do, by the way in which we direct our attention, the actions we take. We judge something complete, and we attend to something else, something "new". Recall James: relations between "parts of experience" are themselves part of experience; and a relation called "completion" arises, becomes part of experience, when we *feel* we have begun something different. "We have *an* experience", Dewey writes, "when the material experienced runs its course to fulfillment" (p. 35). We ascribe to *an* experience a unity, a distinguishing quality or qualities that sets it apart from others; and when that unity or quality changes fundamentally, we declare the experience complete. An experience need not be continuous—cooking a meal might be interrupted by a phone call, or reading a novel halted for days or weeks—but in naming or characterising the experience, such interruptions are disregarded: it is the unifying quality that binds the experience together, not temporal continuity.

But *within* an experience, "every successive part flows freely, without seam and without unfilled blanks, into what ensues" (p. 36). In that sense *an* experience is just a part of life; *an* experience results from "interaction between a live creature and some aspect of the world in which he lives" (p. 44). The interaction proceeds cyclically, in an alternation of states that Dewey characterises as "doing" and "undergoing"; and from this arises the flow, the unity of an experience. "A man *does* something; he lifts, let us say, a stone. In consequence he *undergoes*, suffers, something: the weight, strain, texture of the surface of the thing lifted" (ibid., my italics). The experience consists not merely in the sequence of events but in the binding unity formed by the relationship between them: "An experience ... is not just doing and undergoing in alternation, but consists of them in relationship" (ibid.) (Recall James, again).

But since Dewey situates art not in objects but in experience, it follows that "art... unites the very same relation of doing and undergoing, outgoing and incoming energy, that makes an experience to be an experience" (p. 48). Some experiences, however, have qualities that we designate as "esthetic", and Dewey distinguishes these from others that are "dominantly intellectual or practical" (p. 55). In the latter sort of experience, an end is attained; from the start, the goal is to bring the experience to a useful conclusion, and the experience has no value other than as a means to the end. The experience, we might say, is "theory-driven"; it has no "exploratory" properties, and it includes nothing that has not been anticipated. In an "esthetic" experience, in contrast, doing and undergoing stand in a relationship that is itself of value; the artist and the percipient come to explore a process, not to attain an end. Our interaction with the world, Dewey summarises, "is peculiarly and dominantly esthetic... when the factors that determine anything which can be called *an* experience are lifted high above the threshold of perception and are made manifest for their own sake" (p. 57). Thus an esthetic experience or artistic product is only incidentally about an image, a theme, or a character; at its heart it is about *experience itself*. Hence the value of esthetic experience does not lie in tangible accomplishments, though these may arise in the course of the experience. We may do something new or make a new object; we may learn something about a place, about a person, about form, harmony, design; but above all we learn above all how to experience things esthetically. All experience is recursive in some sense; *an* experience becomes a part of future experience; so for a pragmatist, we might say, esthetic experience is *radically esthetic*.

Education

Having been led to *learning* through art, we are invited into the domain with which Dewey is perhaps most widely associated: educational theory. Dewey's engagement with education persisted throughout his career, beginning with his early, exploratory experiments at the "laboratory school" at the University of Chicago. These culminated in *Democracy and Education* [14], which became a touchstone for "progressive" education in the United States. But Dewey revisited the topic intermittently for the rest of his life, and he wrote a key summary in 1938, just four years after *Art as Experience*. In this—*Experience and Education*—he critiqued the schism between "traditional" and "progressive" education and implicitly linked education with art by means of the term his titles had in common: experience.

Experience and Education is less formal than the earlier volumes, and Dewey assumes an understanding of some concepts that he had previously explained at length. Experience is, as before, an interaction of an organism with its environment; and, as before, interaction includes the two aspects that Dewey called "doing" and "undergoing" in *Art as Experience*. But Dewey's concern is now less with the boundaries of an experience and more with the continuity formed by experience as a whole. It is in this continuity that the basis for a critique of education is found, for "every experience both takes up something from those which have gone before and modifies in some way the qualities of those which come after" [17, p. 27]. People *learn* from experience, and learning is therefore continuous; the question, for Dewey (as for the composition teachers at Stanford), is how to distinguish an educational experience from the learning that happens, willy-nilly, all the time.

Growth—development—also happens continuously, and in some ways would seem to be equivalent to education. But, Dewey observes, some growth enhances future growth, and some hinders it; if the hindrance is too great, the organism withers and eventually dies. Hence, the key question is: "Does [a] particular form of growth create conditions for further growth, or does it set up conditions that shut off the person... from the occasions, stimuli, and opportunities for continuing growth in new directions?" (p. 29). Dewey's answer is unequivocal: "When and *only* when development in a particular line conduces to continuing growth does it answer to the criterion of education" (ibid.; italics original).

This is a *radical* education closely allied to radical empiricism and radical esthetics. In all three domains attention is focused not on objects but relations, not on ends but on processes. These processes are cyclic: experiences—the common basis for all three enquiries—are evaluated by whether they include, enhance, advance future experiences. When they do not—when *experience* is divorced from the environment and situated solely within the self—empiricism becomes idealism, esthetics become formalism, education becomes training. "The most important attitude that can be formed", Dewey writes, "is that of desire to go on learning" (p. 49).

Not all education, not all *experience*, is radical in this sense. We live much of our lives by *habit*, a topic that Dewey [15] and James [21, v. 1 Chap. 4] both treated at length. Habits are necessary and beneficial; they are acquired through experience and they may result in art. But they are not radically esthetic or educational, except insofar as one can speak of acquiring a habit of inquiry or a habit of esthetic engagement. Workaday habits—walking, talking, eating, grooming, resting—do not hinder an organism's growth, but neither do they further it; they merely maintain it. They are beneficial, but they are not educative.

There are, however, experiences that are *not* beneficial—experiences that are mis-educational or *counter*-esthetic. Dewey gives several examples in both domains, and they are remarkably similar. From *Experience and Education* [17]: "A [mis-educative] experience may be such as to engender callousness; it may produce lack of sensitivity and of responsiveness. Then the possibilities of having richer experience in the future are restricted.... [It] may increase a person's automatic skill in a particular direction and yet tend to land him in a groove or rut;... [it] may be immediately enjoyable and yet promote the formation of a slack and careless attitude [that modifies] the quality of subsequent experiences so as to prevent a person from getting out of them what they have to give" (pp. 13–14). And from *Art as Experience* [16]:

"The enemies of the esthetic are neither the practical nor the intellectual. They are the humdrum; slackness of loose ends; submission to convention in practice and intellectual procedure. Rigid abstinence, coerced submission, tightness on one side and dissipation, incoherence and aimless indulgence on the other, are deviations in opposite directions from the unity of an experience" (p. 40).

Stanford Revisited

With rhetoric like this we are led back to the discussion at the Stanford symposium. The problem there—how to teach composition—integrates both of Dewey's concerns, the esthetic and the educational. And certain of Dewey's insights illuminate the points of contention at Stanford. Recall that Dewey rejected an a priori distinction between "fine" art and what might be called the arts of life [33]; how does this affect the Stanford dichotomies between "talent" and "skill", between "art" and "craft"? If "the intelligent mechanic" is artistically engaged, can we not say the same of the intelligent composer writing, say, species counterpoint? When would we *not* want to say that?

For a pragmatist, all depends on the nature of the experience. If the counterpoint is an exercise and *only* an exercise, then the activity does not further growth and is not truly educational; if the experience does not illuminate experience itself, it is not truly esthetic. But the same might be said of a composer determined to display an original "talent", a distinct "voice": if display is the purpose, *experience* is not "lifted high" and the work is not esthetic. It does not matter what kind of work is undertaken; from an educational perspective, the experience is worthy if it enhances the potential for future growth, and from an esthetic perspective, the experience is worthy if elevates the understanding of *experience*. The two amount to very nearly the same thing: an esthetic experience is usually educational; and an educational experience, if concerned with musical composition, will, in general, be esthetic.

Much the same applies to the place of historical repertoire in teaching. Indeed, Dewey addresses exactly this point, more broadly, in *Experience and Education*. First he observes that when one attends to any phenomenon, the attention occurs in the present. Pragmatically speaking, then, there is no necessary difference between encountering a phenomenon that has been encountered before and one that is entirely new; the *experience* can be educational—or esthetic—in either case. Moreover, experience produces "a continuous spiral", Dewey written, an "inescapable linkage of the present with the past", in which *an* experience, when completed, becomes "the ground for further experiences" [17, p. 97]. Again, the educational value resides in the nature of the experience: if one studies Bach in order to have studied Bach (for example, to pass an examination), the experience is not educational; but if the experience of studying Bach leads forward "into an expanding world of subject-matter" (p. 111), it is. A similar argument applies to esthetic value. Shulamit Ran pleads passionately for the study of an historical repertoire "because it is GREAT MUSIC" [27, p. 306]; but the greatness resides not in the object—the "music"—but in the experience. Acquiring technique or knowledge from the study of "Bach" will probably not be esthetically satisfying, and the musical experience will be something other than "great"; but when the experience of Bach's music brings experience itself to the fore, greatness happens.

Finally, the underlying tension between "inner" and "outer" worlds at Stanford also appears in pragmatism, as we have seen. But to a pragmatist there is no reason to take sides; organisms and environment interpenetrate in unending cycles of *doings* and *undergoings*—music "going out" and "coming in", wrote Ives—that constitute *experience*, whether educational or esthetic. This is precisely the model for composition that Roger Reynolds put forth, with the additional suggestion that the educational experience is enhanced if a "thoughtful observer—a mentor—[is] at hand" [29, p. 316]. Again, however, the experience itself must be the touchstone for educational value: if the experience is such that the mentor remains needed for future work, it has restricted growth, and it is mis-educative; but if the experience enables a student to grow in new, unanticipated ways, independent of the mentor, it is profoundly educational.

In sum, it would seem that the deliberations at Stanford invoke and continue a long-standing discourse—now a century old—about "progressive" education. And because the education, in this case, concerns music, they also shed light on the interconnection between esthetics and education that Dewey implied in his two summative volumes. The teachers who gathered at Stanford are engaged in processes of exploratory experiment—in tandem, collectively, and in sequence, individually. They proceed not from an a priori theory by which to teach composition but rather from a theory—a reflective analysis—that develops from experience and that constantly evolves as one experience succeeds another. The teacher grows along with the student; both are educated. The objective is the creation not of *objects* with specific, approved properties but of *experiences*—experiences that are both esthetic and educational, experiences that enhance the qualities of future experiences and that enable growth in "an expanding world". The nature of these experiences cannot be specified in advance without severely limiting their potential; any music that results is, in that sense, and regardless of its surface features, *experimental music*. And the associated education, leading and following, interpenetrating cyclically with the work that is done, is *experimental education*.

As is, I will now claim, the POINT project.

The POINT Project

What is the POINT project, viewed through the lens of pragmatism?

In the final section of this chapter I will argue, first, that the technical roots for POINT extend deep into the bedrock of experimental music, but that POINT can be distinguished from many of its predecessors by the *type* of experiment it manifests. Second, I will suggest that the philosophical roots for POINT reach back to radical empiricism and that it formalises, in a sense, the pragmatic concept of experience, especially as explained by Dewey; because of this, the POINT project is linked with the issues raised at Stanford. Third, I will claim that POINT results in esthetic experiences—music, if you will—that arise in ways consistent with Dewey; in this, too, it serves to illuminate some points of discussion at the Stanford symposium. And finally, I will suggest that POINT is, at its foundation, a pedagogical tool, one that pertains not only to the teaching of composition but also to the agenda of progressive education, one that is of interest not only to the composers and researchers involved but to a much wider community of practitioners.

POINT and Experimental Music

In POINT, algorithms are devised and employed to make musical compositions. One can argue that the foundations of algorithmic composition are lost in history—that a canon, after all, is a kind of algorithm, and that the first notated algorithmic music is "Sumer is icumen in". But the term is more usually associated with a compositional practice that began in the 1950s and that was from the outset associated with the phrase "experimental music". That was the title of Lejaren Hiller's seminal publication [18], recounting the first attempts at computer-assisted composition, and it has persisted to the present in institutions like the University of Illinois's "experimental music studios".

As I have shown elsewhere [5], however, "experimental music" immediately came to mean two different things. On the one hand, in the work of Hiller and other practitioners like Milton Babbitt, the "experiments" conformed to classic "theory-driven" models. A hypothesis was formed and tested by means of an algorithm; this was predicted to produce results possessing specified properties, and if it failed to do so, it was rejected. On the other hand, in the work of John Cage and others, algorithms (for Cage, chance procedures) were employed precisely because their outcomes could *not* be predicted. The result was to be experienced esthetically regardless of its nature.

Cage's experimentation entails "observation", in opposition to Hiller's "tests"; and this, I have said, is similar to the opposition between theory-driven and exploratory experimentation. But the two are not exactly equivalent. In Cage's work, each "experiment"—each "piece", if you will—stands apart from the others; each is sufficient unto itself. There is no need to generalise; in fact, generalisations—theories—are assumed to undermine the value, the individuality, of future experiments. In exploratory experimentation, in contrast, a *series* of experiments are conducted, not quite at hazard and sharing a single field of inquiry (though this may be very vaguely defined). At some point (also vaguely defined) the experimenter brings all the outcomes together to form a theory, a conception of the whole. This theory can then be (but need not be) the basis for subsequent "theory-driven" experiments.

The latter model is used in POINT. An algorithm is devised, applied, and evaluated—but not with the objective of testing or altering the algorithm (though changes may be made) but with the objective of clarifying the objective. "Oh!" the composer says, upon experiencing the outcome, "that's interesting, but I think I must have meant something different". Or, perhaps, "Oh! that's really unexpected; let's see where it goes". The goal to be attained at the end of the series of experiments is, at least in part, determined during the course of the experiments themselves.

From the perspective of the participating composer, the process is therefore one of exploratory experimentation. From the perspective of the researcher, the process is also exploratory, but on a larger scale. No two composers are alike; for each different algorithms are needed, and each composer's own exploratory process is distinctive. In aggregate, the composers form a collection of experiments that share a single field of enquiry—algorithmic composition of music—but do not constitute an orderly route to a pre-determined goal. What *is* the "theory" that unifies all the compositional acts, all the algorithms? We don't know—certainly at the beginning, and quite possibly even at the end. We can be assured only that new relations between "parts of experience" will be observed and that these will themselves be factors in the generation of future experiments.

POINT and Experience

I have just paraphrased William James's characterisation of radical empiricism. POINT is radically empirical because at every stage the relations between its "parts"—whether components of a computer program, musical outputs, or composerly responses—are taken into the experience upon which the next iteration will be based. Indeed, POINT formalises this process; one might even say that it formalises experience in the abstract. *Doing* and *undergoing*—which normally form a fluid continuity, mediated only by the senses—are split by technological intervention: the composer "does" only by means of the computer; the composer "undergoes" only in response to the computer. This split creates clear divisions in the stream of experience: there is *an* experience (an algorithm is devised), then there is *another* (an output results), then a *third* (the composer responds), and so forth. A certain kind of relation—beginning and ending—is elevated, within experience in general, to a dominant position; and at the boundary, the point at which one experience ends and the next begins, there occurs a change in medium: from computer code to musical score to verbal discourse, cycling forward. The participating composer is placed in a *situation* [17, p. 39] that enables self-awareness about the way experience is experienced.

The experience of the researcher embraces a much broader field of interactions, but again these are formalised in a way that creates distinct boundaries between *experiences*. One composer departs, another arrives: the boundary creates two different experiences. Each of these, in turn, can be parsed into distinct units, as above; the whole can be thus, in this instance, structured hierarchically. Think of a visit to an art gallery: "*Les Demoiselles d'Avignon*" names *an* experience, bounded by the frame around the painting, that is contained within *an* experience named "gallery 2", bounded by walls and doorways, that is contained within *an* experience named "the Museum of Modern Art", bounded by entering and exiting a building. But experience need not be nested in this way; we could equally think of a *single*, *unbroken* experience: the journey through the gallery—steps, stairs, pauses, turns. How we experience the experience is determined pragmatically, perhaps arising from the question asked: "Where did I go?" requires something quite different from "What did I see?".

Movement between these experiential poles—the discrete and the continuous (with all possible gradations between)—occurs constantly and is itself part of experience. It occupies a central place in conventional models of composition, such as those put forth at Stanford: the composer is concerned with *materials*, the composer is concerned with *form*; the work is comprised of *parts*, the work constitutes a *whole*; the music is experienced in *isolation*; the music is experienced in *relation* to other music. Self-awareness of one's behaviour, moving forwards and back along these spectra, is assumed to be beneficial, and a teacher serves in part to encourage such self-awareness. By juxtaposing a strictly hierarchical experience with an intuitive, fluid one, POINT provided participating composers with an opportunity for structured self-scrutiny, a *situation* in which they could work upon their method of working.

POINT and the Esthetic

POINT makes no assumptions about the nature of compositional materials. A composer starts from an image; very well, the computer will start from an image. A composer starts from a series of intervals; very well, the computer will generate such a series. Artistic creations—musical works—are not constrained by technical attributes; esthetic experience can occur in the presence of music of any type. Dewey's mechanic, "engaged in his job, interested in doing well and finding satisfaction in his handiwork", would recognise POINT as a useful and adaptable tool.

In that sense, POINT is a kind of work-desk for a composer, or a laboratory for a researcher: it constitutes an "experimental system" that includes equipment, a working method, and designated roles for participants, the whole intertwined with an environment that extends to social and economic concerns. It itself would seem to have no esthetic qualities, other than those experienced in using any well-made tool—a brush, a knife, a skillet. It is in the results that are produced—the "outputs" of the project—that the esthetic will be experienced.

But if this is all that matters for POINT, the project as a whole is simply directed towards the production of "music"; it is goal-directed—the sort of experience that Dewey characterised as "dominantly intellectual or practical". In effect, the project becomes merely a mechanism for composing music, and an obvious question arises: why bother? Composers do perfectly well on their own; what need have they for POINT? If the esthetic experience of a work created in POINT cannot be differentiated from that of a work created elsewhere, from a pragmatic perspective the experience is the same, and POINT is irrelevant.

Clearly something else is at work here; POINT is more than a work-desk, more even than an experimental system. POINT is not of value because it results in compositions that can be esthetically experienced; POINT is an esthetic experience *in itself*. The participating composer interacts with the *experience of composing*, not merely with an environment made up of compositional materials and other familiar entities. In an esthetic experience, Dewey requires [16, p. 57], "the factors that determine anything which can be called *an* experience are lifted high above the threshold of perception and are made manifest for their own sake". And that is precisely what happens in POINT. The esthetic experience forms (informs, transforms, re-forms) future esthetic experiences—because POINT is about *experience*, not about music.

In this respect, POINT helps to explain the underlying unease in Stanford. Most of the teachers gathered there were not entirely comfortable with a pedagogical focus on tools and skills—how to hold the pencil, how to edit the sound-file, how to write invertible counterpoint. Many argued that the skills needed to change, to be reshaped to suit the current situation; but nearly all indicated that instruction in composition should also include something quite different—"talent", "voice", "a 'way'.". But they were often at a loss to explain how to teach this. POINT helps us to understand that the two objectives are not, after all, sides of a single coin; they are wholly different species. In one case, teaching composition is a matter of *production*; it is goal-directed, skill-based, tool-specific. Composers learn how to *make something*. In the other, teaching composition is a matter of *experience*; it is open-ended, exploratory, *experimental*. Composers learn how to *experience making something*. No one denies the necessity for both; but only the latter leads to the esthetic.

POINT and Education

In a practical sense, POINT—like any research project—is concerned with the production of knowledge. New algorithms are written, new compositions come into being, new dialogues are had. New books, chapters, paragraphs are written. The knowledge enters the environment, and (one hopes) people *use* it. In this way POINT enriches a field of study, a domain of interactions.

But from Dewey's perspective an enriched domain is not in itself educational; all depends on what is done with it. Unless the interactions are such that the organism—the student—is enabled to grow further, in "an expanding world", they merely preserve established habits or, worse, impose additional restrictions on behaviour. In particular, an experience that leads to no questions, that simply provides the necessary answers, is not educational. One uses satellite navigation to obtain instructions for travel; these might be informative, but they are not educational. But when one uses a map, there arise possible questions about routes, terrain, and connections; in pursuing these, one obtains not just information but new insights, new possibilities. A cyclist setting out to explore an area wants a map, not a GPS device. The latter tells you "how to"; the former invites "what if". The latter is goal-directed, the former is exploratory.

For that reason, if POINT were *only* its "results", it itself would not be educational even if the results subsequently were used educationally. A participating composer who, through experiencing POINT, arrived at a definitive method for writing music and then used that method religiously for the indefinite future would, in fact, have been *mis*-educated by the experience. POINT is educational precisely to the extent that the circular process that animates it has no destination. The objective is to keep moving, to continue seeking an objective; the "results"—the compositions, the algorithms—are merely breadcrumbs dropped to mark the journey's path.

For the participating composers, then, POINT is educational not because they acquire insights into what they do but rather because they are led to speculations about what they *might* do. Self-knowledge is valuable primarily because it enables one to become other than what one is: it engenders growth. The experiments within POINT enter the field of past experiences and intermingle with their predecessors, establishing new relations between "parts of experience". To the extent that these relations are *exploratory*, in pursuit of an ever-receding theory, new experiences will be educational. And, not incidentally, these experiences will have a potential for producing artistic work, the basis for new *esthetic* experiences to be had by others. For the esthetic—like POINT, like education—is not realised in completed objects; it is found in experiences that fold back on themselves, that enrich the possibility of future esthetic experiences.

For the researcher, too, POINT's educational value does not lie in its accomplishments. Unexpected problems may have arisen and been overcome; new programming solutions may have been devised; new interfaces may have been built. These are well worth reporting and disseminating; they, too, enrich a field of study. But they are also best approached as an incomplete set of exploratory experiments: incomplete, first, because new composers will generate new problems, and there is no end to composers; but incomplete also because the set of relations between the solutions—the new "parts of experience"—and the greater environment raises new enquiries. It is not so much that "further research will follow"—that is merely a platitude—but that further research will be constantly transformed in an interactive cycle that is actually a (re-)application of the model applied in POINT itself. Education is viable to the extent that it reflexively and sustainably generates education; POINT is educative to the extent that it generates a further "POINT" in which it is itself a component.

And so also for the greater community of artists and researchers. POINT is educative to the extent that it invites continuation in new areas and new ways. We learn nothing if we take it as "finished". Future POINTs could simply move to new domains: one can easily imagine analogous projects for writers, for choreographers, for film-makers. But with such extensions POINT veers dangerously close to method, to doctrine; the experimental domain is enlarged, but the work is in some sense replicated. Rather, POINT—as an entity, not a collection of results—needs to enter the domain of discourse about artistic method, artistic research, artistic apprehension. There it can enter into new relations with other projects assembled, not quite arbitrarily, into a field for exploration, for experimentation.

A world of *experience* awaits.

Departure

At the outset I spoke of "the harbour of my being". This was not just a poetic gesture: the metaphor is a true one. I am a harbour. My being—the water of the harbour—interpenetrates with the sea around it. In certain directions (from certain perspectives) the sea and the harbour cannot be clearly distinguished; still, there is another world, outside of the harboured self, with which I interpenetrate. Currents—tides—at times carry that other to me; at other times they carry me to it. There is undergoing and there is doing.

The harbour is, however, bounded; reality has its limits. In both cases, the true boundary is the shore—the change from one state to another—and the world "beyond" is unknowable. The water cannot know the land: it only feels the land's constraint. It can lap gently or rage openly, but it gains access only by small, unnoticeable increments. And when access is gained, it takes the land into itself; water is a universal solvent. Thus I transcend myself, utterly unaware.

The other boundary—the vague, shifting edge between harbor and sea—is open to adventure. I can truly embark; the harbour can empty itself. To engage with the sea, I must abandon my habits: I must open myself to currents, forces, I cannot foresee. The adventure is founded on *faith*, an indefensible belief that understanding will result; but I commit myself—my *self*—thereby to being all at sea. There is no return except by fortune.

References

1. Applebaum M (2012) Existential crises in composition mentorship and the creation of creative agency. Contemp Music Rev 31(4):257–268
2. Bresnick M (2012) Thoughts about teaching composition, or self-portrait as Hamlet and Polonius and Whitman is also there. Contemp Music Rev 31(4):269–276
3. Brooks W (2007) Pragmatics of silence. In: Losseff N, Doctor J (eds) Silence, music, silent music. Ashgate Publishing, Aldershot, pp 97–126
4. Brooks W (2009) Sounds, gamuts, actions: Cage's pluralist universe. In: Essays on and around Freeman Etudes, Fontana Mix. Aria. Orpheus Institute, Ghent, pp 61–95
5. Brooks W (2012) In re: experimental music. Contemp Music Rev 31(1):37–62
6. Brooks W (2013) In re: Re. Unpublished paper, University of Illinois Composers' Forum
7. Brooks W (2014) In re: experimental analysis. Contemp Music Rev (in press)
8. Bruhn CE (2006) Ives's multiverse: the Concord Sonata as American cosmology. PhD thesis. University of New York, New York
9. Campion EJ (2012) Fitting Music Composition Studies for the 21st-Century American University. Contemp Music Rev 31(4):277–282
10. Clarkson A (2001) The intent of the musical moment. In: Bernstein DW, Hatch C (eds) Writings through John Cage's music, poetry, and art. University of Chicago Press, Chicago, pp 62–112
11. Contemporary Music Review (2012), vol 31(4)
12. Crooks EJ (2011) John Cage's entanglement with the ideas of Coomaraswamy. PhD thesis. University of York, York
13. Czernowin C (2012) Teaching that which is not yet there (Stanford version). Contemp Music Rev 31(4):283–289

14. Dewey J (1916) Democracy and education. The Macmillan Company, New York
15. Dewey J (1922) Human nature and conduct. Henry Holt and Company, New York
16. Dewey J (1934) Art as experience. Minton, Balch and Company, New York
17. Dewey J (1938) Experience and education. The Macmillan Company, New York
18. Hiller LA, Isaacson LM (1959) Experimental music: composition with an electronic computer. McGraw-Hill Book Company, New York
19. Ives C (1962) Essays before a Sonata. In: Boatwright H (ed) Essays before a Sonata and other writings. Norton, New York
20. Ives C (1962) Some 'quarter-tone' impressions. In: Boatwright H (ed) Essays before a Sonata and other writings. Norton, New York
21. James W (1890) The principles of psychology. Henry Holt, New York
22. James W (1909) The meaning of truth: a sequel to 'pragmatism'. Longmans, Green, and Co, New York
23. Kostelanetz R (1970) The theatre of mixed means. Pitman, London
24. Lerdahl F (2012) On teaching composition. Contemp Music Rev 31(4):291–296
25. Lindroth S (2012) Teaching composition: artistic growth through confrontation, tact, sympathy, and honesty. Contemp Music Rev 31(4):297–304
26. Perry RS (1974) Charles Ives and the American mind. The Kent State University Press, Kent
27. Ran S (2012) On teaching composition. Contemp Music Rev 31(4):305–312
28. Retallack J (1994) Poethics of a complex realism. In: Perloff M, Junkerman C (eds) John Cage: composed in America. The University of Chicago Press, Chicago, pp 242–273
29. Reynolds R (2012) Thoughts on enabling creative capacity: provocation, invitation, resistance, and challenge. Contemp Music Rev 31(4):313–322
30. Ribe N, Steinle F (2002) Exploratory experimentation: Goethe, Land, and color theory. Phys Today 55(7):43–49
31. Schwab M (ed) (2013) Experimental systems: future knowledge in artistic research. Leuven University Press, Leuven
32. Steinle F (1997) Entering new fields: exploratory uses of experimentation. In: Philosophy of science. Biennial meetings of The Philosophy of Science Association, vol 64(2). University of Chicago Press, Chicago, pp S65–S74
33. Stroud SR (2011) John Dewey and the artful life: pragmatism, aesthetics, and morality. Pennsylvania State University Press, University Park
34. Ulman E (2012) An overview of the symposium. Contemp Music Rev 31(4):249–256

Artistic Research and the Creative Process: The Joys and Perils of Self-analysis

Nicolas Donin

Reflecting on Composition, from the 20th to the 21st Century

We don't live anymore in a world in which composers would secure their posterity by claiming forcefully what the music of the future must be. The age of theoretical treatises, overarching conceptual frameworks, and animated controversies among peers, seems to be gone. Once an important component of (or complement to) composers' creative activity, writing about music was less and less integral to the development of composers born after World War II—up to the point where composers' books, traditionally consisting of collected writings on analysis, theory or aesthetics, were simply replaced with book-length interviews. In the last quarter of the 20th century, the most prominent figures of high modernism declined to go further on the track themselves had set as young revolutionaries in the 1950s: they gradually ceased to theorise and even to write about music, often limiting their interventions to prefatory notes, tributes and autobiographical reflections.

Is this to say that composers stopped reasoning on their art? While the need for grand theories seems to have vanished at least in the official discourse, the reflective side of composition is still viewed by many as an essential component of art music, but it is construed otherwise. On the one hand, composers more often conceive of reflection as a strictly private matter, in contrast to the emphasis on collective theory building and debate that prevailed in the age of extreme intellectualisation of music at Darmstadt, Ivy League Universities, and a handful of studios across the Western world—all of which disseminated the representation of the composer as an intellectual, a researcher, or a scientist of some sort. On the other hand, reflection can take many other public forms than the treatises and manifestos familiar to musicologists: courses, master classes, pre-concert talks, contributions to scientific, artistic, political projects, etc.

N. Donin (✉)
IRCAM STMS Labs, IRCAM-CNRS-UPMC, Paris, France
e-mail: donin@ircam.fr

G. Nierhaus (ed.), *Patterns of Intuition*,
DOI 10.1007/978-94-017-9561-6_16

In line with the previous list, two particular reflection-friendly domains have gained prominence in the last decades and deserve considered attention due to their novelty, as well as their relevance to the present collection of essays: computer programing and artistic research.

Computer programing, be it performed by the composer or by somebody else, implies a formulation of a given musical idea or technique in the language of the computer. Such degree of elicitation at the very beginning of a creative process is neither easy nor welcome. It can prove stimulating to some composers, unhelpful or even frightening to others. This was particularly true before the rise of the personal computer, when specifically designed hardware studios were home to a handful of composers summoned, as it were, to find a common ground between their musical imagination and the programs and procedures made available to them with the mediation of expert users. Boulez reported his own experience, and certainly expressed conventional wisdom of early IRCAM, as he commented on the computer in the following statement: "Above all, the computer forces us to think carefully about the very mechanisms of compositional practice. (…) In any instance of invention, it forces us toward a different pathway, has us look at what's happening from another angle. As a consequence, it disturbs our customs forged by education and practice." [1, p. 46]. Nowadays, computers are more and more pervasive in virtually every domain of our life and we can observe the rise of "digital natives" among young composers—people whose "education and practice" have integrated the computer as well as digital technologies at the deepest level since childhood. These 21st century composers could never conceive of the computer as something external to their world that they would eventually meet, or confront, at some turning point in their creative development. This does not preclude them, however, of adopting a critical approach to the automated component of their compositional practice. While their elders could not escape *reflecting* on computers (they had to choose if and how they would embed computer into their creative world), the newer generations of composers may adopt a more ductile approach: they always use computers at a given rate and then have to choose if and how their use of technology should go beyond business as usual and include a reflective component.

On the other hand, "artistic research" has been continuously developing over the last decades as a practice and, later, as an academic field. I will not discuss artistic research per se (see Darla Crispin's chapter in this book) but put the emphasis on an important dimension of artistic research as we understand it today: each instance of it is an effort to make public something that emerged in the private, fragile and complex realm of a particular artistic practice. In other words, there may be research in every creative process to some extent, but "artistic research" also involves a community of peers of some kind to whom the artistic researcher reports thoughts and acts in an explicit manner. This again does not necessarily imply that the artist/researcher should adopt the traditional forms his ancestors favoured, like treatises or academic lectures. On the contrary, new forms are to be invented as a response to the current social and cultural role of the artist. This is where we come to term with self-analysis—a particular type of discourse that seems to be more and more relevant to artistic research and at the same time raises multiple methodological and epistemological issues.

Defining Self-analysis

What kind of literature emerges from current spaces for critical artistic thinking such as computer programing, artistic research, and so on? In general, "theory" is not the main concern here. "Reporting", "account", "elicitation", "analysis" can probably better describe what the resulting writings consist of. Based on the assumption that post-war compositional theories were unsuccessful in illuminating most of existing music as well as in shaping the subsequent development of musical language, many composers proposed a modest stance: before changing the world by abstractly prescribing what others' music must be, let me first describe how my music itself is developing; and then see if this can be of interest to others. Or, to put it more straightforwardly in the words of Roger Reynolds: "I have no theories. I have ways." [9, p. 1]. Analysing one's own "ways" can be performed as a purely personal process of learning and improvement, but it can also be part of a dialogue with colleagues, students, and listeners. In the latter case, it fulfils a function previously occupied by theoretical writing. Before further discussing the epistemology of self-analytical reports on composition, let us delineate this concept. It is worth distinguishing between different kinds and degrees of self-analysis among the variety of discourses emitted by 20th century composers, often on the fringes of the dominant paradigm of big theories.

First of all, the word "analysis" fits different contexts. It may tacitly refer to "music analysis", or have a broader meaning that virtually connotes any kind of examination methodology used by the self-reflective practitioner. Let us start with the former option, with "self-analysis" meaning "music analysis of one's own piece".

Degree zero of self-analysis occurs when a composer comments for himself/herself on what he/she is writing over the course of the creative process. This is how Joseph N. Straus has described Stravinsky's annotations on manuscript documents pertaining to a particular stage of his late compositional process: "sketching and composing chunks or blocks of material... also involves extensive self-analysis, with Stravinsky carefully notating the serial derivation of each note, particularly in passages where the derivation might not be readily apparent." [11, p. 48]. While Stravinsky did not intend those annotations (rather similar to code comments in current practice of software development) to be available to the public, some of them actually appear in published scores—likely against the composer's will and due to quid pro quos during typesetting of manuscript and proof reading [11, p. 58].

Messiaen and Stockhausen offer radical examples of the exact opposite: they used to write extensive analytical introductions to their works in order to impose meaning to the audience, whatever its time, space and culture might be. That kind of "self-analysis" typically consists of a notice, preface, talk or course in which the basic components and concepts of the piece are presented and illustrated throughout the score. Demonstrating the coherence between the underlying system and its implementation in the work is essential to such didactic writing. The sense of conceptual perfection and aesthetic mastery it attempts at conveying can also contaminate analysis itself. A piece in Stockhausen's catalogue offers perhaps the most conspicuous instance of this tendency: the composer wrote a thorough analytical text about his *Inori* op. 38 (1973–1974) and included it in his official work list as a piece "for a singer" plus "1 transmitter, 2 × 2 loudspeakers, 1 mixing console" and a formal chart to be put up on stage, entitled *Lecture on Hu* (op. 38½, 1971). The lecture must be performed according to instructions no less accurate and constraining than those we can find in other Stockhausen scores.

Previous examples featured a narrow and quite old-fashioned kind of music analysis—a rigid procedure of description aimed at sorting out and labeling bricks of music. Now, what if we rather consider music analysis as a repertory of close reading techniques opening up many possibilities of interpretation and listening? A very different kind of self-analysis then appears. Instead of being paired from the very beginning with the composer's decisions on material and architecture, analysis becomes a means to discover unnoticed aspects of the work in retrospect, to enrich the composer's perception and representation of his own production, and perhaps to generate new conclusions and perspectives out of it. Such an approach is not completely new—it is grounded in the creative use of music analysis that Barraqué, Berio, Boulez, Pousseur and Stockhausen developed in the 1950s based on seminal pieces by Webern, Debussy and other key historical figures—but there is a significant gap between applying close reading to masterworks of the past and to one's own creation. As Klaus Huber put it in his contribution to Wolfgang *Gratzer's Nähe und Distanz* (an edited collection of essays on contemporary pieces confronting, for each piece, the composer's and a musicologist's analyses): "Every composer who starts reflecting on his music after the fact must clearly understand that an adequate analytical approach to one's own work is possible only to a limited extent" [6, p. 250]. Reflecting on his *Der Dichters Pflug* (1989), Huber critically discusses one of the aims he had set for the piece, i.e. writing in third-tones. He highlights en passant an unanticipated consequence of the methods he had devised to develop his pitch material: "the limited selection of pitches and intervals, together with register concentration, brings on considerable "didactic" benefits in order to train the ear to appropriate third-tones" [6, p. 258]. Only in retrospect does he establish the fruitful connection between the nature of his original chromatic space and the rationale of his pitch selection procedure. Elsewhere, he reports how his analytical approach to his past work led him to chart the statistics of duration values used in one section of the piece, which "reveals that *not a single duration is repeated* throughout the sequence" [6, p. 261, Huber's emphasis]—a point that he had overlooked until then, since this was not an intended purpose of the particular compositional techniques in use.

In fact, analysis in Huber's text is not restricted to *music analysis* in the sense of a consideration of the score as an autonomous entity, disconnected from any knowledge of its creative process. More often than not, what the composer analyses in his discussion of *Der Dichters Pflug* are his past intentions, options, trials and errors, over the course of the compositional process; thus the composer's behaviour, emotion and cognition are also an object of the analysis. In other words, the focus of self-analysis is alternatively the score as a product, and score writing as a mental and practical process—none of these two objects being perfectly isolated from the other. Embedding knowledge of both the process and the product, it seems to me, is a defining feature of in-depth self-analysis. The interplay between the work project and the resulting aesthetic object allows the artist to eventually emancipate his/her thinking from the lines that once guided his/her action. Analysis proves instrumental both in recalling the riches of the creative path and in distancing the present and future of the composer from his/her past. Reflection and introspection, turned toward the past as they seem, can also end by feeding prospection (as was the case in the pioneering book of Pierre Schaeffer, *In Search of a Concrete Music* (1952) [10], which is based on the extensive recollection of Schaeffer's early studio experiments in the form of a diary, followed by a theoretical essay charting the future of musical research).

Despite of their diversity, all previously quoted texts have at least two things in common: they relate to a particular work; and they discuss the connection between musical material and structure. The Stravinsky, Messiaen and Stockhausen examples, however, were self-analytical in name only, since they tended to illustrate a theory far more than they shed new light upon the peculiar musical work under scrutiny. In contrast, truly critical self-analysis—be it focused on musical text or its genesis—is capable to affect pre-existing theories as a result of the close examination of the work's ambiguities and potentialities. In my view, self-analysis par excellence approaches the work as (productive or perceptive) experience, process, and singularity—as opposed to a thing, an output, or a particular instance of some general type or overarching system.

In this respect, self-analysis is no less forward looking than theory, but it performs prospective in very different way. Indeed, there is a strong and complex relationship between self-analysis and theory, as far from each other as they are at first sight. One could construe self-analysis as a remedy to theory—this is certainly the way many composers do approach it. According to this view, theories receded because they failed to articulate the particular and the general, the individual and the collective: imperious axioms on the essence of music were, in fact, uncontrolled ad hoc generalization of observations that applied only to the composer's particular style and experience. In contrast, self-analysis relies upon the same events (to some extent at least) but tells the story without any pretension to corral other composers into one's own path toward new music. On the other hand, a partly different conception would insist on the fact that every self-analysis offers observations and hypotheses that could be generalized by confronting them to other composers' findings. Accordingly, self-analysis would be the smartest method to forge general claims firmly rooted into composer's experience. It would be the first step of a long-term, bottom-up process of theorisation.[1]

[1] Donin [4] proposes further discussion of self-analytical accounts by 20th-century composers as a response to theory. For an in-depth discussion of composers' self-commenting in modern music history, see [5].

In brief, self-analysis may well be a partner or a rival to theory in our century, but it isn't anymore a peripheral or exotic reflective practice.

The Challenge of First-Person-Based Research

The relevance of self-analysis as a means to build knowledge, however, is far from obvious. From the perspective of the composer, paying attention to one's past work may just be a waste of time leading to no significant change; and reflecting upon one's own practice may cause more harm than good, since too much reflexivity is expected to inhibit spontaneity, disturb routines, and finally break the very action that it was to illuminate.[2] From the perspective of the researcher, self-analysis is cautious because it seems impossible to assess whether the composer faithfully reported his action and thinking, or not. This argument, however, is not as effective as it might seem. Nobody can go back to the time and place of the creative process and confront the event with the later discourse on it, but this is true, after all, of many past facts attested by a unique testimony—and yet does not preclude the historian for discussing them with a critical eye. Moreover, a significant part of the business of musicology actually implies a critical use of composers' discourse, which must be confronted with evidence of various kinds (e.g. their works, the discourse of their contemporaries, the traces of their creative acts in sketches, drafts, etc.). Finally, there are many different ways of performing self-analysis: while some accounts are deprived of contextual and procedural information, others describe the process of analysis, and then allow for criticism, reenactment or variation by the reader. This is undoubtedly an important criterion for the assessment of self-analytical writings with respect to science: the more they stick to a coherent and explicit methodology, the more they may count as reliable data for musicological or psychological inquiry.

Yet the marginality of self-analysis until recent times is not only a consequence of too much confidence in compositional theory, nor of the composers' doubts about the benefits they could receive from it. It is also part of a ubiquitous suspicion toward first-person-based research in general. Experimental psychology was in no small part edified in response to the introspection-based psychology that had flourished in Germany and elsewhere in the first decades of the century. From the 1950s to the 1990s, a "hard" line prevailed in the delimitation of what was scientific psychology and what was not, preventing mainstream researchers from giving credit to common sense, personal experience, artistic discourse, and any kind of verbal or behavioral traces outside the laboratory. Commonly dismissed as being subjective, non-reproducible and leading to non-falsifiable findings, introspection has only regained some legitimacy in the last two decades, in the wake of post-Artificial Intelligence theories of cognition such as situated action [12], embodied cognition [13] or cognitive anthropology [7]. It is worth quoting a few passages of Varela and Shear's introduction to a seminal issue of the *Journal of Consciousness Studies*, which

[2] I criticized this argument based on psychological and compositional literature in [3].

efficiently addresses some of the epistemological concerns raised by first-person systematic exploration of consciousness. The authors aim at building "a *science of consciousness which includes first-person, subjective experience as an explicit and active component.*" [14, p. 2]. The first-person methods they discuss "share some fundamental common traits or stages": (1) a basic attitude of "suspension and redirection moving from content to mental process", that gives way to (2) "a specific training to pursue the initial suspension into a more full content" (likely with the help of a "mediation or second-person"); finally, (3) "the process of expression and validation will require explicit accounts amenable to intersubjective feedback" [14, p. 11]. As a consequence, "whatever descriptions we can produce through first-person methods are not pure, solid "facts" but potentially valid intersubjective items of knowledge" [14, p. 14]—just as self-analyses were supposed to be in our earlier discussion of a "bottom-up process of theorization". Accordingly, the authors dismiss the view that first-person research should replace third-person research and they express hope that first-person and third-person investigations could illuminate each other in the future. They also suggest that we tend to rely too much on "the apparent familiarity we have with subjective life" and for this reason have still not seriously enough worked on the methodology: "without a sustained examination we actually do not produce phenomenal descriptions that are rich and subtly interconnected enough in comparison to third-person accounts" [14, p. 2]. This is also true, obviously, of the study of music composition. To my knowledge, only a handful of composers have attempted at producing "phenomenal descriptions" of their activity, and they have done so at a very modest scale.[3] Studies involving "second-person" mediation are less uncommon, but still scarce. In this case, the composer's reflection is made possible by a tight collaboration with a partner, who may help him/her to define the object and the method at the outset, and then nurture, monitor and record the process (see a short literature review in [3]).

In the last page of their Introduction, Varela and Shear added en passant a remark that also has significant epistemological resonances: "Human experience, they write, is not a fixed, predelineated domain. Instead, it is changing, changeable and fluid. If one undergoes a disciplined training in musical performance, the newly acquired skills of distinction of sounds, of sensitivity to musical phrasing and ensemble playing, are undeniable. But this means that experience is explored and modified with such disciplined procedures in non-arbitrary ways." [14, p. 14]. Researchers thus shouldn't shy away from interacting with the domain of experience they study by means of a first-person approach: reflexivity or feedback are better understood as a parameter, or dimension, of the research process than as a source of artefacts affecting the purity of the experiment. As a matter of fact, Varela and Shear referred to the practices of introspection, phenomenological reduction, and Buddhist meditation that were discussed in the same volume that they introduced. But their point is generally true of any disciplined effort toward self-understanding, and artistic research,

[3] The earliest attempt might be Janáček's short essay on his stream of ideas and thoughts as he wrote a passage of a cantata during a night [8].

for example, could add new elements to the list in the years to come—as the authors themselves probably guessed by mentioning music performance as an example.

The POINT project offers an interesting variety of the idea of second-person-aided introspection as a transformative process. It has composers express their aesthetics at the outset, receive theoretical and technical feedback from the very definition of the musical project until the production of the final piece, and end at a point to which neither them nor the researchers could have done in isolation. Here, "analysis" is not performed in the form of the composer's disciplined reflection on her/his own work. It rather pertains to the investigators, who select from the composers' discourse a given issue or problem that they can articulate within a formalised, analytical model; the comparison between the output of the model and the output of the composer's activity then leads to further evaluation, discussion and implementation. This dialectics between intuition and formalisation is certainly present in every compositional process to some extent, but it is never as explicit as in this case characterized by the maximal formalisation of a real-world intuition—an experimental instance of dialogical, in vivo self-analysis.

Conclusion

Self-analysis is a crucial component of the current rebuilding of musical thinking along the lines of "artistic research". It includes a series of methods, still in progress, for tracking the aesthetical and psychological stakes of composition, both as process and product. For composers, it represents an opportunity to extract knowledge from the intense experience of creating music, and also to feed a dynamic of creativity and transformation. For scientists, it is one of the most promising prospects for documenting and analysing the compositional process in unprecedented ways, leading to a better understanding of creativity and human cognition, with respect to psychology, as well as musical style and methods of composition, with respect to musicology. But there is still a lot to invent and to experiment with, before we can rely upon truthful methods and fix all epistemological flaws and challenges. We must gather evidence from the past, devise new projects relevant to artistic and scientific research at the same time, and confront our findings with those from a myriad of partly shared, partly different endeavors that address first-person-based research and the creative process. At the moment, we as a research community experience the perils of self-analysis no less frequently than its joys.

POINT is at the cross roads of many issues addressed in the previous pages. If computer programing and artistic research are two strategic places in which reflection on composition develops, then POINT is more strategic than anything as it mixes both. Moreover, the project generated an authentic dialogue between composition and science, maintaining a delicate balance between the individual and the collective, intuition and formalisation, first-person claims and third-person perspectives. It also raises the crucial issue of comparison of the incomparable which is at the heart of generalisation in artistic research: can we elaborate a model, or several models, of the creative process out of dialogical accounts of creative processes and aesthetical worlds as different as those gathered in the present book? Can we compare such models with those sketched in former comparative studies in the process of composition?[4] These questions will become ever more challenging if artistic research, social sciences and the humanities are to join their forces in a truly interdisciplinary effort—as I reasonably hope they will.

References

1. Boulez P (1981) L'in(dé)fini et l'instant. In: Le compositeur et l'ordinateur. IRCAM. Paris, pp 46–47
2. Delalande F (2007) Towards an analysis of compositional strategies. Circuit: Musiques Contemp 17(1):11–26
3. Donin N (2012) Empirical and historical musicologies of creative processes: towards a cross-fertilization. In: Collins D (ed) The act of musical composition: studies in the creative process. Ashgate, Farnham, pp 1–26
4. Donin N (2013) L'auto-analyse, une alternative à la théorisation? In: Donin N, Feneyrou L (eds) Théories de la composition musicale au XXe siècle, vol 2. Symétrie, Lyons, pp 1629–1664
5. Gratzer W (2003) Komponistenkommentare: Beiträge zu einer Geschichte der Eigeninterpretation. Böhlau, Weimar
6. Huber K (1997) Des Dichters Pflug. In: Gratzer W (ed) Nähe und Distanz. Nachgedachte Musik der Gegenwart. Wolke, Hofheim, pp 249–263
7. Hutchins E (1995) Cognition in the wild. MIT Press, Cambridge
8. Janáček L (1993) How ideas came about. In: Zemanová M (ed) Janáček's uncollected essays. Marion Boyars Publishers, London, pp 69–75
9. Reynolds R (2002) Form and method, composing music: the Rothschild essays. Routledge, New York
10. Schaeffer P (2012) In search of a concrete music (trans: North C, Dack J). University of California Press, Berkeley
11. Straus JN (2001) Stravinsky's late music. Cambridge University Press, Cambridge
12. Suchman LA (1987) Plans and situated actions: the problem of human-machine communication. Cambridge University Press, New York
13. Varela FJ, Rosch E, Thompson E (1991) The embodied mind: cognitive science and human experience. MIT Press, Cambridge
14. Varela FJ, Shear J (1999) First-person methodologies: what, why, how? In: Varela FJ, Shear J (eds) The view from within. Special issue of J Conscious Stud 6(2/3):1–14

[4] For example the Germinal project [2] and the MuTeC project [3].

Musicking Beyond Algorithms

Sandeep Bhagwati

Intelligence Amplifiers

> *Haben Sie sich schon einmal klargemacht, dass nahezu alles, was die Menschheit heutigen Tages noch denkt, Denken nennt, bereits von Maschinen gedacht werden kann, hergestellt von der Cybernetik, der neuen Schöpfungswissenschaft? Und diese Maschinen übertrumpfen gleich den Menschen, ihre Ventile sind präziser, die Sicherungen stabiler als in unseren zerklafterten Wracks.*[1]
>
> Gottfried Benn aus "Der Radardenker" (1949)

In 1917, Sigmund Freud[2] introduced a powerful cultural trope that excites our imagination unto this day: he identified three narcissistic insults for humanity. At first, Copernicus had showed earth to be just another planet. Then Darwin had classified humans as just another animal. And, finally, Freud himself declared that we "are not masters in our own home", i.e. that we do not have any meaningful control over our decisions and impulsions. A century later, we can add at least one further narcissistic insult: not even our intellect seems to be unique. It is a well-known conundrum of artificial intelligence research that once highly valued 'brainy' activities (such as playing chess, solving complicated equations, identifying a writer through stylistic analysis) are easier to perform by computer systems than real-world problems that were hitherto not thought to require any elevated intellectual capacity (such as tidying a house, being unpredictably violent, or changing bandages). Computers, it seems,

[1] "Do you realize that almost anything that humanity thinks today, or calls thinking, can already be thought by machines, made by cybernetics, the new science of creation? And these machines instantly trump mankind, their valves are more precise, their fuses more stable than in our disintegrating wrecks.", from: "The Radar Thinker (1949) [2, p. 71], passage translated by Sandeep Bhagwati.

[2] "Eine Schwierigkeit der Psychoanalyse" [4, pp. 1–7].

S. Bhagwati (✉)
Faculty of Fine Arts, Concordia University Montréal, Montréal, Canada
e-mail: sandeep.bhagwati@concordia.ca

G. Nierhaus (ed.), *Patterns of Intuition*,
DOI 10.1007/978-94-017-9561-6_17

will outcompete intellectuals sooner than they will replace housekeepers, dictators and nurses. Are humans hence distinctly human only when they manipulate other humans?

In his 1964 book "Summa Technologiae" [6, p. 159], Stanisław Lem already prognosticated that machine intelligence will one day surpass human intelligence—and that we will need such superhuman "intellectronics", as he called it, because the social and geophysical problems of mankind may soon become so complex that they will overtax the abilities of human flesh-and-blood intellects. Fifty years later, Lem's book still reads as a remarkably fresh and pertinent take on urgent problems—don't we ourselves increasingly suspect that human intelligence alone may be not enough to manage our global messes? One of Lem's most provocative concepts in "Summa" is the so-called "intelligence amplifier":

> The intelligence amplifier ... is the exact equivalent, in the domain of intellectual activity, to the amplifier of physical strength, i.e. every human-controlled machine. Cars, excavators, cranes, machine tools, are amplifiers of strength—as well as every device to which man "is attached to" not as a source of strength, but of control ... Would an intelligence amplifier multiply intelligence by about the same factor by which a normal machine amplifies the physical strength of its human operator, it could attain an I.Q. of about 10 000. The potential for devising such an amplifier is no less real than that of making a machine that is a hundred times stronger than a human.[3]

While such an intelligence amplifier is not yet available as a universal tool, specialised approaches[4] to such an amplification of humanity's intellectual capabilities have steadily gained ground in all fields of human endeavour—even the humanities, and other humanocentric activities[5] that formerly did not use computation. Sifting and parsing the world's ever-growing data streams, all its problematic applications and ramifications aside, has offered us intellectual resources (and, yes, intelligence, in every sense of the word) that were simply inaccessible even a decade ago. In some cases, the results of Big Data crunching look as if they could be the first steps towards an intelligence amplifier—for example in climate science, a field that could not have produced tangible results if its analyses and models relied on human brains alone.[6]

---

[3] *Summa Technologiae* [6, p. 159], passage translated by Sandeep Bhagwati.

[4] In fact, it would be more reasonable to assume that intelligence amplifiers would be rather specialised tools—just as we have no general physical robot that can amplify each of our many physical actions: instead we have cars for faster locomotion, power tools for better penetration, forklifts for better lifting, vibrators for better stimulation etc. If we think about it in this way, there is a case to be made that we already now routinely use Lemian intelligence amplifiers—from shopping 'genies' to route planners, from commodity usage predictors to high speed securities traders.

[5] Such as medicine, baseball team management, predicting the outcome of elections or fashion design—and, obviously, music composition.

[6] Of course, as Jaron Lanier [5] has pointed out, we cannot assume that this amplified intelligence will be beneficial to our societies and way of life—will it not almost necessarily re-create hierarchical, feudalist, societies where some lording corporations and their employees control the flow of information (and thus, the amplified intelligence) and the others simply will be their dumb 'material'—a society that does not need the kind of middle class that currently earns its living solely by its specialised use of intelligence?

If fully realised, such an intelligence amplifier would be the ultimate blow for the narcissistic eurocentric self-image of humanity. Already, its humble beginnings gnaw at the last vestiges of a carefully upheld creed that humanity's multi-versal intelligence somehow is a unique phenomenon. Once more, we will be forced to redefine our role as humans not only in the affairs of the universe—but also within our societies! What aspect of being human would we perceive as being crucial for living, mating, educating? Perhaps, some would propose, our creativity in the arts? And if so—how would we protect it from the intelligence amplifiers? Or would we (and could we), quite to the contrary, use them to make "better" arts—and music?

Generative Paradises

> *We can discern a secret parentage between areas of human life that usually are not thought together: sleep and stupidity, the oldest retreats of the unworldly, come into contact with the cultures of drugs, meditation, speculation—and music, the graceful art that, as they say, removes us from the greyness of hours into a better world. They follow each other like links of an immune system to ward off an infectious, overly demanding world...*
>
> Peter Sloterdijk, Where are we when we listen to music?[7]

At this point it must be noted that, at the time of writing, most of what people listen to as music every day could indeed already be (and sometimes already is) generated or emulated competently by computers, both as structure and as sound. Already today, having live human musicians make pop, or free jazz, gamelan, techno or dhrupad is more a question of aesthetic, social (and hence: commercial) *choice*, than one of technological necessity—because all these styles (and many more) can already be convincingly emulated by software systems. You nowadays need humans to *perform* (and sell) such music to other humans, but you would not strictly need them any more to actually *compose* it—especially if the preferred future mode of reception remains similar to today's predominant mode: "listening to loudspeakers/headphones while doing other things".

Even the need to have humans perform and sell music may change: technologically and socially, it would not be far-fetched to imagine a future world where 99 % of music in people's lives is indeed algorithmically generated on the spot, and intelligently embedded into the situations they are in, heightening the emotive experience of being alive.[8] Every life would have its personalised soundtrack, similar to what is already routine in the controlled environment of digital games. In such a world only 1 % of

[7] Wo sind wir wenn wir Musik hören? [10, pp. 294–325], passage translated by Sandeep Bhagwati.

[8] Similar in emotional impact to—but not as physically hazardous as—other drugs and alcohol. The ubiquitous "soma" drug of Huxley's "Brave New World" could, in such a scenario, turn out to be not a pill or a fluid at all—but rather such bespoke live-generated music—exactly what the Sloterdijk quote, too, seems to imply.

music would still be made for conscious, focused listening—and its audience would expect it to be composed, at least in part, by humans.

The music research program reported in this book, however, is interested not in further understanding these already mostly well-understood mainstreams of current musical life. Rather, it looks at the algorithmic exploration and simulation of highly specialised musical outliers and their aesthetics. It looks for algorithmic representations of some kinds of music that we do not already know very well, musics from the much-less-than-1 % of musical life. These are musics that have gained traction only within a comparatively small community of experimental and improvised eurological musickers. In essence, this project wanted to find out which methods of analysis and modelisation would be appropriate to which of these specialised musical aesthetics and creative techniques, and how close these models can come to re-simulating a convincing instance of these musics—convincing, that is, for the small community they exist in or, at least, their chosen representative, the composer/improviser.[9]

Why is it interesting to research and analyse such minuscule music communities? Moralistic arguments aside, we could turn to the logic of capitalism for a possible explanation. The current functional relationship between musical experimentalism (i.e. new music) and musical mainstream (established music formats, from pop to classical music) in capitalisation-driven societies could be likened to the corporate nexus between research and development and sales departments—the former creates, identifies and develops future profit opportunities for the latter. It is not random coincidence that Stockhausen, Reich and Ferrari have become such revered figures in techno circles, or that Partch has been such an inspiration for Tom Waits and his arrangers. In this perspective, the algorithmic analysis of outlier music seems at first sight a kind of fundamental long-term research without immediate commercial resonance: societies fund such research principally because of the you-can-never-know-where-this-might-come-in-handy reasoning.

This is cultural evolution at work: produce as many variations on a theme and let society validate them. Yet even the imagined future paradise for algorithmic music described above could not sustain itself indefinitely without human input—humans, being humans, always yearn for change. Mostly, it is true, they will desire only incremental change—but sometimes they actually do realise that the time has come for ground-breaking change. Computer creativity could conceivably provide the former, but the latter would still need to draw on the quirkiness of flesh and blood brains. Such change usually comes from the margins—the innovators, the experimentators. In an evolutionary scenario, these outliers would become the providers of genetic mutations, of weird organisms and ideas that were never developed with validation and survival in mind—and yet might just mean cultural survival in a changed environment, by profiting from on conjunctions of social, aesthetical and musical undercurrents impossible to foresee—and thus, to encapsulate in algorithms.

[9] It is evident from the composers' statements from this project that some of them quickly subverted this intention—either by limiting their descriptors to the analytic team or by accepting the resultant software compositions as "raw material" for subsequent work. In this, they used some of the strategies for meta-composition and meta-listening that I will describe later.

However, even the most extreme experimental aesthetic would first need to be represented as an algorithmic model. Only then can it duly fulfill its role as an aesthetic mutation in any future ecology of generative music: hence, the research programme discussed in this book could be seen as a step towards such a future cultural industry. Such an industry would, of course, need to identify effective methods and strategic approaches to algorithmically represent even the most maddeningly idiosyncratic of musical languages. For human aesthetic inventiveness to retain its decisive role also in future musical endeavours, we must be able to translate individual flesh-and-blood-based musical creativity into formalised computer language—as a genetic mutation pool for the continued evolution and continual emergence of new musical styles.

Meta-phenomena

Some people create with words or with music or with a brush and paints. I like to make something beautiful when I run. I like to make people stop and say, 'I've never seen anyone run like that before. 'It's more than just a race, it's a style.

Steve Prefontaine[10].

Style—for the purposes of this text, this term is used not in the usual sense of an "historical" or "genre" style.[11] Rather, it is more widely defined as: *a dynamic emergent meta-phenomenon that allows us to bind into one perceptual/aesthetic model a number of not necessarily related, but salient features that we can perceive as common to several separate instances or samples*. Such an emergent style can, for example, be discovered in the way a person dresses over years and on many different occasions, or in the specific way themes are developed in all of Beethoven's compositions—or by comparing many different musicians who have played a North Indian *raag* (a musicological concept that in itself is remarkably similar to the above definition of "style").

One of the important aspects of this idea of style is that it usually is a meta-feature that emerges from being exposed to many different instances of a phenomenon. Such an emergent style cannot be intentionally "composed", "decreed" or "designed"—it arises through the cumulation and distillation of instances and can only be distilled through retrospective analysis. However, once distilled and thus defined[12] such a style then can eventually become prescriptive and predictive, i.e. musicians can now compose or improvise "in a style", or even create an algorithmic model of it. Any such stylistic model would need to be based on a comparatively large dataset of previous examples and may require unique analytic strategies appropriate to the specific style

[10] Steve Prefontaine was a runner who once held the American record in seven different running events, from 2,000 to 10,000 m. The quote is attributed to him (without bibliographical data) see [13]

[11] Although both obviously would be specific instances of the proposed definition!

[12] And this definition by necessity must be open-ended, in order to leave room for new instances.

under scrutiny. Style thus always must be the result of an observation a posteriori, while *stylistic models* can become prescriptive, even generative.

Used in this sense, *style* therefore seems like a perfect aesthetic category for a world beginning to deal with Big Data. Some of the tools developed for the analysis of datastreams, as well as most neural networks, amplify our own real time data-processing ability to an extent we could not have accessed before. They allow us to discern salient features and emergent behaviours at speeds that we can work with—through their emulations of complex processes we can "see" or "hear" stylistic features in real time—and thus make sense of a datastream that would otherwise be an impenetrable and overwhelming jungle of sensory impressions. Such big data analytic algorithms speed up the process of perceptual learning i.e. they give us, in a compressed format, all the information we need to understand a style. Here lies an interesting paradox: because style is a necessarily aesthetic[13] category, such systems cannot themselves "define" a style. Yet, they can successfully emulate and imitate one, in real time. Stylistic modelling does not even need to be based only on styles that are already well-understood. For most purposes in generative sonic music, mathematical descriptors that simply emerge from the analytic process are good enough—even if there is no musicological term for them.

In classical music, *stylistic variation* is a type of variation that re-imagines melody not through motivic or other technical variation modes but through embedding it in a different stylistic context. Such stylistic variations were more often than not seen merely as harmless salon travesties[14]—but in the late 20th and 21st century they have increasingly gained ground as serious aesthetic devices, for example in the music of Wilhelm Killmayer, Alfred Schnittke, Luciano Berio or John Zorn. In Frederic Rzewski's Variations on *The People United Will Never Be Defeated* (1975), conventional compositional procedures such as transposition, permutation, reduction etc. are used to create emergent stylistic simulacra: music that distinctly evokes a specific style, but actually is the result of a rigorous, almost algorithmic development of compositional elements. This aesthetic method was further—and now algorithmically—explored by Clarence Barlow. In his computer-assisted scores of the early 1980s, which he called "musica derivata", Barlow often used specific styles as attractors to his generative algorithms: e.g. in his piano trio 1981 the music feels as if tossed in a force field generated by three different stylistic attractors (Clementi, Schumann and Ravel), being near to one at one moment, and evoking one of the others in the next moment. But these moments of memory and recognition are fleeting—most of the time, Barlow's algorithms traverse a myriad of interpolated emergent musics that at almost each moment have everything that would qualify them as a style: solely their transient nature seems to bar them from being canonised by a proper name.[15]

---

[13] "Aesthetic" in the sense of "pertaining to how we humans evaluate our perceptions".

[14] Such as the popular variations in the styles of composers from Bach to Brahms, written by 19th century composer Siegfried Ochs on the children's song "Kommt ein Vogel geflogen", and many similar variation cycles in that vein, even in popular music and comedy.

[15] Similarly, almost no stylistic innovation that may come up during an improvisation has a name: only when it is re-visited often enough by the inventor or others (and/or recorded and distributed) will

In such experiments, style was acknowledged as much more than an intangible personal idiosyncrasy, fit to be used by others only as a parody. Stylistic modeling became an important creative tool in music, used to convey aesthetically different moods, much in the same way different chords had been used in the 19th century. In the string quartets of Alfred Schnittke and in the complex musical set-ups of John Zorn, composing with style became as, or even sometimes more, important than composing with harmony or with duration.

Listening to this kind of music is not possible without a conceptual frame of mind: a style always evokes the environment that originally provided us with the raw data that made us understand it as a style. A style is never a neutral technical parameter (as are duration or pitch)—it is, in fact, the most semantic of all musical parameters: when we hear a style appear in a piece of music, it always carries with it a wealth of meaning *beyond* music. This is already noticeable in mono-stylistic music—and must be even more so in complex poly-stylistic compositions where style becomes a structural element of a composition, where a change of style means as much or more for the overall form of a composition as a change of tonality in 19th century music.

As such, style forces us to reconsider the most familiar trope of conventional modernist listening: the focus on sound as the only acceptable aesthetic material, the affectation of our aesthetic perception through listening alone. Stylistic listening re-defines the role of the sonic in musicking. But in order to understand this new role, we must first understand how 20th century musicians and sound artists thought about sound.

Celebrating Materials

Composers combine notes, that's all.
Igor Stravinsky[16]

Today, in eurological[17] art music contexts, any sound, and even silence, can be listened to as music. Yet sounds in themselves, of course, are not yet aesthetic. Making music and sonic art is a process of framing, ordering, contextualising sound:

(Footnote 15 continued)
it be understood as a new aesthetic entity—and thus accrue cultural relevance. This poses a fundamental aesthetic problem for algorithmic practices such as live-coding and computer-improvisation: their *practice* is often understood to be culturally relevant, while their *sonic* result mostly is not—a paternalistic attitude we normally apply only to amateur work, not to high-level intellectual pursuits. At the time of writing, it is not yet clear whether this situation points to a problem in the resultant music or in our cultural prejudices around music.

[16] Igor Stravinsky, Robert Craft, *Dialogues* [11, p. 52].

[17] I employ this term, coined by Lewis [7] in reference to certain forms of jazz practiced in Europe to designate music practiced around the world that is based on the European heritage of musicking, composition, and discourse. I find it more adequate within a global perspective than the terms more conventionally used, such as "Western Classical Music", "Western Art Music", or the falsely universalist term used for this musical tradition by most musicologists in Europe and North America, namely "music".

we all know that one and the same constellation of sonic events can equally well be declared sublime music—or unwanted noise. As John Cage is famously reputed to have said: "If you celebrate it, it is art, if you don't, it isn't.".

Mostly, it is communities that create a framework of spaces and times where sounds are 'celebrated': whether humming while working or whether listening to concerts in halls, whether participating in carefully timed rituals in sacred places, whether going on exploratory soundwalks or whether chanting around the campfire—we always need a context to actually "hear/understand"[18] a sequence of sounds as "musical" or "artistic". In a context that makes us perceive it so, any sound sequence can be music, regardless of whether it was organised with intent or whether it just quasi-randomly emerged from a complex sonic environment.[19]

One of the many processes used in contextualising sound is the process called "composing". Composing is a very peculiar way of contextualising sound. For it relies on a tightly regulated behaviour towards music—namely, the expectation of re-listening and re-production. In other words, composition needs faithful interpreter-performers and, vitally, an audience.[20]

From this follows, obviously, that composing, even if it quintessentially is the act of establishing *context-independent* relationships between sounds, must nevertheless always take into account what performers and listeners can know, what they all can expect and what they therefore can learn to play and/or perceive. While composing, the composer must thus engage with the cultural acoustics and the social setting of the listening situation and must understand how musical elements conceived and ordered independently will ultimately transform within any specific context. Scores must be

[18] The French word *entendre* means both "to hear" and "to understand" [and even: "to signify"]—a homonymy well exploited in French music aesthetics. It indeed seems to embody a useful correlation of concepts for the purpose of my argument.

[19] Conversely, one cannot even hear a Beethoven symphony as music—unless one learns how to listen to it. When my father was a teenager in 1950 s newly post-colonial Bombay, he loved music. One day, a schoolmate gave him a LP record of Beethoven's 5th Symphony. My father was eager to listen to this new treat—but when he lowered the needle onto the disc, he did not hear/understand anything musical: just an impenetrable, loud, messy and noisy jungle of nonsensical sounds. The next day, he complained to his friend about this puzzling cacophony. The friend told him about Beethoven and his high status in the West, and asked my father to listen again and again—even in the background while doing homework or while playing. Although his family had been high-level activists in Gandhi's movement and political prisoners of the British Empire, my father probably did still consider it important to know the music of the not-really-former colonisers. He followed his friend's advice. And, indeed, after a few days of constant re-listening and exposure, the music finally jumped out from the jungle: he could hear/understand it.

[20] The proudly defiant statement, often uttered by avantgarde composers working in the concert/festival circuit, that "they do not write for the audience", however truthful as a description of their mental state while writing, and however revealing as a socio-political commentary on the relationship between audiences and composers, obviously cannot be literally true: even the most arcane piano/orchestra/ensemble etc. composition is still conceived for a concert audience, it plays a defined role in the cultural politics of institutions promoting this music and most importantly, *it usually wants to be listened to in silence—preferably several times*. In that expectation, it relies on a context that exists before itself—in a trivial, but not irrelevant sense, all compositions intended for concert performance or audio publication are just fluctuations (content) within a larger context of public presentation of music, fulfilling its situational, social, political—and yes, even aesthetical expectations.

performed, and in writing the score, it can be vital to know what type of performance will await it: an acousmatic concert, a chamber setting with a few friends, a symphony hall, or a microphone and a savvy editor who concocts a stereo soundfile that will sound good on iPod headphones?

Furthermore, in order to re-produce a composed sequence of sounds, composing, especially as it has evolved in eurological art music, must also deeply engage with the physicality and the history of its sonic technologies. Eurological music, as an art, has always been fairly technology-driven: musical instruments, from the bone flute to brass instruments, from the church organ via the piano to the Theremin, and finally all the recording technologies from wax cylinder to digital devices are not just utilitarian technologies that (re)produce music in the way that a stove produces heat—every instrument and sound device is in itself already an reified aesthetic statement and determines the way music is (and can be) made with it, including all Lachenmannian subversions. Every instrument thus embodies a specific cultural approach to a community and its value systems: a guitar implies another socio-aesthetic context than a harpsichord or an acousmatic loudspeaker.

At this point, though, it must be noted that such a comprehensive perspective on composition has not been—and often still is not—the dominant inner game of eurological composition. All too many musickers, and especially those composers who celebrate it as art, believe that music is a largely self-contained entity, unperturbed in its essence by mere social context and the trivial limitations of physicality. Composing is taken to be an idealistic action that establishes a meta-physical, immaterial architecture of evolving time-sensitive relationships—more often than not it is definitely not seen as an intervention in a real physical and social environment.

The reasons for this are manifold—but one major reason could be that, unlike paint, canvas or stone, the material substrate of music was intangible and difficult to pin down. Voices and instruments have always been elusive and transient sources of sound. Until the late 20th century, the relationship between scores and sounds tended to be fickle and unreliable. Variances in sonic character between two realisations, from time to time and from place to place, might well be aesthetically more relevant than the internal variances within a composition. Rather than notate every detail of their envisioned sonic world, many composers of occidental music opted to not think about precise sound as a major component in their compositional process at all: for by abstracting, reducing and transcribing complex sonic events into separate conceptual (and thus: under-descriptive) objects such as notes and chords, they were able devise and manipulate musical architectures without actually engaging with the physicality of sound.

This is a principal reason why, for a very long time, and often even today, most eurological composers preferred writing for a traditional instrumentation, such as a string quartet, a choir, or a piano etc. Such time-tested instrumentations, and the clearly delineated social settings they had grown out of, afforded composers the same benefits that a controlled environment affords experimental scientists: they could focus on the inherently musical and structural questions at hand. By the middle of the 20th century, eurological scores had developed rich and complex sonic classifications, as well as evolutionary and administrative systems that could in theory work well with any arbitrary catalogue of sounds.[21]

It is mainly for this reason that eurological art music has not played a leading role in the proliferation and exploration of the kind of conceptual art practices that have transformed so many other eurological art forms over the last 50 years—because eurological music had long ago established its own strong, but explicitly music-immanent, theory and practice of conceptualism.[22]

Thus, while conceptual art forms in the visual arts increasingly focused on socio-aesthetic contexts *beyond* or via the materiality of art-making, aiming at immaterial constructions of the aesthetic, 20th century music and sonic art largely took the opposite direction: given that all they could rely on was this elusive, immaterial relation between movements of air molecules and electrochemical nerve messages that together give rise to the phenomenon of sound, with its ephemeral and intangible architectures, prevailing trends in avantgarde music and the sonic arts have, over the last 100 years, tended to first focus on substantiating the counter-intuitive claim that, indeed, *sound is material*—and then on working *with* and *through* this materiality that they had just proclaimed.

Ironically, however, even the so-called materiality of music is much more abstract than in the visual arts: on the one hand, emerging digital audio technologies seemed to encourage working with sound itself as an intuitively sculptable and thus as a 'material object'. But in reality, the practices both of sculptural transformation and direct manipulation of sounds heavily relied on what one, in reference to Walter Ong, could perhaps call "secondary"[23] materiality: a virtual materiality enabled solely by mathematical representation and processing.

Yet, even such heavily mathematised digital representations of sound were still too varied to be subjected to algorithmic analysis and processing. Their mathematical representations were too manifold: while representing sound through a more

[21] It should be pointed out, however, that abstracting musicking from the sonic is not the only possible reaction to the ephemerality of sound. Other music traditions display a large variety of approaches in coping with the fickleness of the sonic. On the one hand, some East Asian (Chinese, Japanese) art music traditions seem to address the matter head-on—they consider the timbre of a note to be even more important than its place in time. They therefore must precisely notate instrumental playing technique rather than, say, duration and precise rhythm. Others deal with the issue by considering the actual sound of musicking as an aesthetically rather marginal element. Hindustani music philosophy, for example, posits a spiritual ideal of sound, *dhvani*, that must necessarily always remain unmatched by actual sonic events—and, like qin aesthetics, it also knows the concept of *anahata*: inaudible sound.

[22] Only recently can we observe a conceptual turn in composition that mirrors the context-conscious conceptualism of the visual arts. The author of this text feels a certain creative affinity with the dispersed and heterogeneous non-movement of conceptualist composers that might be inferred from individual aesthetics as diverse as those of Peter Ablinger, Mark Applebaum, John Oswald, Clarence Barlow, Johannes Kreidler, Sergej Newski, Chris Newman, Hannes Seidl, Martin Schüttler, Alexander Schubert, John Zorn, etc.

[23] In his book *Orality and Literacy* [9], Ong develops the concept of a "secondary orality" that manifests itself in highly technological (and hence literate) societies—e.g. in TV talkshows or the telephone. Similarly, one could define a "secondary materiality" as one that relies on virtual de- and re-constructions of its "materials".

than 25-dimensional, time-sensitive and context dependent parameter space might be manageable for real-time sonic resynthesis in neural networks,[24] algorithmic musicking needed less detail and more abstraction—it needed musicological concepts.

Thankfully, computer musicians could draw on the other, the reductionist tradition of materiality in music. Since the beginning of the 20th century, composers had increasingly begun to view and designate as 'material' those under-descriptive symbols of their usual craft: their notes, chords, scales, rhythms. In many different theoretical moves[25] they prised them out of their usual contexts, isolated their different functions and broke down complex *clangs*[26] until they were left with three basic materials: pitch (hertz), duration (milliseconds) and amplitude (dB). From these, at least in theory, all musicological concepts could now be re-constructed bottom up through detailed layering—for example, a sonic timbre was conceptualised as a set of coordinated pitches with correlated sets of durations and amplitudes. This reductionist approach to a re-ordering of the sonic realm finally paved the way for advanced algorithmic processing within a well-manageable parameter space and with the possibility of creative inventions of alternate musical entities, ranging from novel clang-objects, as e.g. in Tenney's compositions, to novel stylistic models, as e.g. in Barlow's music.

While both approaches, the intuitive secondary materiality of electroacoustics as well as the counter-intuitive symbolic materiality of computer music were, of course, deeply imprinted with the millennium of eurological music theory and cultural acoustics that had preceded them, they both also opened up a remarkable new terrain for creative exploration: the decontextualisation and parameterisation of these 'sonic atoms', as well as the conceptual dissociation of sound from both the flow of perceived time and the physical limitations of the musician's bodies, afforded composers unfettered expeditions into new sonic and dramaturgical experiences. On the one hand, they could now build novel types of harmony, rhythm and melody from scratch, types that may have been implicit in older musics, too, but could not be controlled and realised with earlier music technologies. Even though these new, additively developed musical elements were still mostly devised for and filtered through the existing sonic affordances of traditional instruments, their collateral effect was to introduce new sound qualities into the landscape of composable music.

On the other hand, the possibility for composers to concoct sequences and superpositions of musical 'atoms' without heeding the established conventions of instrumental performance forced musicians to discover new playing techniques and develop new performance skills. These often palpably virtuosic techniques again and again provoked moments of pure wonder in the attentive listener—split-second sleight-of-hands which often served to divert the audience's ears away from emotional

---

[24] As, for example, in the live-computer improvisation project *Native Alien* discussed in Section "Through the veil".

[25] Non-diatonic and non-identical scales, alternate definitions of consonance in chords (e.g. triads based on the fourth), dodecaphony in its various guises, diverse takes on microtonality, synaesthesias of various kinds, etc.

[26] James Tenney's term for perceivable sonic building blocks in complex compositions [12].

and structural architectures, and primed them to pay more attention to the adventurous life of sounds.

In the 1950s, Karlheinz Stockhausen had fed back his studio experience around non-dramaturgical, explorative, in-depth encounters with sound into his instrumental concert music, and had called it "moment-form". This formal idea has audibly become the dominant standard music architecture of the late 20th century—new compositions, and especially those using algorithmic processing, often appear to be nothing more than seemingly random walks through exquisite soundscapes untrammeled by dramaturgical or narrative flow.

All these trends point to a significant change in the way new music deals with time. Philosopher Albrecht Wellmer expresses this novel aesthetic concern when he writes:

> Time is not everything, not even in music. Music often rebels against it in the name of a present which gathers and concentrates its dimensions within itself.[27]

To do justice to this 'rebellion' in temporal processing of sound, a new kind of listening seems necessary—indeed, one could ask if "listening", in the traditional sense, is still what we [should] do with music.

Through the Veil

> *If aestheticised art—oblivious of boundaries as art can be—begins to draw our entire reality into the dream and intoxication of art and in some sense replaces reality by art: then this is not only aesthetisation of art, but aesthetisation of reality. This is not really good, because aesthetisation of reality means anaesthesia of humans. Thus aesthetica—dangerously—become anaesthetica.*
>
> Odo Marquard, Aesthetica and Anaesthetica[28]

Much of what this article discusses is informed not only by the research project that is the subject of this book, but also by my own research into live-improvising software architectures, especially the experiences and insights afforded by my recent research-creation project *Native Alien*.[29] In this project, we developed a response-driven creative sonic environment for an improvising solo musician, a co-improvising computer system that would feed back both generative sounds and score information to the musician, so that both of them would co-evolve in musicking within a style-model-driven architecture.[30]

---

[27] Albrecht Wellmer in: Identity and Difference [14, p. 71].

[28] See [8, pp. 12–13], passage translated by Sandeep Bhagwati.

[29] There is no comprehensive article on *Native Alien* yet, as the project is still in evolution and testing, but some glimpses [video, audio and text] may be found on the project website http://matralab.hexagram.ca/projects/native-alien/ [accessed on July 1, 2014].

[30] In this case, the style-models in question are not based on pre-existing musical styles: rather, a major part of the compositional act was to closely define each of the nine style-models invented for this piece—a comprovisation technique I call working with "encapsulated traditions", see [3].

Our interest here was the evolution of stylistic features and musical material over time—and in particular, as we could see in the Barlow example cited above, the emergence of new, stylistic entities somewhere in the no man's land *between* the defined style-models written into the meta-score of this work. Using the live situation to both explore these emergent styles, both via a performer's intuition and via computational layerings of different algorithms, has become one of the most exciting aspects of *Native Alien* performances.[31] And at the same time, this work opened up many new questions around listening, both from the performer's and from the audience's side.

In traditional music reception, theories of listening mostly view listening as an act that is somewhat similar to understanding a message. An attentive and educated listener (supposedly as opposed to a "normal" listener) is able to decode all the *intentional* information contained in the music. Adorno has described such an ideal listener in his *Introduction to the Sociology of Music* [1]:

> The *expert listener*... can be defined as one who listens adequately. He would be the fully aware listener, who hardly misses anything and who, at every moment, can give an account of his listening. Who, for example, confronted for the very first time with a largely liquefied work devoid of architectonic supports such as the second movement of Webern's String Trio, could name its formal elements, would, for the time being, be a satisfactory example of this type. While spontaneously following even complex music, he hears the sequence of past, presence and future moments together in such a way that a meaningful context is crystallized. He can distinctly grasp even the convoluted intricacies of simultaneous, complex harmonies and their polyphony...

Even Adorno concedes that this ideal is rather utopian, and somewhat totalitarian in its expectations—but he nevertheless posits it as a worthy goal to strive towards. The systemic problem with this model, however, is that Adorno expects a differentiation through mere perception that can only be sustained in a perfectly coherent environment: one can only become an expert listener *within* a clear stylistic framework. If one knows the rules one can understand how they are fulfilled and/or broken. But what if one does not know the rules—or if the rules change while the music plays? An expert listener for Webern might well flounder if confronted with the aesthetic universe of José Maceda—or with the genre once called *Intelligent Dance Music*. If playing with stylistic modeling becomes an aesthetic tool not only for musical material, but also for musical dramaturgy, if perceptible musical structures are organised, demarcated, and sequenced in ways alien to eurological music, even an expert listener of Adornian description would find it difficult to identify the salient structural properties of any given sonic flux solely by listening. Such a listener would be even less able to penetrate aurally into algorithmically composed music, where generative processes, the rules of the music, are interactively layered and nested within each other in a myriad of ways.

---

[31] In April 2013, *Native Alien* was presented at Mumuth Graz in a concert performance with Mike Svoboda (trombone) and Navid Navab (computer). Navab is a co-creator of *Native Alien*, and his role is similar to that of a conductor: he does not actively "play" or "trigger" any of the sounds that emerge, but watches over the musical evolution of the entire orchestra of algorithmic improvisers. Recently, bass trombonist Felix del Tredici has been the main performer with the system.

Thus, when styles become a feature of algorithmic composition, they seem to demand another approach to the role of listening in making sense of music and sound: an approach that understands how a style can emerge from the interaction of complex systems and massive datastreams—and therefore can penetrate beyond both the sonic surface and the obvious musical structure and dramaturgy. Listening to algorithmically generated music confronts us with some of the most nagging questions of music-making today: what does music signify when algorithms are responsible not only for its sound and its material structure, but also for its composition and style? Is this a meta-implementation of John Cage's idea of *non-intentionality* in music—a systematic, mechanical non-intentionality that does not even result from a conscious artistic engagement with the non-intentional? And how can such a non-intentionally non-intentional music mean anything at all, subjective likes and dislikes aside?

In his book *Aesthetisches Denken* [15], Wolfgang Welsch sketches a situation that perfectly illustrates different modes of listening: When you hear two people debate a point, you can try to understand their explicit arguments—this would be listening to their *intended* meaning. Or, you can additionally try to listen to the tones of their argument, trying to determine whether they are angry at each other, whether they think they are losing the argument, whether they both believe what they say etc.—and this would be meta-listening: the act of carefully scrutinising the layers of *implied* and *contextual* meaning in any given situation.[32]

If composers would indeed only combine notes, as the famous quote by Igor Stravinsky claims, we would need no meta-listening. We would just need to analyse the notes, and unpack their intended message, openly manifest in the way they are combined. This was indeed, for a long time, the premise of musicological analysis, especially in the German speaking countries, where the term *werkimmanente Musikanalyse* (work-immanent music analysis) stood for a noble devotion to the musical score alone, an analysis untainted by biographical anecdotes and other 'incidental' facts—or, indeed, by the sordid vicissitudes of an actual performance.

Such an immanent focus on the material substrate, and the strong belief in its power to anchor and decisively determine the meaning of music also underlies, in many ways, the practice of algorithmic analysis, and also the recent musicological methodology called "Analysis by Composition"[33]—a methodology that has some technical parallels to the present project, even if their goals and intentions are different. If notes and their combinations are indeed all we want to think about in considering music, such approaches can hold some promise of success. But if we look at musicking in a more comprehensive perspective, we need to look at more than the reductionist,

---

[32] Welsch calls this act "aesthetic perception", a term which makes more sense for verbal or even visual situations where you can separate *semantic* perception and *aesthetic* (or formal) perception more neatly than in music, a non-semantic formal mode of communication. Hence my term 'meta-listening'.

[33] "Analysis by composition" as an emergent musicological field heavily relies on algorithmic music technologies, both on offline algorithmic composition tools and on a variety of live-improvisation softwares already in regular concert use such as *OMAX* (IRCAM Paris), *Voyager* (Columbia University New York), *Prosthesis* (Goldsmiths College London) and *Native Alien* (matralab, Concordia University Montreal) and more.

but enveloping 'architextures' of musical materiality or of sonic reality. Like *maya*, the elaborate veil of illusion in old Indian thought, the combinations of notes and the many richly sensual sounds present in a music piece may actually be nothing more than a smokescreen that hides from us the rich and manifold reality of musicking.

If all we think about in listening to music is its reality of sound and notes and the immediate emotions they evoke, we may indeed become, as Marquard writes, anaesthesised by their immersive aesthetical reality. We therefore need to practise listening as an exploration of the implied and contextual meanings present in the sonic surface of any music we hear. This, of course, is desirable both for non-algorithm-based and for algorithm-based creations. But while non-algorithmically composed music allows us to speculate about the composer's intentions—and to listen (and analyse) as if our task were to discover or decode these through the sonic structure alone—we cannot assume that there is any decodable intent to a event we perceive in algorithmically composed sonic art. We may well assume an arcane justification for its presence, namely the calculations that led to it—but this justification is not any sufficient indication of conscious artistic intent—except if the composer already *expects* us to meta-listen, i.e. if the meaning of the musical act we hear does not primarily reside in its sonic surface, but in its context.[34]

In engaging with algorithmic music, we seem to be well-advised to listen not only to its immersive sensuality or even to its audible dramaturgy—for we are invited to hear something beyond these phenomena: to discern when and how a style appears and disappears, how a texture emerges from hidden processes, how a modified rhythm affects our sense of time, etc. At all times, it seems, we must be aware of the context from which these calculated sonic structures emerge: what social, moral, conceptual, aesthetic function does the music we hear have for us, the listeners, as well as for the meta-composers of the algorithms—and, thereby, for all the other musickers at this location and at this moment in time. So that, even as the algorithms absent-mindedly churn out their potentially infinite progression of notes and sounds, we are still able to listen beyond the 'anaesthetic' environment that this *maya* of notes and sounds creates for us—and to again experience, through meta-listening, those precious moments of epiphanous aesthetic perception that we so crave in our restless entanglements with art.

Amplifying Intelligence

> *Thus, all thinking moves to the robots who will take care of the basic necessities. What remains are the rudiments of an earlier volcanic existence, and wherever they show up, they already appear to us as inhuman, and a mess.*
>
> Gottfried Benn, The Radar Thinker (1949)

In this quote, Gottfried Benn expresses a typical Western intellectual's fear and self-loathing when confronted with the idea of amplified intelligence. Similar fears

[34] As is the case in many of the works of the conceptual composers mentioned in footnote 22.

are palpable in many conversations with musicians and composers whenever the subject of computer-aided composition arises. Even if they hesitatingly should accept the 'remote possibility' that computers might someday indeed compose something that an educated, regular listener would accept as 'good' music (and surprisingly many resist the very notion), they will immediately voice concerns over the demise of composition as an art form.[35]

Such concerns echo the worries of naturalist and portrait painters who once fumed at the invention of photography. Depicting the visible world as mimetically as possible in two dimensions had over generations become a finely honed skill and a complex aesthetical style, one that demanded years of devotion and study, trial and error. Now, it suddenly looked to be replaced by a mechanical and chemical process accessible even to laymen with no artistic credentials. But, as we all know, the visual arts have survived the aesthetic (and economic) shock of photography quite well—and some of their creative survival histories can be seen as strategies that already now imbue the art of composition with new relevance, and make intelligent and artistic use of intelligence amplification.

Diversification

Grammar aside, music does not exist in the singular: there is no reason why algorithmic music should be subject to aesthetical concerns relevant to earlier traditions of musicking. The project-at-hand aims at emulating human-made music and the resultant scores/musics are then evaluated by musicians—but it also shows that these evaluations need not always be pertinent: it is almost like asking a late-1800s portrait painter to evaluate the similarity of an early black-and-white photo portrait to the person he himself has also just painted—one would obtain an interesting opinion, but certainly not conclusive, evaluative evidence. In fact, such aesthetic transfers and legacies may even muddy the issue, just as early "artistically valuable" staged photographs ignored what has turned out to be photography's major aesthetic innovation: the possibility to work in an aesthetic of the plausible snapshot, a new kind of meta-naturalism.

Algorithmic music thinking, both in composition and in analysis, had already developed aesthetic schisms before computers entered the stage, opening up musicking to novel kinds of meta-sonic architectures (serial, spectral, randomized)—but its real aesthetic contribution to musicking may yet lie in time- and context-sensitive re-fashionings and re-thinkings of musical architecture—a potential many projects and compositions have hinted and worked at, but which has to date not developed any distinct artistic ecology or aesthetic school of thought.

---

[35] Indeed, should the 'paradise' evoked in section "Generative paradises" one day become reality, the case for teaching sonic composition as more than a software skill for corporate use probably may become hard to make.

Conceptualisation

Question: If technologies (or indeed even human improvisers and creative musicians) can take over all the skill and materials part of making a score, what on earth could there be left for a composer to do? Answer: Almost everything of real artistic interest! For example, work on the score itself could begin in earnest: what can a score be, what are its particular functions, which score parameters can become aesthetically meaningful, in which setup etc.? Or one could work on the internal ecology of a score: how to regulate the inner relationships as they flow, how to make artistic statements using energetical, textural, emotive meta-structures? Or one could focus on non-musical elements such as new forms of using text, light, space, movement etc. while the computer in the background composes music that fits these artistic decisions. An invitation to listen and look beyond the nitty-gritty may not be an entirely bad thing for new music, given the seemingly endless parade of etude-like pieces that one can hear at new music and sonic arts festivals—pieces plodding through the technical variants and details of one single sonic obsession! Of course, learning skills in real depth will always be useful to a beginning composer: the process itself can serve to hone and sublimate your mind. But do we really still need to insist that the skills required in studying composition must necessarily center around writing scores and mastering eurological music theory? New types of ear training for the sonic arts have already be implemented in daring music schools—what if they started to teach listening into algorithms, too...

Re-appropriation

More than 20 years ago, in my first in-depth collaborations with an algorithmic computer program I did not use the software patches I wrote to actually compose music—I used them to develop new styles. I would fiddle around with them, letting them make audible music all the time, trying to get a visceral sense of what a parameter change or a logical change would shape the music into: which styles would emerge? What could be an interesting turn, structure, texture etc.—and what stylistic behaviour would fit to the aesthetic idea I had in mind for the pieces I was writing? Once I had found that stylistic setting, I would let the computer compose many instances within that style, listening to the (still very awkward) playback again and again—until I felt I had an intuitive sense of how this music worked. Then I moved to my desk, quit the software—and started to compose quasi-freely in the style I had now absorbed, armed only with pencil, paper, and my algorithmically primed inner ear.

The modes of re-appropriating the particular aesthetic of a specific algorithmic music (not its inner mechanics but its, one may call it, 'sound') can be manifold and must by necessity be highly idiosyncratic—but such a non-devotional attitude towards algorithmic music can offer a wide range of hitherto impossible compositional approaches, including new ideas about the relationship between sketch and composition...

Contextualisation

This is an aspect that has come up repeatedly over the course of this text: to think of music as an event in a social context—and to use this situation artistically, in the manner of conceptual art practices prevalent in the visual arts since the 1960s. The emphasis on (and the near-inevitability of) specific skills, practices and social institutions in making music has hitherto prevented conceptual art to really blossom in music. But these skill-based aspects can now be dealt with by algorithms—working conceptually in music does no more face any technical hurdles. Of course, one then still needs a compelling and well-structured idea and concept...

De-centering

Obviously, algorithmic musicking has up to now always had a strong eurological art music bias: many of its underlying ideas not only about the basic elements (parameters and relevant structures) of music making, but also about how, when and in what social context music should be presented, come from eurological art music practices, i.e. from traditional or experimental forms indebted to the so-called 'classical' art music of Europe. Opening up to aesthetic ideals or artistic practices, social contexts and musical tools from musical traditions or genres beyond the eurological parish might expand and enrich our musical life not only with new techniques or new concepts, but also with new musical contexts and new kinds of practitioners: there is a case to be made, for example, for the insight that interactive live-improvised computer music may yet find its strongest practitioners in Carnatic or Hindustani art musicians, because the way in which they think about music while improvising it is much more compatible with the conceptualisation in a software than are the inner games in eurological music practices. In another example from Hindustani art music, the cultural association of musical structures to times of day could spawn an entire algorithmic aesthetic of context- and situation-based real time composition. Such kinds of intimate and entangled engagement with global aesthetical ideas and practices of sound and music will thus almost certainly throw open a wide field of research into the aesthetical impacts and affordances of algorithmic composing and music fabrication.

* * *

All these strategies have one thing in common: they rely on and at the same time subvert the specialized kind of intelligence amplifiers that algorithmic music softwares have steadily become over the past decades. Whether used offline or live, they already determine the way we make music—and will even more determine the way we will listen to music in the future. In the olden, golden days, composing music used to be a secretive and highly-regarded cultural activity. Already, over the course of the last century, being a composer has become a proposition with significantly less cultural prestige. This slide down the social and cultural slope will continue as music alchemisms turn into sonic algorithms. Maybe in the future, just like chess players and marathon runners today, we composers will continue to cook up music not because there is any real need for it, but purely because we happen to like it and have a talent for it, pursuing a healthy, absorbing pastime that has become irrelevant to society at large.

Or we can invent a new *raison d'être*, a new social justification to use our musical imagination—a motivation rooted in a deep engagement with the world around us, a way to amplify our musical intelligence through new contexts and tools, leveraging the magmatic forces of our earlier 'volcanic existence' that Benn still had found so embarrassing—and to erupt anew, forging musical realities that the algorithms we invented, for all their intelligence, will not yet have foretold.

References

1. Adorno TW (1962) Einleitung in die Musiksoziologie. Suhrkamp, Frankfurt am Main
2. Benn G (1991) Der Radardenker. In: Sämtliche Werke, Prosa 3. 1949, Stuttgart
3. Bhagwati S (2014) La superposition de traditions encapsulées dans les partitions comprovisées. In: Ayari M (ed) Penser l'improvisation. Editions Delatour, Sampzon in print
4. Freud S (1917) Eine Schwierigkeit der Psychoanalyse. In: Imago. Zeitschrift für Anwendung der Psychoanalyse auf die Geisteswissenschaften V. pp 1–7
5. Lanier J (2013) Who owns the future? Simon and Schuster, New York
6. Lem S (1981) Summa technologiae (trans: Griese F 1964). Suhrkamp, Frankfurt am Main
7. Lewis GE (2004) Improvised music after 1950: afrological and eurological perspectives, (1996). In: Fischlin D, Heble A (eds) The other side of nowhere. Wesleyan University Press, Middletown
8. Marquard O (2003) Aesthetica und Anaesthetica. Wilhelm Fink, München
9. Ong W (1982) Orality and literacy. Routledge, London
10. Sloterdijk P (1993) Wo sind wir wenn wir Musik hören? In: Weltfremdheit. Suhrkamp, Frankfurt am Main
11. Stravinsky I, Craft R (1982) Dialogues. University of California Press, Los Angeles
12. Tenney J (1988) Meta (+) Hodos: a phenomenology of twentieth-century musical materials and an approach to the study of form. 1961/1986. Frog Peak Publications, Oakland
13. URL: http://running.about.com/od/runninghumor/a/prefontainequotes.htm (visited on 01/07/2014)
14. Wellmer A (2004) On music and language. In: Cross J (ed) Identity and difference: essays on music, language, and time. Leuven University Press, Leuven
15. Welsch W (1990) Ästhetisches Denken. Reclam, Stuttgart

Boulez's Creative Analysis: An Arcane Compositional Strategy in the Light of Mathematical Music Theory

Guerino Mazzola

We investigate part I of the famous composition *Structures pour deux pianos* by Pierre Boulez with regard to their mathematical construction principles and interpret the analytical results in order to obtain computational schemes for generalized compositions following Boulez's approach and also in the lines of Boulez's principle of creative analysis. These generalized schemes are then implemented in rubettes of the software Rubato and yield corresponding compositions. Our analysis confirms the visionary force of Boulez's innovation in that his matrix methods for part I turn out to be in complete congruence with the category-theoretical situation created by generally addressed points in the spirit of the Yoneda lemma and then systematically used by Alexander Grothendieck.

Boulez's Arcane Idea of a Creative Analysis

In [5], Pierre Boulez describes a compositional strategy, which he calls "analyse créatrice". It is opposed to what he calls "sterile academic" analysis in that the analytical results are used as germs to create new compositions. It is not only analysis, but more a strategy of being creative in composition. His procedure transcends the purely analytical or compositional activities: He proposes a coherent double activity, which includes both, analysis *and* composition, and doing this in a creative way. We will therefore deal with both, analysis *and* composition, and the latter will more specifically be realized by use of the music composition software Rubato [11] for reasons that intrinsically relate to Boulez's computational approach to serialism.

G. Mazzola (✉)
School of Music, University of Minnesota, Minneapolis, USA
e-mail: mazzola@umn.edu

G. Nierhaus (ed.), *Patterns of Intuition*,
DOI 10.1007/978-94-017-9561-6_18

Let us explain the immediate consequences of Boulez's strategy[1]: Anne Boissière [2] has given a concise summary of Boulez's ideas on creative analysis, which comprise these core items:

- The analysis focuses on the limits of the given composition and may neglect historical adequacy. These limits open up what has not been said, what was omitted or overlooked by that composer. In this sense, it is a genuine creative activity in the sense of creativity theory, as exposed in [14]. This hermeneutic work does not aim at a new composition qua special case of what has been recognized (deduction), nor is it meant to help create the new composition by a transition from the particular to the general (induction). Referring to Gilbert Simondon's philosophical reflections [16], the creative movement consists in the opening of a topological neighborhood of the given analysis in a space of analytical parameters. In this space, analytical structures topologically similar to the given one are chosen and used as initial points for the construction of new compositions. Simondon coins this "horizontal" movement a "transduction".
- In this transduction process, Boulez calls the composer's *gesture* the movement towards new compositions, which share precisely those analytical structures similar to the given analysis. I.e. we use the analysis of the given work and make a some "small" value changes for the analytical parameters. For example, if we have exhibited a set of retrograde symmetries that govern the given work, we may extend that set and include also pitch inversion symmetries. Or if we have recognized that an instrumental voice is derived by some systematic procedure, e.g., time expansion (dilation) plus transpositions from the leading voice, then we may add more instruments and apply the same procedure, e.g., further dilations plus transpositions to define these new voices.
  This creative gesture—building new works from the transgression of the analytical structure discovered in the given work—is what Boissière calls a *detonation*. It is precisely this act of breaking the given structures and stepping into unknown neighborhoods, which characterizes Boulez's concept of an *open work*.

To our knowledge, these ambitious claims have not been reified by concrete examples: How should and would such a strategy work in detail? This is what we have accomplished in a formal (mathematical) setup and on the level of computer-aided composition. And this is what we want to discuss in this contribution. In view of Boulez's rather poetical text, such an enterprise cannot be more than a initial proposal. But we believe that it could open a fruitful discussion about the dialectic between analysis and composition. It is therefore completely logical to pursue the trajectory to its completion, namely to the construction of a full-fledged composition[2] (see Sections "Implementing the First Creative Analysis on Rubato Composer" and "A Second More Creative Analysis and Reconstruction").

---

[1] For a more philosophical discussion of this approach, we refer to [13, Chap. 7].

[2] It should however be noticed that such a creative analysis had been applied in the case of Beethoven's op. 106 [10] before we knew about Boulez's idea. The present approach is somewhat more dramatic since we shall now apply Boulez's idea to two his own works, namely *Structures* [3, 4].

Creative Analysis, Quantum Mechanics, and Topos Theory

If we qualify Boulez's description of a creative analysis as arcane, dark, mysterious, it is so not so much because of its abstract idea, in fact, Boissière's above summary is quite precise on that level. The dark side is rather located in the question of how to create a conceptual framework to enable a topological space that would yield a natural transduction of given data, a creative extension of "boxes" defining the given composition. Is there any chance to have a natural extension of such box spaces? The situation Boulez is invoking requires that, while acting on the given composition analytically, similar to a physicist making an experiment, the object in question acts back on the analyst and evokes a detonation towards a new composition. In physics, this intimate interaction is well known from quantum mechanics, where the measurement of an observable hits back to the experimenter's and the object's state. *One could view creative analysis in music as an interaction that carries over quantum mechanical approaches and problems to music.*

It is in the vein of this analogy that we believe being addressed to rethink Boulez's proposal in terms of the type of mathematical music theory we have been developing since the early 80s, and which has been exposed in [12] as a theory constructed upon Alexander Grothendieck's topos theory, see [9] for a thorough reference. Why? Because quantum mechanics and topos theory share deep commonalities of conceptualization, refer to John Baez's[3] fascinating work [1], for example. These commonalities touch upon the crucial question of what is reality. It is a well-known quandary in quantum mechanics that it has redefined reality in a rather spectacular way, in particular embedding the duality of particles and waves in a formalism where those apparent conflicts disappear. Without delving into technicalities, one can summarize this solution saying that the physical reality of quantum mechanics is rooted in observables A, which are linear operators on a Hilbert (state) space. These operators cannot be experienced as such, but only via their measurement actions upon states M, yielding probability measures with expectation value $\langle A|M\rangle$. Although A is not visible as such, its action on all states M delimits A from every observable $B \neq A$.

Topos Theory shares this phenomenology with Quantum Mechanics in the following sense. It has one of its roots (besides more concrete ones from Algebraic Geometry and Mathematical Logic) in the fact that a mathematical category, one of the most important structures in mathematics, introduced by Sounder MacLane and Samuel Eilenberg in 1945, is an awfully abstract thing. A category C consists of so-called objects $X, Y, \ldots$ and morphisms $f : X \to Y$, together with some conditions allowing for composition of morphisms. There is nothing more concrete here. Objects and morphisms are encapsulated things, you cannot find out what is "within" an object or morphism. This is similar to the abstract nature of observables. In mathematics, there is a very powerful solution to this problem, namely the celebrated lemma of the Japanese computer scientist (!) Nobuo Yoneda. He explained it 1954 in Paris to Eilenberg before his return to Japan. The lemma goes as follows: Instead of looking at an

[3] And yes, he is the singer Joan Baez's cousin, music matters...

object X in a category $\mathcal{C}$, Yoneda considers the functor $@X$. It associates with every object Y of $\mathcal{C}$ the set $Y@X = \{f : Y \to X\}$ of morphisms from Y to X. Then, every morphism $g : Z \to Y$ maps $Y@X$ to $Z@X$ by the assignment $f \mapsto f \cdot g$. This new structure $@X$ is a special case of a functor, i.e. a datum F that assigns a set $Y@F$ for every object Y of $\mathcal{C}$, and a map of sets $g@F : Y@F \to Z@F$ fore every morphism $g : Z \to Y$ (with some technical conditions which don't matter for now). Yoneda's lemma states that an object X of $\mathcal{C}$ is uniquely determined by its functor $@X$. This means that the abstract object X can be "realized" by the system of classical sets $Y@X$ its functor defines. Intuitively, this means that X is determined as an abstract object by its behavior when "observed" from all the "addresses" Y. This is the remarkable parallel to Quantum Mechanics: even in the most abstract categories, their objects are "understood" as soon as we know how they behave when being acted upon from all addresses. The functor $@X$ plays the role of an observable's measurement potential.

Why is this so important for Boulez's creative analysis? Because it suggests that in such an analysis, one should try to construct the "functor" $@X$ of a composition X and thereby reveal the full identity of X in the sense that $@X$ opens X to its variety of unseen perspectives from new "musical addresses" Y. This is exactly what we propose as a general methodology to realize Boulez's program and then as a concrete procedure when performing the creative analysis of Boulez's *Structures*. *Generating "the functor of a composition" is what we propose as a general methodology for the reification of Boulez's ideas.* Why is mathematics the right scientific framework for such a creative program? Because to extend boxes means to generate a generalization that is natural with respect to the given data. Mathematical conceptualization does exactly this, it captures the essential features of a special situation and makes evident their abstract general background. The latter is where we then may start with extended contexts, where the power of the abstract mechanism automatically enables more general operations. This has been the driving force of Grothendieck's revolutionary achievements in Algebraic Geometry: Reduce the problem to its intrinsic essence. This is the opposite of what in mathematics is known as "abstract nonsense": The abstract is the key to the concrete. We shall relate to this methodology in the course of the following analysis.

Why Performing a Creative Analysis of Boulez's Structures*?*

Our choice of Boulez's *Structures* is not by case. It relates to the prominent role which these compositions have played in the development of serialism. This is also reflected in the fact that György Ligeti has published a careful analysis of *Structures*, part Ia. Ligeti's investigation [7] is, on the one hand, neutral and precise, on the other, it abounds of strong judgements on the work's compositional and aesthetic qualities.[4] The very success (or failure) of the serial method has been related to

[4] We shall come back to this point in the course of our own analytical work using modern mathematics instead of plain combinatorics.

this composition, which was not only one of Boulez's successes, but also a turning point in his compositional development. In view of Boulez's principle of creative analysis, when applied to the *Structures*, one is immediately led to the question: Would it be possible to write a world of new music on the principle of serialism or was it just a radical experiment without too much long range effects? This *is* an important question when taking the idea of creative analysis for serious, and not only as a recipe for fabricating yet another work.

In our case, the *Structures*, the Boulezian gesture of opening a work's limits is a doubly critical and difficult one: On the one hand, it should help determine whether the huge calculations that lead to the composition are at any rate worth being reused with aesthetic success. On the other hand, the method of serialism also marks the computational limits of humans to compose music.

We must understand here how to integrate computational power in creative works of music. And on what level of creation this can or should be done. Boulez's *Structures* is an excellent testbed to learn this lesson. It teaches us that the control of laborious computational processes cannot be systematically delegated to very limited human calculation power. To paraphrase Schoenberg (and to make clear that there is a life beyond strictly human composition) *somebody had to be Boulez*.

Of course, computers are widely used by modern composers, but it is a common belief that creativity is delineated from such procedures, it terminates when the big ideas are set, and computers are just doing the mean calculations. Apart from being classically wrong we shall see that this is not realistic. In fact, no composer would contest the creative contribution of trying out a new composition on the piano, listening to its acoustical realization, and playing it on the keys, which may give a strong feedback for the creative dynamics, even on the gestural level of one's hands, as is testified by Ligeti and other composers [8]. Here, Marshall McLuhan is wrong: the medium is not the message, it however gives the message's germ the necessary mould and resonance to grow into a full-fledged composition.

Before delving into the technical details we should address the question whether not only computational computer power is necessary or advantageous for modern compositions, but also conceptual mathematical power. Isn't musical composition *anyway* sufficiently controlled by plain combinatorial devices: permutations, recombinations, enumerations, and the like? The question is in some sense parallel to the question whether it is sufficient to control a computer's behavior on the level of binary chains, or, say on the level of machine language. Or else the question whether it is not sufficient to perform a composition for piano by controlling the mechanical finger movements and forgetting about all that psycho-physicological "illusions" such as gestures.

The parallelism lies in the fact that all these activities are shaped by high-level concepts that create the coherence of low-level tokens in order to express thoughts and not just juxtapose atomic units. Of course can one write a computer program in machine language, but only after having understood the high-level architecture of one's ideas. The artistic performance of a complex composition only succeeds when it is shaped on the high mental level of powerful gestures. And the composition of computationally complex musical works needs comprehensive and structurally powerful concepts.

Combinatorics is just one of the machine languages of mathematical thinking. We shall see in the following analysis that it was precisely Ligeti's combinatorial limitation which hindered him to understand the real yoga of Boulez's constructions. You can do combinatorics, but only if you know what is the steering idea. Much as you can write the single notes of Beethoven's "Hammerklavier Sonata", if you know the high-level ideas. The mathematics deployed in the modern mathematical music theory is precisely the tool for such an enterprise. It is not by chance that traditional music analysis is so poor for the composition of advanced music: Its conceptual power is far too weak for precise complex constructions, let alone for their computer-aided implementation.

Reviewing Ligeti's Analysis

Ligeti's analysis [7] of *structures Ia* exhibits a totality of four rows dominating this part of the composition. It starts from the given serial rows S_P for pitch classes and S_D for durations (the primary parameters), as well as S_L for loudness and S_A for attack (the secondary parameters). It then discusses that central 12×12-matrix[5] $Q = (Q_{i,j})$ which gives rise to all row permutations for the four parameters. Whereas the construction of Q is relatively natural, its subsequent permutations for the primary parameters, and still more radically those for the secondary parameters seem to be fairly combinatorial. Ligeti calls these constructions a combinatorial fetishism.[6] This is even aggravated when it comes to the secondary parameters, where Boulez applies what Ligeti calls chess board knight paths, a procedure which in Ligeti's understanding qualifies as numerological game without any musical signification.

This disqualification is confirmed in Ligeti's final remarks on the new ways of hearing which are enforced by this compositional technique. He compares the result to the flashing neon lights of a big city which, although being driven by a precise machinery, generate an effect of statistical sound swarms. He concludes that with such a radical elimination of expressivity, still present in Webern's compositions, the composition finds its beauty in the opening of pure structures. And Boulez—we follow Ligeti's wording—in such a "nearly obsessive-compulsive neurosis, strains himself at the leash and will only be freed by his colored sensual feline world of 'Marteau'".

Ligeti's main critique of Boulez's approach is that he abstracts from the parameters and plays an empty game of meaningless numbers instead. We do contradict this verdict and on the contrary show that in the language of modern mathematics, topos theory, to be precise, Boulez's strategy is perfectly natural, and in fact, the only reasonable when dealing with such diverse parameters such as pitch classes, durations, loudnesses, and attacks. When we say "natural", we mean mathematically natural,

---

[5] Ligeti names it R, but we change the symbol since R is reserved for retrograde in our notation.

[6] "... schliesslich die Tabellen fetischartig als Mass für Dauernqualitäten angewandt ...".

but the fact that a musical construction is only understood by advanced mathematical conceptualization, and not by naive combinatorial music theory, proves that mathematical naturality effectively hits the musical point. A fact that will later also be confirmed by the possibility to implement our findings in the music software Rubato in order to comply with the creative part of Boulez's principle. Music theorists have to learn that from time to time, conceptual innovations may even enlighten their ossified domains. It is not the music's fault if they are "dark to themselves".[7]

A Creative Analysis of *Structure Ia* Based on Yoneda's Lemma

The initial problem in Boulez's construction is that there is no intrinsic reason to transfer the 12 pitch class framework to the other parameters. Although the number 12 is natural in pitch classes, its transfer to other parameters is a tricky business. How can this be performed without artificial constructs?

Let us first analyze Boulez's matrix Q construction. It yields one pitch class row for every row. The ideas run as follows. We get off ground by a modern interpretation of what is a dodecaphonic pitch class series S_P. Naively speaking, S_P is a sequence of 12 pitch classes: $S_P = (S_{P,1}, S_{P,2}, \ldots, S_{P,12})$. More mathematically speaking, it is an affine morphism[8] $S_P : \mathbb{Z}^{11} \to \mathbb{Z}_{12}$, whose values are determined by the 12 values on $e_1 = 0$ and the 11 basis vectors $e_i = (0, \ldots, 1, \ldots, 0), i = 2, 3, \ldots, 12$, where the single 1 stands on position $i - 1$ of that sequence. This reinterpretation yields $S_{P,i} = S_P(e_i)$. The condition that a series hits all 12 pitch classes means that the images $S_P(e_i), S_P(e_j)$ are different for $i \neq j$.

This reinterpretation of a dodecaphonic series means that it is viewed as a $\mathbb{Z}^{11}$-addressed point of the pitch class space $\mathbb{Z}_{12}$ in the language of topos theory of music [12]. This language views the series as a point in the space $\mathbb{Z}_{12}$, but just from the perspective of a particular domain, or address, namely $\mathbb{Z}^{11}$. In topos theory of music, a space $\mathbb{Z}_{12}$ is replaced by its functor $@\mathbb{Z}_{12}$, which at any given address B, i.e., module over a specific ring, evaluates to the set $B@\mathbb{Z}_{12}$ of affine module morphisms $f : B \to \mathbb{Z}_{12}$. This means that the address B is a variable and that our dodecaphonic series is just a point at a specific address among all possible addresses. It is, to come back to our initial discussion of Quantum Mechanics and Topos Theory, the "measurement" of $@\mathbb{Z}_{12}$ for the "state" or address B. It is remarkable to understand here that to understand $\mathbb{Z}_{12}$, it is not sufficient to look at the set or even module structure of $\mathbb{Z}_{12}$. We have to consider all possible "measurements" of its functor. And Boulez, when considering his 12-tone rows, just positions his perspective on the address $\mathbb{Z}^{11}$, among an infinity of potential other addresses. In this context, the

---

[7] Title of a Cecil Taylor LP.

[8] An affine morphism $f : M \to N$ between modules M, N over a commutative ring R is by definition the composition $f = T^t \cdot g$ of a R-linear homomorphism $g : M \to N$ and a translation $T^t : N \to N : n \mapsto t + n$. Affine morphisms are well known in music theory, see [12].

change of address is completely natural. What does this mean? Suppose that we have a module morphism $g : C \to B$ between address modules. Then we obtain a natural map $B@\mathbb{Z}_{12} \to C@\mathbb{Z}_{12}$, mapping $f : B \to \mathbb{Z}_{12}$ to the composed arrow $f \cdot g : C \to \mathbb{Z}_{12}$. For example, if we take $B = C = \mathbb{Z}^{11}$, and if $g(e_i) = e_{12-i+1}$, then the new series $S_P \cdot g$ is the retrograde $R(S_P)$ of the original series.

Our claim, to be proved in the following analysis, is that *all of Boulez's constructions are simply such address change maps,* and as such follow a very systematic construction. So the combinatoriality is viewed as a particular technique from topos theory. Evidently, Boulez did not know this, since topos theory was not even invented at that time, and Yoneda's lemma was only published in 1954, 1 year after the publication of *Structure I*. But this makes his approach even more remarkable; one could even state that in view of this temporal coincidence, *Boulez's structures are the Yoneda lemma in music.*

Functorial Address Changes Replace Parameter Transformations

Boulez uses the following functorial trick to get rid of the unnatural association of different parameters with the serial setup stemming from pitch classes. One observes that for any (invertible) transformation $T : \mathbb{Z}_{12} \to \mathbb{Z}_{12}$, we have a new pitch class series, namely the composition $T \cdot S_P$. For a transposition $T = T^n$, we get the n-fold transposed series. For an inversion $U(x) = u - x$, we get the inverted series, etc. Evidently one may also obtain this effect by an address change as follows. If $T : \mathbb{Z}_{12} \to \mathbb{Z}_{12}$ is any (invertible) affine transformation, then there is precisely one address change $C(T) : \mathbb{Z}^{11} \to \mathbb{Z}^{11}$ by a base vector permutation such that the diagram

$$\begin{array}{ccc} \mathbb{Z}^{11} & \xrightarrow{C(T)} & \mathbb{Z}^{11} \\ {\scriptstyle S_P}\downarrow & & \downarrow{\scriptstyle S_P} \\ \mathbb{Z}_{12} & \xrightarrow{T} & \mathbb{Z}_{12} \end{array} \tag{1}$$

commutes. Instead of performing a parameter transformation on the codomain of the pitch class row, we may perform an address change on the domain $\mathbb{Z}^{11}$. Note however that the address change $C(T)$ is also a function of the underlying series S_P.

What is the advantage of such a restatement of transformations? We now have simulated the parameter-specific transformation on the level of the universal domain $\mathbb{Z}^{11}$, which is common to all parameter-specific series. This enables a transfer of the transformation actions on one parameter space (the pitch classes in the above case) to all other parameter spaces, just by prepending for any series the corresponding address change. So we take the transformation T on $\mathbb{Z}_{12}$, replace it by the address change $C(T)$ on $\mathbb{Z}^{11}$ and then apply this one to all other series, i.e., building $S_D \cdot C(T)$, $S_L \cdot C(T)$, $S_A \cdot C(T)$. This means that we have now a completely natural understanding of the derivation of parameter series from address changes, which act as mediators between pitch class transformations and transformations on other parameter spaces. This is the only natural way of carrying over these operations between intrinsically incompatible parameter spaces. We replace the spaces by their functors and act on the common addresses. This is quite the opposite of purely combinatorial gaming. It is functoriality at its best. Without this functorial restatement of what is a series and how transformations operate, no unified understanding of Boulez's simultaneous operations on different parameter spaces would be possible. What can be done on the functorial level cannot be done on the parameter spaces which have incompatible ontologies. For a deeper understanding of Boulez's selection of his duration series, also relating to Webern's compositions, we refer to [6].

The System of Address Changes for the Primary Parameters

From the functorial point of view, nearly everything in Boulez's construction of part Ia is canonical. The most important address change is the matrix Q. It is constructed as follows. Its ith row $Q(i,-)$ is the base change $C(T^{S_P(i)-S_P(1)})$ associated with the transposition by the difference of the pitch class series at position i and 1. The natural[9] number $Q(i,j)$ in the matrix is therefore $Q(i,j) = S_P(i)+S_P(j)-S_P(1)$, a symmetrical expression in i and j. Moreover, we now see immediately from the definition of the operator C in the above commutative diagram that the composition of two permutations (rows) of the matrix is again such a permutation row, in fact, the transpositions they represent are the group of all transpositions. We may now view Q as an address change $Q : \mathbb{Z}^{11} \boxtimes \mathbb{Z}^{11} \to \mathbb{Z}^{11}$ on the affine tensor product $\mathbb{Z}^{11} \boxtimes \mathbb{Z}^{11}$, see [12, E.3.3], defined on the affine basis $(e_i \boxtimes e_j)$ by $Q(e_i \boxtimes e_j) = e_{Q(i,j)}$. For any such address change $X : \mathbb{Z}^{11} \boxtimes \mathbb{Z}^{11} \to \mathbb{Z}^{11}$, and any parameter Z series $S_Z : \mathbb{Z}^{11} \to \mathit{ParamSpace}$ with values in parameter space *ParamSpace*, we obtain 12 series in that space by address change $S_Z \cdot X$ of the series, and then restricted to the ith rows of X, or equivalently, prepending the address change (!) $row_i : \mathbb{Z}^{11} \to \mathbb{Z}^{11} \boxtimes \mathbb{Z}^{11}$ defined by $row_i(e_j) = e_i \boxtimes e_j$.

Given any such address change matrix $X : \mathbb{Z}^{11} \boxtimes \mathbb{Z}^{11} \to \mathbb{Z}^{11}$, we therefore get 12 series in every given parameter space. So we are now dealing with the construction of specific matrix address changes and the entire procedure is settled. The general idea is this: One gives two address changes $g, h : \mathbb{Z}^{11} \to \mathbb{Z}^{11}$ with $g(e_i) = e_{g(i)}, h(e_i) = e_{h(i)}$ and then deduces a canonical address change $g \boxtimes h : \mathbb{Z}^{11} \boxtimes \mathbb{Z}^{11} \to \mathbb{Z}^{11} \boxtimes \mathbb{Z}^{11}$ by the formula $g \boxtimes h(e_i \boxtimes e_j) = e_{g(i)} \boxtimes e_{h(j)}$. So, when X is given, we obtain a new address change of the same type by building the composed address change $X \cdot g \boxtimes h$. For example, the retrograde matrix in Ligeti's terminology is just the matrix $Q \cdot Id \boxtimes R$ deduced from Q by the address change $Id \boxtimes R$. And Ligeti's U-matrix is deduced from Q by $U \boxtimes U$, where U is the address change associated with the inversion at $e_\flat$, i.e., it is the composite $Q \cdot U \boxtimes U$.

---

9 We represent elements $x \in \mathbb{Z}_{12}$ by natural numbers $0 \leq x \leq 11$.

Now everything is easy: for the first piano, for the primary parameters pitch class P and duration D, and for parts A and B Boulez creates one matrix $Q^1_{P,A}, Q^1_{D,A}, Q^1_{P,B}, Q^1_{D,B}$ address change each, all deduced from Q by the above composition with product address changes $T^1_{P,A} = U \boxtimes Id, T^1_{D,A} = U \cdot R \boxtimes U \cdot R, T^1_{P,B} = U \cdot R \boxtimes U \cdot R, T^1_{D,B} = R \boxtimes U$ via

$$Q^1_{P,A} = Q \cdot T^1_{P,A}, Q^1_{D,A} = Q \cdot T^1_{D,A}, Q^1_{P,B} = Q \cdot T^1_{P,B}, Q^1_{D,B} = Q \cdot T^1_{D,B}. \quad (2)$$

This is quite systematic, moreover, the second piano is completely straightforward, in fact the product address changes of this instrument differ just by one single product address change, namely $U \boxtimes U$:

$$\begin{aligned} T^2_{P,A} &= U \boxtimes U \cdot T^1_{P,A}, T^2_{D,A} = U \boxtimes U \cdot T^1_{D,A}, \\ T^2_{P,B} &= U \boxtimes U \cdot T^1_{P,B}, T^2_{D,B} = U \boxtimes U \cdot T^1_{D,B}. \end{aligned} \quad (3)$$

How to Change Addresses for the Secondary Parameters

For the secondary parameters, loudness and attack, Boulez takes one such value per series—deduced from the given series S_L, S_A—that was derived for the primary parameters. Intuitively, for each row in one of the above matrixes, we want to get one loudness and one attack value.

For loudness, we start with the Q matrix address change for piano 1 and with the U matrix for piano 2. We then take an address change $a : \mathbb{Z}^{11} \to \mathbb{Z}^{11} \boxtimes \mathbb{Z}^{11}$ for part A, and another $c : \mathbb{Z}^{11} \to \mathbb{Z}^{11} \boxtimes \mathbb{Z}^{11}$ for part B. These address changes are very natural paths in the given matrix. Path a is just the co-diagonal of the matrix, i.e., $a(e_i) = e_{12-i} \boxtimes e_i$, and path c is the path shown in Fig. 1.

What about Ligeti's verdict that these paths are simple numerological choices? They are both closed paths if one identifies the boundaries of the matrix. Path a is a closed path on the torus deduced from Q by identifying the horizontal and vertical boundary lines, respectively. And path c is closed on the sphere obtained by identifying the adjacent left and upper, and right and lower boundary lines, respectively. The torus structure is completely natural, if one recalls that pitch classes are identified exactly like the horizontal torus construction, while the vertical one is a periodicity in time, also a canonical identification. The sphere construction is obtained by the parameter exchange (diagonal reflection!) and the identification of boundary lines induced by this exchange, see Fig. 2.

For the attack paths, one has a similar construction, only that the paths a and c are rotated by 90° clockwise and yield paths α and γ. Again, piano 1 takes its values on Q, while piano 2 takes its values on the U matrix. So apart from that rotation everything is the same as for loudness.

In summary, we need only one product address change given by the U transformation for the primary parameters in order to go from piano 1 to piano 2, whereas

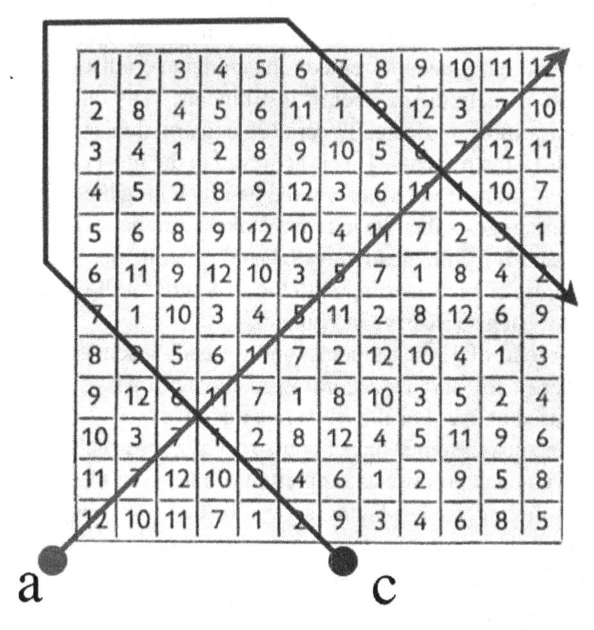

Fig. 1 The two paths a, c for loudness, part A and part B, in Ligeti's Q matrix for piano 1; same paths in the U matrix for piano 2

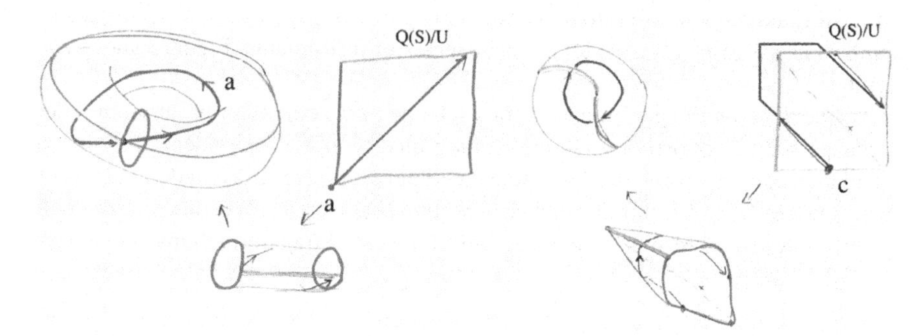

Fig. 2 The *geometric shapes* (*left* torus for a, *right* sphere for c) explaining the two paths a, c

one rotation by 90° suffices to switch to the secondary parameter paths. Observe that this rotation is just the address change on the matrix space $\mathbb{Z}^{11} \boxtimes \mathbb{Z}^{11}$ induced by a retrograde on each factor! It could not be more functorial, and barely more beautiful.

Our First Choice of a Creative Analysis

In view of the functorial setup which we have constructed, a first transduction is now immediate. Of course, there are many ways to shift from the given analytical data to neighboring data in the space of analytical data. A first way is evident, and it is also the one which we urgently need to remedy the evident serialist imperfection of the given construction, namely the number of instruments. Why only two instruments? In order to obtain a more intrinsically serialist construction, one should not work with two, but with 12 instruments. This is achieved in the most evident way: We had seen that the second piano is derived from the first by taking the U matrix instead of the Q matrix. This suggests that we may now take a number of 12 address changes $U_i : \mathbb{Z}^{11} \to \mathbb{Z}^{11}$, starting with the identity Id, and generate one instrumental variant for each such address change, starting with the structure for the first piano, and then adding variants for each successive instrument.

This enables a total of 12 instruments, and for each a sequence of 12 series for part A and 12 series for part B, according to the 12 rows of the matrix address changes as discussed above. For the ith series this gives us 12 instruments playing their row simultaneously. Boulez has of course not realized such a military arrangement of series. We hence propose a completion of the serial idea in the selection of the numbers of simultaneously playing series. To this end, observe that the series S_P of pitch classes has a unique inner symmetry which exchanges the first and second hexachord, namely the inversion $I = T^7. -1$ between e and $e_\flat$, i.e. the series defines the strong dichotomy No. 71 in the sense of mathematical counterpoint theory [12, Chap. 30]. In part A, we now select the instrument $S_P(i)$ from below and then take $I(S_P(i))$ successive instruments in ascending order (and using the circle identification for excessive instrument numbers). For part B we take the I-transformed sequence of initial instrumental numbers and attach the original serial numbers as successively ascending occupancies of instruments. Figure 3 shows the result.

The next step will be to transform this scheme into a computer program in order to realize such compositions and to test their quality. It is now evident, that such a calculation cannot be executed by a human without excessive efforts and a high risk of erroneous calculations. It is also not clear whether such creative reconstructions will produce interesting results, or perhaps only for special transformational sequences $U_1, U_2, \ldots, U_{12}$.

Implementing the First Creative Analysis on Rubato Composer

As mentioned in the introduction, the realization of a variety of creative analyses in terms of notes is beyond human calculation power, or at least beyond the patience and reliability of a composer. Therefore, we have implemented the above mathematical procedure on the music software Rubato Composer [15]. This comprises seven new rubettes (Rubato PlugIns), specifically programmed for our procedure, see also Fig. 4: BoulezInput, BoulezMartix, Transformation, BaseChange, Chess, SerialSystem, and Boulez2Macro. We call them *boulettes* in order to distinguish them from general purpose rubettes.

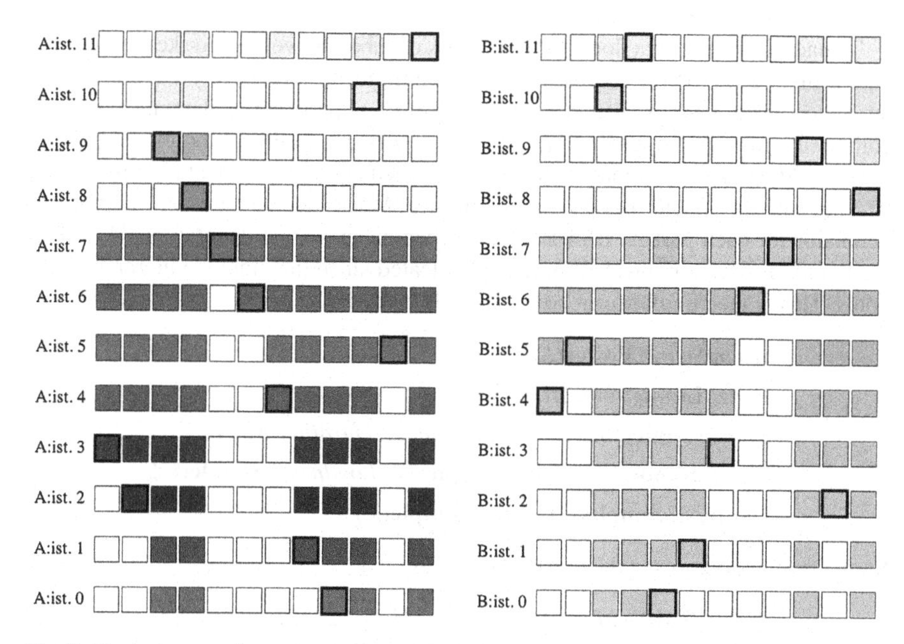

Fig. 3 The instrumental occupancies in part A, B, following the autocomplementarity symmetry $I = T^7.-1$ of the original pitch class series. The lowest instruments are taken according to the series while the occupancies are chosen according to the I-transformed values. E.g. for the first column, we have the serial value 3, and its I transformed is 4, so we add 4 increasingly positioned instruments

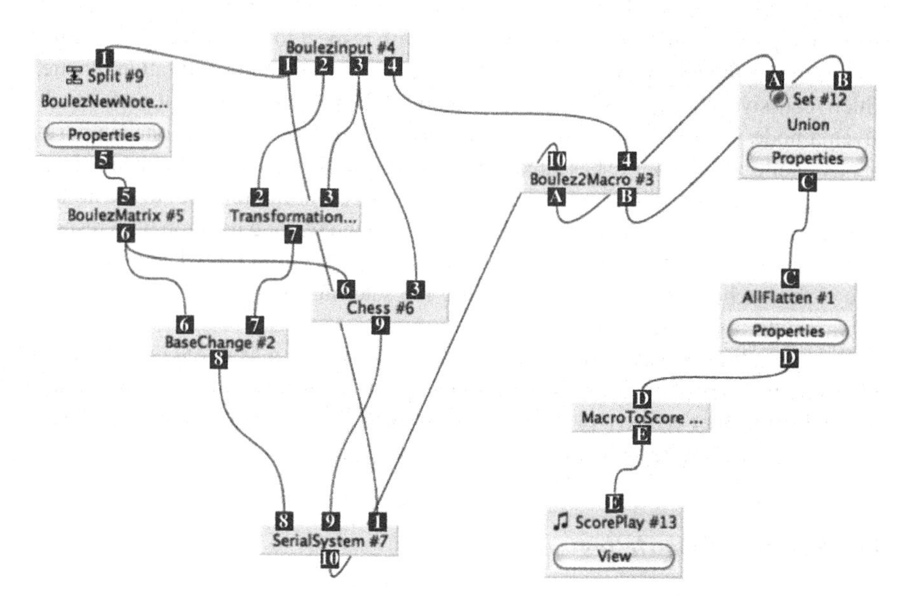

Fig. 4 The Rubato network generating midi files (played the ScorePlay rubette) with arbitrary input from the creative analysis that is encoded in the BoulezInput rubette

To understand the data flow of this network of rubettes, we have to sketch the used data format, see [15] for details. Rubettes communicate exclusively via transfer of denotators. These are instances of forms, a type of generalized mathematical spaces comprising universal constructions, such as powersets, limits, and colimits, which are derived from the functors associated via Yoneda to mathematical modules.

The outputs A and B of boulette Boulez2Macro create one zero-addressed denotator for each part A, B of Boulez's scheme. These two denotators, M_A, M_B, are not plain sets of notes, but more sophisticated since they include hierarchies of notes. This is the circular form comprising M_A, M_B:

$$\begin{aligned}
&MacroScore{:}.\mathbf{Power}(Node)\\
&Node{:}.\mathbf{Limit}(Note, MacroScore)\\
&Note{:}.\mathbf{Simple}(Onset, Pitch, Loudness, Duration, Voice)\\
&Onset{:}.\mathbf{Simple}(\mathbb{R}), Pitch{:}.\mathbf{Simple}(\mathbb{Z}), Loudness{:}.\mathbf{Simple}(\mathbb{Q})\\
&Duration{:}.\mathbf{Simple}(\mathbb{R}), Voice{:}.\mathbf{Simple}(\mathbb{Z})
\end{aligned}$$

So the formal notation of the two denotators is

$$\begin{aligned}
&M_A{:}0@MacroScore(M_{A,0}, M_{A,1}, \ldots, M_{A,m}),\\
&M_B{:}0@MacroScore(M_{B,0}, M_{B,1}, \ldots, M_{B,n})
\end{aligned}$$

with the nodes $M_{A,i}, M_{B,j}$, respectively. Each node has a note, its *anchor note*, and *satellites*, its MacroScore set denotator. Observe that the concept of an anchor with satellites is *grano cum salis* also the approach taken by Boulez in his multiplication of chords, where the anchor is the distinguished note, and where the satellites are represented by the intervals of the other notes with respect to the anchor. This output A, B, is then united in the Set rubette and its output C is sent to the AllFlatten rubette, which recursively "opens" all the nodes' satellite MacroScore. How is it performed? Given a node with empty satellite set, one just cuts off the set. Else, one supposes that its satellite MacroScore has already recursively performed the flattening process, resulting in a set of notes. Then one adds these notes (coordinate-wise) to the node's anchor note. This means that the satellites are given the relative position with respect to their anchor note. A trill is a typical example of such a structure: The trill's main note is the anchor, while the trill notes are the satellites, denoted by their relative position with respect to the anchor note.

From the Boulez2Macro boulette the output is given as a MacroScore denotator for strong reasons: We want to work on the output and want to take it as a primary material for further creative processing in the spirit of Boulez, a processing, which, as we shall see, requires a hierarchical representation.

The System of Boulettes

But let us see first how the Boulez composition is calculated. We are given the following input data:

Outlet 1 in the BoulezInput boulette contains the series for all parameters as a denotator *Series*:@$\mathbb{Z}^{11}$*BoulezSeries*(S_P, S_D, S_L, S_A) of the form *BoulezSeries*:. lim(P, L, D, A) with the factor forms P:.**Simple**$(\mathbb{Z}_{12})$, D:.**Simple**$(\mathbb{R})$, L:.**Simple** $(\mathbb{Z})$, A:.**Simple**$(\mathbb{R}^3)$. The attack form A has values in the real 3-space, where the first coordinate measures fraction of increase of nominal loudness, the second the articulatory fraction of increase in nominal duration, and the third the fraction of shift in onset defined by the attack type. For example, a sforzato attack (*sfz*) would increase nominal loudness by factor 1.3, shorten duration to a staccato by 0.6, and the third would add to the nominal onset a delay of $-0.2 \times$ nominal duration. As discussed in section "Reviewing Ligeti's Analysis", the address $\mathbb{Z}^{11}$ yields the parameterization by the 12 indices required for a serial sequence of parameters. For example, the pitch class series is the factor denotator S_P:$\mathbb{Z}^{11}$@$P(3, 2, 9, 8, 7, 6, 4, 1, 0, 10, 5, 11)$.

Outlet 2 contains the two address changes for retrograde R and inversion U. They are encoded as denotators R:$\mathbb{Z}^{11}$@*Index*$(R_1, R_2, \ldots, R_{12})$, U:$\mathbb{Z}^{11}$@*Index* $(U_1, U_2, \ldots, U_{12})$ in the simple form *Index*:.**Simple**$(\mathbb{Z})$, which indicates the indices R_i of the affine basis vectors, which are the images of the basis vectors e_i $C(R)$ and inversion $C(U)$.

Outlet 3 encodes the above sequence $U. = (U_i)$ of address changes for the instrumental sequence, i.e. a denotator $U.$:$\mathbb{Z}^{11}$@*Sequ*$(U_1, U_2, \ldots, U_{12})$ with U_i:$\mathbb{Z}^{11}$ @*Index*$(U_{i,1}, U_{i,2}, \ldots, U_{i,12})$, where we use the list type form *Sequ*:.**List**(*Index*). In fact this also works for any number of instruments, but we restrict our example to 12 instruments of our choice.

Outlet 4 encodes the registers, which must be defined in order to transform the pitch classes into real pitches. This information is given in the same form as the address change sequence, where the coordinates for the ith sequence are the octave numbers, where the pitches of the respective pitch class series in the corresponding instrument are positioned. Octaves are numbered starting from octave 0 at pitch 60 in midi format. This information is also used to position the pitches according to an instrumental range.

The Split rubette takes the input series *Series* and sends its pitch class factor S_P to outlet 5. This denotator is taken as input of the BoulezMatrix boulette and yields the famous matrix Q at outlet 6, which we interpret as a denotator Q:$\mathbb{Z}^{11} \boxtimes \mathbb{Z}^{11}$@*Index*$(Q_{i,j}, i, j = 1, 2, \ldots, 12)$.

The boulette BaseChange is devoted to the calculation of the address changes on outlet 8 for the primary parameters described in the four formula groups (2) and (3) for the primary parameters of the 12 instruments and take as input the matrix Q from outlet 6 and the sequence $U.$ of instrumental address changes.

The boulette Chess is devoted to the calculation of the corresponding address changes on outlet 9 for the secondary parameters loudness and attack, as described by the chess board paths.

From the address change systems on input 8 and 9 of the SerialSystem boulette, and taking as a third input the total series from outlet 1, the total system of series is calculated according to our formulas described in sections "The System of Address Changes for the Primary Parameters" and "How to Change Addresses for the Secondary Parameters". This produce outlet 10, which is finally added as input,

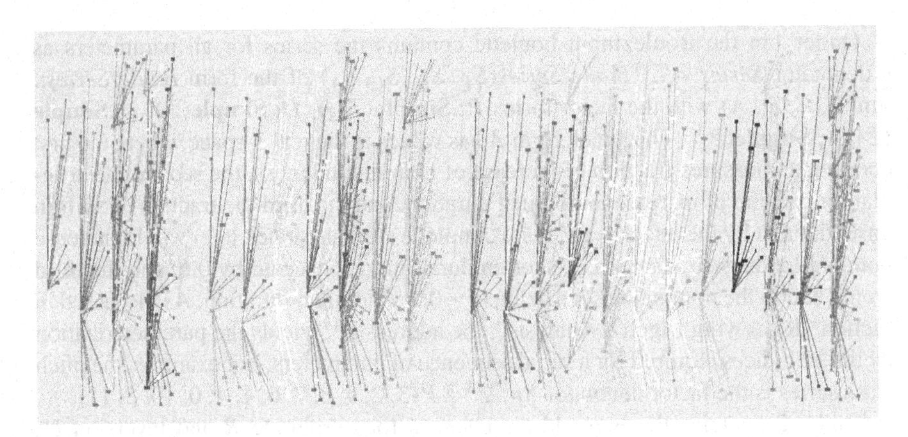

Fig. 5 The final "raw" material for 12 instruments. Instruments are distinguished by grayscales. Satellites pertaining to a given anchor note are connected by rays to that anchor note

together with the input 4 of octaves, to calculate the effective parameters. The pitches, nominal durations, and loudnesses are now given, the nominal onsets are calculated to produce the rectangular scheme shown in Fig. 3. The attack data is used to transform the nominal values into the attack-specific deformations, and we obtain the required outputs A and B.

This double output is a denotator of form *MacroScore*. Its nodes in part A and B are 144 series each. The anchor note of each serial node is taken to be the first note in the series. The satellites of this node are the remaining 11 notes with their relative position with respect to the anchor note. Moreover, the output denotators at A and B have one instrumental voice number for each instrument. Taking the union of these parts in outlet C, we obtain a large *MacroScore* denotator $M_C = M_A \cup M_B$. Selecting from this system the series as shown in Fig. 3 yields the final "raw" material. This one will now be used to generate more involved creative constructions in section "A Second More Creative Analysis and Reconstruction". The system as calculated by Rubato is shown in Fig. 5. The graphical representation is realized on the BigBang rubette for geometric composition. The input to this rubette is the denotator M_C, while the selection of the instruments according to the selection shown in Fig. 3 is made by direct graphically interactive editing. The functionality of the BigBang rubette is discussed in section "The BigBang Rubette for Computational Composition".

A Second More Creative Analysis and Reconstruction

One of the most creative extensions of techniques in musical composition is the opening of the transformational concept. This was already a crucial argument in Boulez's own construction of derived Q matrixes, where he invented that ingenious tool of address change in order to extend pitch class transformations to parameters, where such operations would not apply in a natural way. Our extension of Boulez's approach was presented above and implemented in Rubato's boulettes, yielding the denotator M_C:0@ *MacroScore*().

In this section, we shall add other extensions of the given transformations and apply them to the construction of huge extensions starting from the present "raw material" M_C. There are two threads of extensions, which we shall expose: The first is the conceptual extension, i.e., conceiving new types of transformations, while the second deals with the associated concrete manipulation of compositions on the level of graphically interactive gestures.

The background of this double strategy is the following general idea: The formulaic rendition of compositional tools, when implemented in software, pertains to what is somewhat vaguely called algorithmic composition. This is what happens in Rubato's boulettes. The drawback of such an implementation is that the result is "precooked" in the cuisine of the code and cannot be inspected but as a *res facta*. A composer would prefer being able to influence his/her processes in the making, not only when it is (too) late.

This is, why we have now realized a different strategy: The transformations, which are enabled by the BigBang rubette [15, Chap. 17], are immediately visible when being defined, and they can be heard without delay. The general idea backing this approach is that conversely, any algorithm should be transmuted into a graphically interactive gestural interface, where its processes would be managed on the flight, gesturally, and while they happen (!). Why should I wait until rotation of musical parameters is calculated? I want to generate it and while I actually rotate the system by increasing angles, I would like to see the resulting rotated set of note events, and also hear how that sounds, and then decide upon the success or failure of that rotation.

The Conceptual Extensions

The conceptual extension of transformations has two components: the extension of the transformations as such and the application of such transformations as a function of the hierarchical structure of the *MacroScore* form. The serial transformations on the note parameters of a composition usually comprise the affine transformations generated by inversion (pitch reflection), retrograde (onset reflection), transposition (pitch translation), and time shift (onset translation). But it also includes the construction of assemblies of iterated transformations, not just one transformed note set, but the union of successively applied transformations. The latter is typically realized by regular patterns in time, when rhythmical structures are constructed. So we have these two constructions: Given a set of notes M and a transformation f, one either considers one transformed set $f(M)$ or else the union $\bigcup_{i=0,1,\ldots,k} f^i(M)$. The latter is well known as a rhythmical frieze construction if f is a translation in time. If we generalize frieze constructions to two dimensions, using two translations f, g in the plane, we obtain a wallpaper $\bigcup_{i=0,1,\ldots,k,j=0,1,\ldots,l} f^i g^j(M)$.

Extensions of Single Transformations

The natural generalization of such transformational constructions is to include not only those very special transformations, but any n-dimensional non-singular affine transformation f in the group $\overrightarrow{GL}_n(\mathbb{R})$, whose elements are all functions of shape $f = T^t \cdot h$, where h is an element of the group $GL_n(\mathbb{R})$ and where $T^t(x) = t + x$ is the translation by $t \in \mathbb{R}^n$. This generalization is again a consequence of the mathematical strategy that extracts the essential features of structures and uses their fundamental properties instead of sticking to historically grown special cases whose specialities don't really matter. It is however well known from mathematical music theory [12, Chap. 8.3] that any such transformation can be decomposed as a concatenation of musical standard transformations, which, each, involve only one or two of the n dimensions. In view of this result, we have chosen the generalization of the above transformations to these special cases in 2-space: (1) translations T^t, (2) reflections Ref_L at a line L, (3) rotations Rot_α by angle α, (4) dilation $Dil_{L,\lambda}$ vertical to the line L by factor $\lambda > 0$, (5) shearing $Sh_{L,\alpha}$ along the line L and by angle α. These are operations on real vector spaces, while we have mixed coefficients in the *MacroScore* form. The present (and quite brute) solution of this problem consists in first embedding all coefficients in the real numbers, to perform the transformations and then to recast the results to the subdomains, respectively.

Given the group $\overrightarrow{GL}_n(\mathbb{R})$ of transformations (generated by the above two-dimensional prototypes), we now have to manage the hierarchical structure of denotators in the *MacroScore* form. How can transformations be applied to such objects? To this end, recall that a *MacroScore* denotator is[10] a set M of nodes $N = (A_N, S_N)$, which have two components: an anchor note A_N from the (essentially) five-dimensional form *Note* and a *MacroScore* formed satellite set S_N. Common notes are represented by nodes having empty satellite sets. Given a transformation $f \in \overrightarrow{GL}_n(\mathbb{R})$ and a *MacroScore* denotator M, a first operation of f upon M is defined by anchor note action:

$$f \cdot M = \{(f \cdot A_N, S_N) | N \in M\} \tag{4}$$

This type of action is very useful if we want to transform just the anchors and leave the relative positions of the satellite notes invariant. For example, if the satellites encode an embellishment, such as a trill, then this is the right operation in order to transform a trill into another trill.

It is straightforward to generalize this operation to any set S of nodes in the tree of *MacroScore* denotator M such that no two of them are hierarchically related (one being in the satellite tree of the other). The above situation of formula (4) refers to the top level anchors. Suppose that S consist of nodes N. For non-satellite nodes, we have the above function. Suppose next that such a node N is a satellite pertaining to a well-defined anchor note $A(N)$. Thinking of that anchor note as a local coordinate origin,

---

[10] All denotators in this discussion will be zero-addressed.

we may apply a transformation $f \in \overrightarrow{GL}_n(\mathbb{R})$ to all selected satellite nodes of $A(N)$ by the above formula (4), yielding a transformed set of satellites of the same anchor note. We may apply this operation to each set of satellites of given anchors occurring in S. Since there are no hierarchical dependencies, no contradiction or ambiguity appears, i.e. no note will be transformed together with one of its direct of iterated satellite notes. This means that we are simultaneously applying f to all satellite sets of S. This means that we take the disjoint union $S = \bigsqcup S_k$ of satellite sets S_k pertaining to specific anchor notes A_k and then apply a simultaneous transformation f to each of these S_k. Denote this operation by $f \odot S$.

There is a further operation, which may be applied to a set S sharing the above properties. This one takes not the relative positions of S-elements, but their flattened position and then applies the transformation f to these flattened notes. It is the operation one would apply in a hierarchical context, such as a Schenker-type grouping, but without further signification of the hierarchy for the transformational actions. After the transformation, each of these transformed flattened notes is taken back to its original anchor note. For example, if $s = 1$, and if $N = (A_N, S_N)$ is a satellite of level zero anchor note $A(N)$, then we first flatten the note (once), which means that we take $N' = (A(N) + A_N, S_N)$, we then apply f to its new anchor $A(N) + A_N$, yielding $N'' = (f(A(N) + A_N), S_N)$, and we finally subtract the original anchor, yielding the new satellite $N''' = (f(A(N) + A_N) - A(N), S_N)$ of $A(N)$. This operation is again denoted like the above operation, i.e., by $f \cdot S$.

Extensions of Wallpapers

Let us now review the construction of wallpapers in view of a possible creative extension. Mathematically speaking, a wallpaper is a structure that is produced by repeated application of a sequence of translations $T^{\cdot} = (T^{t_1}, T^{t_2}, \ldots, T^{t_r})$ acting on a given motif M of notes. Each T^{t_i} of these translations is repeatedly performed in the interval numbers of the sequence $I. = (I_i = [a_i, b_i]), a_i \leq b_i$, of integers, what means that the total wallpaper is defined by

$$W(T^{\cdot}, I.)(M) = \bigcup_{a_i \leq \lambda_i \leq b_i} T^{\lambda_1 t_1} T^{\lambda_2 t_2} \ldots T^{\lambda_r t_r}(M) \tag{5}$$

This formula has nothing particular regarding the special nature of the different powers of translations. Therefore the formula could be generalized without restrictions to describe grids of any sequence of transformations $f. = (f_1, f_2, \ldots, f_r)$ for $f_i \in \overrightarrow{GL}_n(\mathbb{R})$, thus yielding the generalized wallpaper formula

$$W(f., I.)(M) = \bigcup_{a_i \leq \lambda_i \leq b_i} f_1^{\lambda_1} \circ f_2^{\lambda_2} \circ \cdots f_r^{\lambda_r}(M) \tag{6}$$

which also works for negative powers of the transformations, since they are all invertible. In our context, the motif M will no longer be a set of common notes, but a denotator of *MacroScore* form. Therefore we may replace the naive application of transformations to a set of notes by the action of transformations on such denotators as discussed above. This entails that—mutatis mutandis—we have two transformation wallpaper for a set S of nodes of a *MacroScore* denotator with the above hierarchical independency property, the relative one

$$W(f., I.) \odot S = \bigcup_{a_i \leq \lambda_i \leq b_i} f_1^{\lambda_1} \circ f_2^{\lambda_2} \circ \cdots f_r^{\lambda_r} \odot S \tag{7}$$

or else the absolute one:

$$W(f., I.) \cdot S = \bigcup_{a_i \leq \lambda_i \leq b_i} f_1^{\lambda_1} \circ f_2^{\lambda_2} \circ \cdots f_r^{\lambda_r} \cdot S \tag{8}$$

This strategy generalizes the transformations and the motives in question. A last generalization is evident, when looking at the range of powers of the intervening transformations. To the date, these powers are taken within the hypercube $D = \prod_i I_i$ of sequences of exponents. However, nothing changes if we admit more generally any finite "domain" set $D \subset \mathbb{Z}^r$ and make the union according to the sequences of exponents appearing in D:

$$W(f., D) \odot S = \bigcup_{(\lambda_1, \lambda_2, \ldots, \lambda_r) \in D} f_1^{\lambda_1} \circ f_2^{\lambda_2} \circ \cdots f_r^{\lambda_r} \odot S \tag{9}$$

or else the absolute one:

$$W(f., D) \cdot S = \bigcup_{(\lambda_1, \lambda_2, \ldots, \lambda_r) \in D} f_1^{\lambda_1} \circ f_2^{\lambda_2} \circ \cdots f_r^{\lambda_r} \cdot S \tag{10}$$

The above constructions were not specified with regard to the addresses involved in these denotators. In the following, we have not yet implemented this functorial point of view in the composition tools of the BigBang rubette, but this is by no means problematic since Florian Thalmann has implemented functorial wallpaper construction in an earlier work [17].

The BigBang Rubette for Computational Composition

The BigBang rubette was implemented during a research visit of Thalmann at the School of Music of the University of Minnesota. It allows for graphically interactive gestural actions for transformations and wallpapers on *ScoreForm* denotators. We shall not describe all transformations in detail, but show the typical gestural action to be taken for a rotation of a denotator, see Figs. 6 and 7.

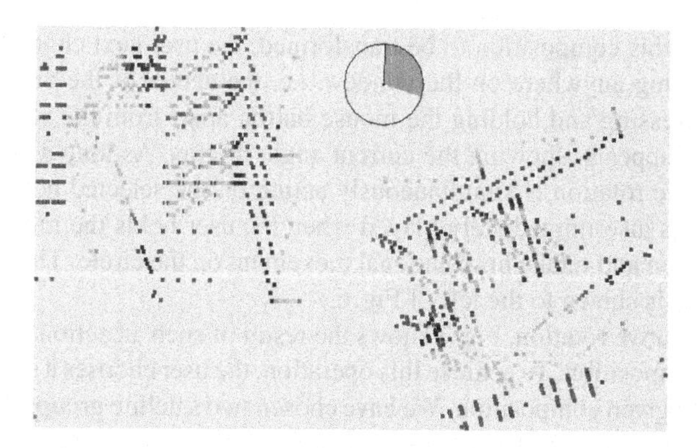

Fig. 6 Rotation of the first bars of Beethoven's op. 106, Allegro (*left*). The rotation *circle* shows the mouse movement on its periphery, the original is also shown

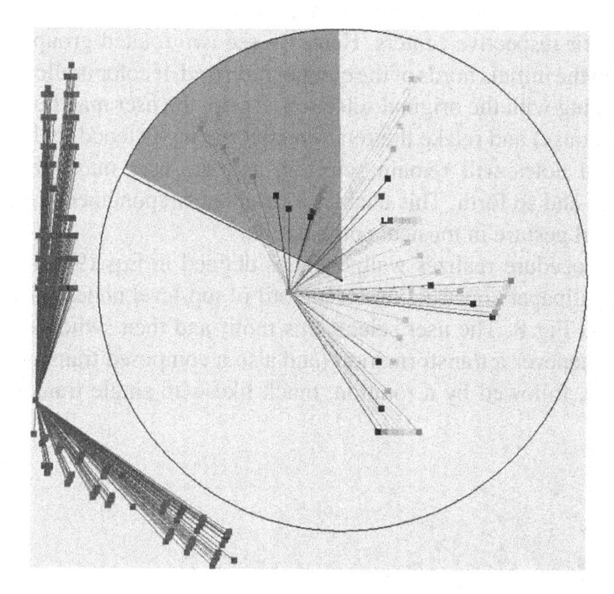

Fig. 7 Here, a relative rotation is performed on the two satellite sets, with their two anchor notes at the rays' centers. The original positions are also shown

The user loads (or draws) a composition (a denotator in *MacroScore* form) M. This is shown in the left half of Fig. 6, the example is the first bars of Beethoven's op. 106, Allegro. This composition is shown in the plane of onset (abscissa) and pitch (ordinate), but the user may choose any two of the five axes corresponding to the note parameters and perform all transformations on the corresponding plane. After having selected with the mouse[11] (drawn rectangles around the critical note groups)

[11] In a more recent development, the mouse-driven input is being replaced by a direct gestural input using the hand recognition software Leap Motion.

the notes from this composition to be transformed, the user next chooses a rotation center by clicking anywhere on the window, i.e. the center of the circle on top of Fig. 6. Then pressing and holding the mouse button apart from the selected center, a rotation tool appears, showing the current angle in gray. As long as the mouse is not released, the rotation is simultaneously acting on the selected note group. The rotated music is also immediately played when the user holds the mouse still. The user may hold on and retake his rotational movement on the circle. The visual result in our example is shown to the left of Fig. 6.

As to the relative rotation, Fig. 7 shows the result of such an action, together with the original composition. To achieve this operation, the user chooses a set of satellites throughout the given composition. We have chosen two satellite groups derived from the composition in Fig. 6. Then the user chooses one anchor note and defines the center of rotation relative to that anchor. Here, our center was chosen near the anchor of right satellite group. Then the same gestures are performed as in the previous rotation. The circle is shown in Fig. 7, and all chosen satellite notes are rotated relatively to their respective centers. Here, we see two rotated groups: the left one stemming from the initial chords of the composition (red, if color is allowed), the right one is overlapping with the original selection. Again, the user may hold on (without releasing the mouse) and retake the rotation after having listened to the result.

The selected notes will remain selected, and the user may then add a next transformation, and so forth. This enables a completely spontaneous and delay-less transformational gesture in musical composition.

A similar procedure realizes wallpapers as defined in Eqs. (9) and (10). Let us illustrate the wallpaper construction for a motif of top-level nodes, as shown by the darkened set on Fig. 8. The user selects this motif and then switches to wallpaper mode. Now, whenever a transformation (and also a composed transformation, such as a translation, followed by a rotation, much like with single transformations) is

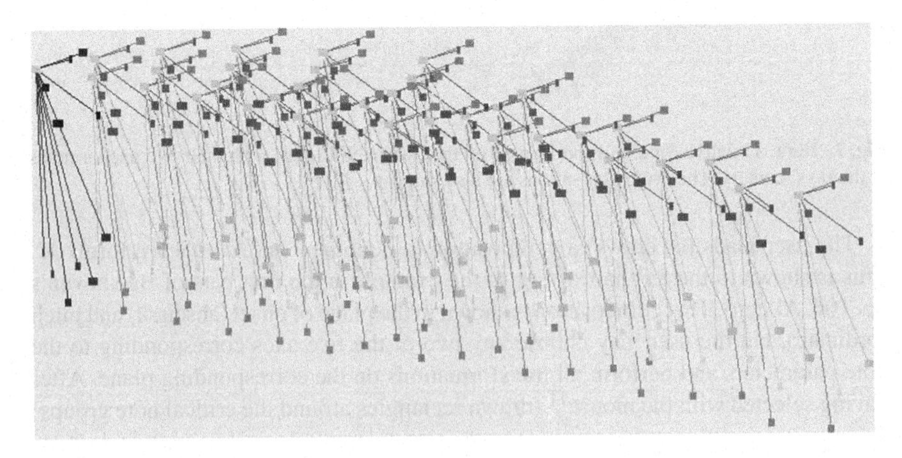

Fig. 8 A wallpaper is built from a motif (*darkened*). Two transformations are used, both are translations, followed by rotations, and shrinking dilation

defined by the previous gestural action, the union of all iterated transformations of the motif is simultaneously shown (and heard). The range of iteration (the powers of that transformation) can be set at will. For a second transformation, the wallpaper mode is clicked again and allows the user to perform a second transformation, and a third, and fourth, etc. The user may also switch to another parameter plane when adding new transformations, and thereby create wallpaper structures in less evident, but musically precious parameters, such as loudness and voice. The example in Fig. 8 has two transformations, each of them being a translation, followed by a rotation and then a dilation.

The BigBang rubette also allows for multidimensional alterations and morphing. These are deformation operations, which alter given notes (on specified levels of the macroscore hierarchy) in the direction of another composition, which might be anything, or just a single point of attraction. We do not discuss this technique further here and refer to [17] for details.

A Composition Using the BigBang Rubette and the Boulettes

Here is a composition, logically named *restructures*, which Guerino Mazzola and Schuyler Tsuda co-composed by use of the above techniques, starting from the raw material M_C shown in Fig. 5. We however also applied the alteration techniques implemented in the BigBang rubette, but will not discuss this technique further here (the composition can be downloaded[12]).

This composition has four movements. Each movement is transformed according to a specific geometric BigBang rubette technique, which we describe in the following paragraphs. After executing these operations, the 12 voices of each movement, which are available as 12 separate MIDI files, were elaborated by adequate orchestrations. This was realized by Schuyler Tsuda, who is an expert in sound design. He orchestrated and attributed the MIDI files to specific sounds in order to transform the abstract events into an expressive body of sound.

The first movement (Expansion/Compression) takes a copy of M_C, then "pinches" the satellites (but not the anchors!) of part A in the sense that the first (in onset) satellites are altered 100 % in pitch direction only to a defined pitch, whereas the last satellites are left as they were (0 % alteration). The satellites in-between are pinched by linear interpolation. The same procedure is applied to part B, however this time the pinching is 100 % at the end and 0 % at the start. This is shown in Fig. 9.

Instrumentation 1: Voice 1 = grand piano, voice 2 = scraped, bowed, rolled, and struck suspended cymbals, voice 3 = electronic mallets, voice 4 = solo cello, voice 5 = pizzicato strings, voice 6 = electronic space strings, voice 7 = plucked e-bass, voice 8 = grand piano, voice 9 = electronic percussion, voice 10 = timpani, voice 11 = electronic toms, voice 12 = electronic bells (Fig. 10).

[12] http://www.encyclospace.org/special/restructures.mp3, accessed 4 Jul 2014.

Fig. 9 First movement: variable pinching the satellite onsets. 100 % pinching at start and end onsets, no pinching in the meet of end of part A and start of part B

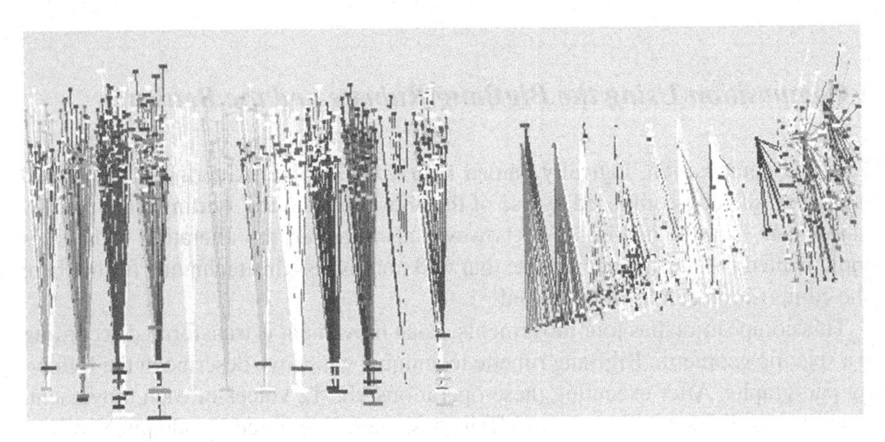

Fig. 10 Fourth movement: sucking down the anchors, expanding their durations, lifting the satellites in part A, then progressive pinching of notes in part B

For the second movement (Space-Time) then take another copy of M_C and expand the onsets and the durations of the anchors of the second appearance to the double, which yields the situation shown in Fig. 11.

Instrumentation 2: Voice 1 = strings, voice 2 = flute and horn, voice 3 = grand piano, voice 4 = sine waves, voice 5 = electronic voice, voice 6 = grand piano, voice

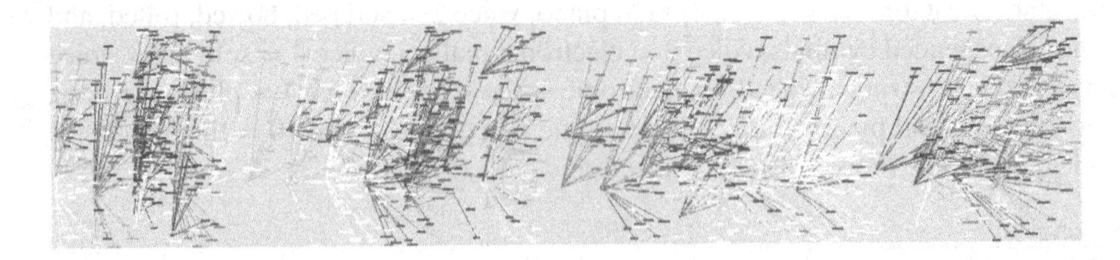

Fig. 11 Second movement: expanding onsets and durations

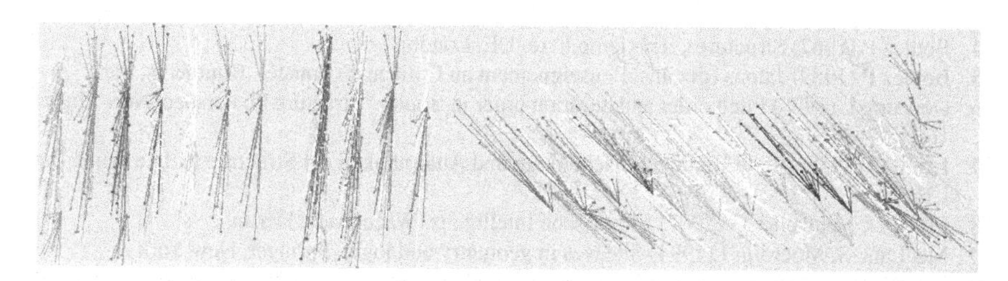

Fig. 12 Third movement: retrograde inversion of anchors and satellites in part A, rotation of satellites in part B

7 = trombone and tuba, voice 8 = electronic strings, voice 9 = triangle and finger cymbals, voice 10 = bowed piano, voice 11 = clarinets, voice 12 = electronic bells.

For the third movement (Rotations), taking again a copy of M_C, and focusing first on part A, we apply a retrograde inversion to the anchors, and then in a second operation also to all satellites relative to their anchors. We then take part B and apply a rotation of all satellites, relative to their anchors, by 45° in counterclockwise direction. The result shown in Fig. 12.

Instrumentation 3: Voice 1 = sine waves, voice 2 = oboe and bassoon, voice 3 = pizzicato strings, voice 4 = marimba, voice 5 = horns, voice 6 = electronic mallets, voice 7 = temple blocks and tam-tam, voice 8 = grand piano, voice 9 = electronic percussion, voice 10 = sine waves, voice 11 = trombones, voice 12 = electronic bells.

Finally, for the fourth movement (Coherence/Opposition), taking again a copy of M_C, we take part A, and pinch to low pitches the anchors, and dilate their durations, whereas the satellites are pinched to hich pitches. In part B, we also operation such separation of pitch of satellites from anchors, but we also execute a progressive pinching of the pitches towards a fixed pitch towards the end of the composition. The result is shown in Fig. 10.

Instrumentation 4: Voice 1 = glockenspiel and electronic noise, voice 2 = glockenspiel and electronic noise, voice 3 = grand piano and electronic noise, voice 4 = harp, electronic noise, and pizzicato strings, voice 5 = sine waves, voice 6 = finger cymbals and timpani rolls, voice 7 = electronic bells, voice 8 = grand piano, voice 9 = Chinese opera gong and low and high gongs, voice 10 = bowed cymbals, voice 11 = triangle and bass drum rolls, voice 12 = triangle and bass drum rolls.

References

1. Baez C (2006) Quantum quandaries: a category-theoretical perspective. In: French S et al (eds) Foundations of quantum gravity. Oxford University Press, Oxford
2. Boissiere A (2002) Geste, interprétation, invention selon Pierre Boulez. In: Revue DEMéter, Lille-3 University
3. Boulez P (1953) Structures, premier livre. UE, London

4. Boulez P (1967) Structures, deuxième livre. UE, London
5. Boulez P (1989) Jalons (dix ans d'enseignement au Collège de France). Bourgeois, Paris
6. Deyoung L (1978) Pitch order and duration order in Boulez' Structure Ia. Perspect New Music 16(2):27–34
7. Ligeti G (1958) Pierre Boulez: Entscheidung und Automatik in der Structure Ia. Die Reihe IV, UE, Wien
8. Ligeti G, Neuweiler G (2007) Motorische Intelligenz. Wagenbach, Berlin
9. MacLane S, Moerdijk I (1994) Sheaves in geometry and logic. Springer, New York
10. Mazzola G (1985) Gruppen und Kategorien in der Musik. Heldermann, Berlin
11. Mazzola G et al (1996) RUBATO on the internet. http://www.rubato.org. Visited on 24 Aug 2014
12. Mazzola G et al (2002) The topos of music—geometric logic of concepts, theory, and performance. Birkhäuser, Basel
13. Mazzola G (2007) La vérité du beau dans la musique. Delatour, Paris
14. Mazzola G et al (2011) Musical creativity. Springer, Heidelberg
15. Milmeister G (2009) The Rubato Composer software. Springer, Heidelberg
16. Simondon G (1989) Du mode d'existence des objets techniques. Aubier, Paris
17. Thalmann F (2007) Musical composition with grid diagrams of transformations. MA thesis, University of Bern, Bern

Algorithmic Music Composition

David Cope

Introduction

I am currently in the process of completing a book titled *The Algorithmic Universe*. My premise for that book is that everything in the universe, all matter and energy, derives from algorithms. Interestingly, I find myself somewhat embarrassed that I haven't included music between its covers, only so much space allowed. Therefore, when asked to write a chapter for the current volume on exploring compositional strategies using algorithmic techniques, it seemed a perfect fit to make the thesis for my own book complete. Thus my premise in this chapter is simple—all composers use algorithms while composing whether aware of doing so or not. To analyze and understand the works of these composers requires the discovery of those algorithms. Once discovered, awareness of these algorithms will greatly increase the appreciation of these works for listeners. Therefore, I begin this chapter with the simple statement: *all* composers are *algorithmic* composers, not just those who profess it so, and I here attempt to prove it.

Definitions

As most readers of this chapter already know, an algorithm is a recipe, a series of instructions on how to solve a problem. In our case, this problem means composing a work of music. Like recipes, these instructions can be highly restricted, vague, or somewhere in between these extremes. This represents an important distinction, since many assume that algorithms, principally known for their use in computer technology, require precision.

D. Cope (✉)
University of California, Santa Cruz, USA
e-mail: howell@ucsc.edu

G. Nierhaus (ed.), *Patterns of Intuition*,
DOI 10.1007/978-94-017-9561-6_19

Fig. 1 J.C.F. Fischer's fugue no. 1 (1702)

As an initial example, John Cage in his *Music of Changes* (1951) selected pitches, durations, tempi, and dynamics by using the I-Ching, an ancient Chinese collection of 64 hexagrams that prescribe methods for arriving at divine or random results, depending on your point of view. Cage regarded his *Music of Changes* as indeterminate in regard to composition, but determinate in regard to performance, because the work is fixed from one such performance to the next, see [1]. Clearly Cage's approach represents and example of a highly restricted algorithmic category, since every aspect of its score results from a precise set of instructions.

Figure 1, presents a less restrictive approach popular in the baroque period of music history—the fugue.

While little is known about Fischer [2], his precise and often interesting fugues demonstrate the basic characteristics that Bach polished to perfection. The alternate tonic-subdominant relations of its four entrances (the exposition) are followed by a brief series of episodes and subject entries (mm. 5–10) ending with a varied repeat of the exposition. This work clearly represents an algorithm in use. There are, of course, many other rules for fugues I need not mention here, but provide the fugue with the harmony, counterpoint, style, and modulations necessary to produce a moderately restrictive popular Baroque form. One could further argue the theme itself results from an algorithm, but doing so is not necessary to make the point here.

In contrast, Fig. 2 presents an example of a very loose algorithmic composition.

This work provides precious little for the orchestra by way of instructions as to how it is to be performed (there are no program notes or other words regarding performance). One might suspect, therefore, that no two performances will likely

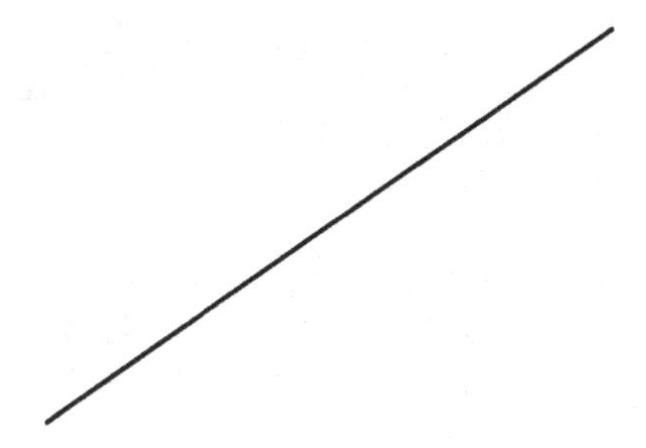

Fig. 2 *A Straight Line* for orchestra. Paul Ignace, 1948

resemble one another much. In fact, the opposite is usually true, as the direction of the straight line and its singular nature, typically create performances that move upwards slowly in pitch, get louder in dynamics, and proceed using faster and faster rhythms.

These works by Fischer, Cage, and Ignace all clearly result from algorithms. Each also points out differences between implicit and explicit algorithmic composition. In the cases of Cage and Fischer, the two scores provide little indication of an algorithm at work. In the case of *Music of Changes*, for example, the standard notation used in the score completely hides the algorithm at work. Most listeners would consider the fugue by Fischer, while clearly the result of a fairly strict algorithm both in terms of form and style, simply rules of the period in which he lived and not an indication that such forms result from algorithms. The explicit nature of the Ignace graphic score firmly indicates a loose algorithm at work, though audiences without access to the score might assume that the music is precisely notated.

Thus, I argue that all three of these scores represent algorithmic approaches to composition, regardless of the fact that all three result from quite different approaches, styles, and degree of strictness. If I've been clear in my approach here, the question now becomes, how do composers compose *without* using algorithms? And my answer is simple; *they don't*. All composers use algorithms whether they know it, admit it, or deny it. And thus begins my argument that it is the use of computers that many people fault in computer composition rather than algorithms; however, since computers are simply tools composers use for their already extant algorithms, they in no way represent the cause for works composed by them to be considered foreign to the traditional concepts associated with human creativity and composition.

Common Practice Music

Classical music, as we currently describe it, has existed for well over a millennium. It began, according to at least many musicologists, as Gregorian chant (single line melodies consisting of mostly stepwise motion) sometime roughly in the 9th century AD, and has continued to the early 20th century. Of course, anything as broad as this description is bound to be imprecise, but one can adhere to the basic idea and still allow for exceptions. During this time, composing algorithms of all the types

Fig. 3 K.222 by Mozart (1775)

Fig. 4 Beethoven's ninth symphony, movement 4 (1824)

mentioned earlier abounded. Whether they knew it or not, composers followed strict and not so strict rules that gave listeners the ability to predict, even on first hearing, what was coming next with a high degree of accuracy.

The single-line melody of chant lasted for centuries in the Catholic Church before developing textures of counterpoint and harmony. The singular nature of these lines, their algorithm, involved the laziest possible motion from one pitch to the next. Motion by major and minor seconds belongs to scales. Occasional small leaps occurred, but rarely so, at least at the beginning. This prevalence on small intervals meant that the mostly amateur singers involved in performances could sing them more accurately. So ingrained were these mostly stepwise lines that they continued as a basic attribute of classical styles throughout what is now referred to as the Common Practice Period (roughly 1600–1900).

Figures 3 and 4 presents a good example of this type of algorithm at work in two melodies from the Classical period (Mozart and Beethoven).

While the two themes are not precisely the same, they are clearly recognizable to anyone who knows both works. And yet there is little or no evidence that Beethoven ever heard the Mozart. The two themes derived from a common algorithm in the classical aether used by many composers, rather than from plagiarism.

Note that both themes in Figs. 3 and 4 are fundamentally stepwise. That is, they contain almost nothing but major and minor seconds (the Mozart has one third, but it separates a repetition). This simple movement derives from the same Gregorian chant previously mentioned. Seconds are much easier to perform than leaps. The term 'voice-leading' is used to describe these motions, and, as I will soon discuss, of even instrumental parts in music.

The music in Figs. 5, 6 and 7 demonstrates how voice leading extends into harmonic motions as well as melodies. In Fig. 5, Bach typically uses seconds in each of his choral lines (bass parts are typically somewhat exempt from this rule), particularly in his chorales. In Fig. 6, Mozart seems to deviate from this norm in his Piano Sonata 16. However, looking at the structure from a voice-leading perspective proves that (Fig. 7), with the exception of the outlines of triads in the upper melodic part, the other 'voices' move by seconds.

Thus, even when composers used instruments, making leaps of many different sizes easy to play, these seconds persisted, often becoming superstructures to yet

Fig. 5 Beginning of Bach Chorale 34 (*Erbarm dich mein, o Herre Gott*, Riemenschneider); BWV 305

Fig. 6 Mozart sonata 16, beginning, K. 545

Fig. 7 Voice-leading structure in Mozart's sonata 16, beginning, K. 545

another aspect of the growing complexity of classical music: harmonic progression. As this latter idea became the norm (several notes occurring at the same time), a kind of syntax arose: some groups of notes (chords) moving to others were common, some rare, and some forbidden. Without going into details, this syntax became as algorithmic as the step-wise motion between melodic notes. By the time the Classical Period (roughly 1750–1825) arrived, listeners could as easily predict the next chord as they'd previously predicted the next melodic note.

As example of such prediction, I often give my students dictation by playing a musical example but one time only. I call these one-shots. When I find a class initially intimidated, I give them a no-shot dictation. In other words, they hear nothing, but

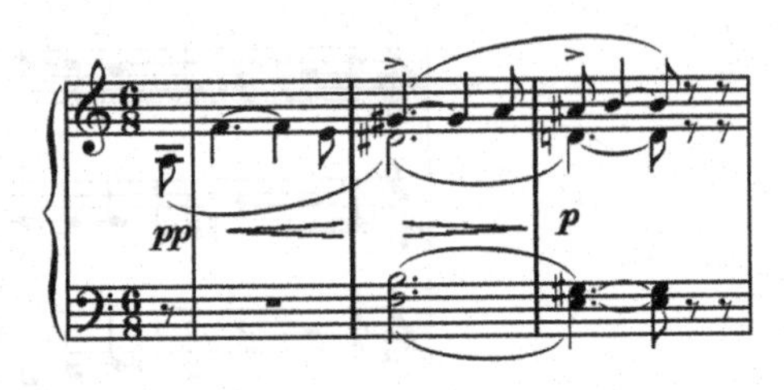

Fig. 8 Opening of the *Tristan und Isolde* by Richard Wagner (1859)

Fig. 9 Louis Spohr's earlier opera *Der Alchymist* (1829)

have to figure out what I might have played. To help them along, I give them the number of chords I didn't play, and the mode: major or minor. With few exceptions, most of my classes get 75 % or more of these progressions correct. The reason is that harmonic progressions in classical music result from algorithms. Once begun, they generally follow predictable patterns to the cadence.

Even when classical music extends itself into realms unknown, one finds examples of new chords developing over time among many different composers, rather than springing whole body from one iconoclast: in other words, a developing algorithm rather than a strikingly new one.

Figures 8 and 9 show the famous Tristan chord, supposedly originated by Wagner in his opera *Tristan und Isolde*, followed by a transposed but almost identical earlier version by Louis Spohr.

There are many other pre-Wagner examples of this chord in the music of Beethoven, Schumann, and Liszt, a good friend of Wagner. Also note here the movement by seconds in the instrumental lines, a clear vestige of traditional voice leading.

And this is just the beginning of a full discussion on classical music algorithms. Instruments have limited ranges requiring composing rules to avoid exceeding them. Performers also face fingering problems on their instruments that often cannot be surmounted, balance problems which, if not algorithmically cared for, may result in their sound becoming lost in the overall balance, a nearly infinite number of rhythms that at certain speeds cannot be executed, and so on. Each limitation requires an algorithm that composers share, whether they know it or not.

While no one could predict an entire piece or even an entire phrase from an initial idea, the idea of relatively strict algorithmic processes in use is difficult to argue. Most if not many of the decisions that composers make subconsciously or by inspiration derive from algorithms. All this is clear to any serious music theorist, and to any conscientious listener of classical music.

The same could easily be said of popular music, rock-and-roll, blues, country and western, jazz, non-western music, and so on. While the algorithms may differ among styles, the fact that algorithms play important and even critical roles is undeniable. Music of random choices, while possibly interesting to some, does not serve as good listening for most. And, as we've seen, even random music is algorithmic by its very nature.

Thus, without going into great detail here, composing music of any style is algorithmic. And in many cases, some of which may be surprising, algorithms reside at the base of even the most striking examples of music that seem to defy analysis, at least the analysis of the times in which it was composed (analysis, of course, being the search for algorithms).

Therefore, no matter that many consider composing a creative enterprise and composers having complete freedom in what they choose, music of every type is constricted by algorithms that themselves dictate choices.

Examples from the 20th Century

Perhaps no other composer in the history of music exemplifies the apparent conflict between implicit and explicit algorithmic composition than Arnold Schoenberg. His music ranges from late-romantic music influenced by Richard Wagner (see *Gurre-Lieder*, for example) to more atonal later works (see his own *Variations for Orchestra*). Figure 10 presents another few measures from this highly influential opera by Wagner, *Tristan und Isolde*, and a work that Schoenberg highly admired.

Here we see the characteristic chromaticism (all 12 pitches present), long-held chords (all measures basically E-Major dominant in nature), and non-harmonic pitches with delayed resolution to add to the dissonance present (though always eventually resolved). As well, however, the primary intervals present in this example and in the entire opera as well, are seconds and thirds, a clear evolution from common-practice voice leading.

Fig. 10 Richard Wagner, *Tristan und Isolde*, Prelude (1859), mm. 64–68

Fig. 11 From Schoenberg's *Verklärte Nacht* (Transfigured Night, 1899), piano arrangement mm. 126–130

Figure 11 then presents a piano arrangement of the string sextet by Schoenberg (1899) *Verklärte Nacht*.

Here we see a somewhat more advanced level of chromaticism than in Wagner's work, much more variety in chord choices and motions, seventh chord dissonance (where some of the chords that might have appeared dissonant in Wagner are here consonant), and yet the music remains tonal (see the dominant tonic in B Major on the second and third beats of measure 128). However, the majority of intervals, certainly melodically as well as harmonically, remain primarily seconds, and to a lesser degree thirds. Schoenberg continued this style up until roughly 1908 and his lush romantic style would produce many astounding works (again see *Gurre-Lieder* for huge orchestra, chorus, and vocal soloists).

Figure 12, the opening measures of Schoenberg's Opus 19, *Six Little Piano Pieces*, provides a transition work in which the chromaticism of the previous two examples continue, but the triadic basis and tonality is lost or at least blurred to the point of disappearing (a controversial issue with Schoenberg who maintained that tonality continued in his music, but more with individual notes than with conventional major and minor scales).

For many, like me, this so-called atonal music lacks the firm algorithms of Figs. 10 and 11, thus leading Schoenberg to firmer ground as presented in Fig. 13, the beginnings of his more formal serial period.

This work presents a twelve-tone row in its original form in the right hand (as E, F, G, Db, Gb, Eb, Ab, D, B, C, A, Bb) and the transposition of that row at the augmented fourth in the left hand (note that the last four notes of this version appear ahead of time in the bottom voice in combination with the intervening notes of pitches 5–8). Schoenberg defines his serial process in his book *Style and Idea* [3], a book that contains several other of his papers. Repeated pitches abound, with seconds in

Fig. 12 From Schoenberg's Opus 19, *Sechs Kleine Klavierstücke* (1911), No. 1, mm. 1–5

Fig. 13 From Schoenberg's Opus 25, *Suite für Klavier*, Präludium (1921), mm. 1–5

clear abundance. Clearly, however, this music separates itself more clearly from the seconds-based voice leading of the previous examples.

Figure 14 presents five measures of Anton Webern's Piano Variations that extend Schoenberg's ideas one step further.

Here we see serialism applied to articulations, durations, dynamics, and even range, so that once described in algorithmic terms, one can imagine that the work might have been created by an assistant or a computer (not actually true, but it bears stating nonetheless). Interestingly, however, even in this pointillist environment, the row used—with the octaves reduced to pitch classes—contains primarily seconds, less thirds and one augmented fourth, an interesting correlation with traditional voice leading of classical common-practice algorithms.

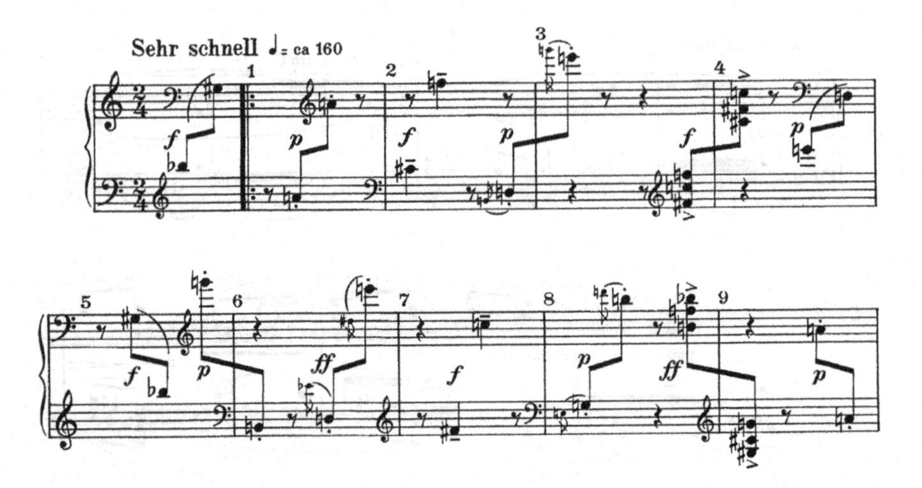

Fig. 14 Anton Webern, *Piano Variations, II* (1936), mm. 1–9

Computers and Music Composition

The first use of computers in music involved using them to create unusual musical timbres rather than as compositional tools. As an extension of earlier analog electronic music, digital processes allowed composers incredible new freedom over the use of their materials. Interestingly, while many composers felt this apparent freedom gave them more choices than composers in previous centuries, the very materials they used—synthesis of new instruments, temporal flexibility, and so on—required algorithms. In fact, all of the programs that composers created and were used by others were called algorithms. And, of course, the end results—which output they preferred most—were the results of the most interesting algorithm of all: the human composer.

For this, now, I must digress slightly to make an important idea clear. Up to this point in the chapter, I've included algorithmic processes as under the control of human composers: i.e., as tools to use on paper, rather than as separate processes defined by machines. With the advent of computational technology, the human composer begins, apparently to share responsibility for the ultimate composition. And this concept often has repercussions with audiences that require explanation.

We, as humans, rely on non-computational algorithms every second of our lives. Even in the womb, for it is there that DNA, an algorithm that determines our hair color, gender, eye color, the color of our skin, and a huge number of other aspects about ourselves that determine who we'll become. As well, the world around us, the plants, the weather, other people, affect our algorithmic brains in ways that make us even more special and different from one another. In short, we couldn't exist without these algorithms; nor could the universe around us. And our music, too, reflects these in our biases (as we've seen).

Thus, the use of computers in the composing process is a natural continuation of the historical use of algorithms. Nothing has changed except that a tool for extending our creative processes has given us possibilities that no one prior to the mid 20th century could possibly have imagined. Thus, computer composing is an extraordinary addition of power and not a transfer of it to a lifeless amalgam of metals and chemicals. There is nothing but good that will arise from our use of computers in this way.

Computer Algorithms

For decades, I and many other composers have used computers for creating music explicitly in our composing. For many of us, doing so seems natural because we'd been using paper algorithms for many decades previous. It was a difficult transition, however, possibly because computers require codes that need to be followed precisely in order for the hardware and software to work correctly. In other words, it required learning a computer language. Paper algorithms, on the other hand, do not require such specific codes. Ultimately, however, both paper and computer algorithms serve exactly the same purpose: to inform someone or something on how to proceed with a recipe like that defined at the beginning of this chapter.

This process of using computers to actually compose music still seems foreign to some, as if computer programs are somehow beings in themselves, capable of making decisions without our permission; as if the output were not ours, but theirs. This is, of course, ridiculous. Computers are not intelligent. They may be confusing, complex, and occasionally unpredictable, but they are certainly not intelligent. As well, like many of my colleagues, I have severe doubts that computers, at least as we know them now, will ever become intelligent. Intelligence requires a complex link to the world of the physical and development through evolution. Regardless, however, the only attributes that computers currently bring to algorithmic composition are speed and accuracy.

Projects and Compositions

I have greatly enjoyed reading about the various projects described in this book. Each of these projects, whether explicitly stated or not, represents an example of algorithmic composition. Whether based on personal aesthetics, intuition, formalization, or curation of output, they carry with them premises based on planning, prediction, rules, and definitions that exemplify what it means to be algorithmic. I applaud their efforts and look forward to appreciating their final results when I can hear them performed.

References

1. Cage J (1961) Composition: to describe the process of composition used in music of changes and imaginary landscape no. 4. In: Silence: lectures and writings. Wesleyan University Press, Middletown, pp 57–59
2. Rudolf W (1990) Johann Caspar Ferdinand Fischer, Hofkapellmeister der Markgrafen von Baden, vol 18. Quellen und Studien zur Musikgeschichte von der Antike bis in die Gegenwart. Peter Lang, Frankfurt am Main
3. Schönberg A (1950) Style and idea. Philosophical Library, New York

Contributing Researchers

Sandeep Bhagwati

Sandeep Bhagwati is a multiple award-winning composer, theatre director and media artist. He studied at Mozarteum Salzburg (Austria), Institut de Coordination Acoustique/Musique IRCAM Paris (France) and graduated with a Diplom in Composition from Hochschule für Musik und Theater München (Germany) His compositions and comprovisations in all genres (including 6 operas) have been performed by leading performers at leading venues and festivals worldwide. He has directed international music festivals and intercultural exchange projects with Indian and Chinese musicians and leading new music ensembles. He was a Professor of Composition at Karlsruhe Music University, and Composer-in-Residence at the IRCAM Paris, ZKM Center for Arts and Media Karlsruhe, Beethoven Orchestra Bonn, Institute for Electronic Music Graz, CalArts Los Angeles, Heidelberg University and Tchaikovsky Conservatory Moscow. He also was a guest professor at Heidelberg University in 2009 and a visiting research fellow to the University of Arts Berlin in 2013/14. From 2008 to 2011, he was the director of Hexagram Concordia, a centre for research-creation in media arts with a faculty of 45 artist-researchers and extensive state-of-the-art facilities. Since 2013, he is the artistic and musical director of Ensemble Extrakte Berlin, a poly-traditional ensemble for post-exotistic glocal musicking.

As Canada Research Chair for Inter-X Arts at Concordia University Montréal since 2006 he currently directs **matra**lab, a research/creation center for intercultural and interdisciplinary arts. His current work centers on comprovisation, inter-traditional aesthetics, the aesthetics of interdisciplinarity, gestural theatre, sonic theatre and interactive visual and non-visual scores.

G. Nierhaus (ed.), *Patterns of Intuition*,
DOI 10.1007/978-94-017-9561-6

William Brooks

William Brooks is Professor of Music at the University of York and a Senior Fellow at the Orpheus Institute, Ghent. He is also Emeritus Professor at the University of Illinois. A scholar as well as a composer, he has written extensively about American music (especially Charles Ives and John Cage) and about experimentalism. He is noted for his compositions for voice and has received commissions from the British Arts Council, the Gulbenkian Foundation, the Arts Council of Ireland, and other agencies. His music is published by Frog Peak.

David Cope

David Cope is Professor Emeritus at the University of California at Santa Cruz, and teaches regularly in the annual Workshop in Algorithmic Computer Music (WACM) held in June-July at UC Santa Cruz. His books on the intersection of music and computer science include *Computers and Musical Style, Experiments in Musical Intelligence, The Algorithmic Composer, Virtual Music, Computer Models of Musical Creativity* and *Hidden Structure* and describe the computer program Experiments in Musical Intelligence which he created in 1981. Experiments in Musical Intelligence's works are published by Spectrum Press and include Horizons for orchestra, three operas in the style of, and with librettos consisting of, letters by the composers Mozart, Schumann, and Mahler, and a symphony and piano concerto in the style of Mozart, and a seventh Brandenburg Concerto in the style of Bach. Experiments in Musical Intelligence's works are available on five Centaur Records' albums (Bach by Design, Classical Music Composed by Computer, Virtual Mozart, Virtual Bach, Virtual Rachmaninoff). Works composed in his own style include ten symphonies, ten string quartets, several chamber orchestra pieces, and a host of other works, most of which have been performed around the world.

Darla Crispin

Darla Crispin is an Associate Professor of Musicology at the Norwegian Academy of Music (NMH), Oslo. She works there as a member of a research team of pianists, composers and musicologists exploring collaborative and interdisciplinary research techniques and their application to both newly-composed and canonical piano repertoire.

A Canadian pianist and scholar with a Concert Recital Diploma from the Guildhall School of Music and Drama, London and a Ph.D. in Historical Musicology from King's College, London, Darla specialises in musical modernity, and especially in the music of the Second Viennese School. Her most recent work examines this repertoire through the prism of artistic research in music, a process which has been reinforced through her work as a Research Fellow at the Orpheus Research Centre in Music from 2008–2013.

As well as developing her own research, Darla has been responsible for the development of innovative postgraduate programmes in two leading UK Conservatoires: the Guildhall School of Music & Drama and, from 2002–2008 the Royal College of Music, where she was the founding Head of the College's Graduate School, overseeing both Masters and Doctoral programmes. She is an Honorary Member of the RCM, a Fellow of the Royal Society of Arts and a member of the Advisory Board for the Platform for Artistic Research (PARSE) for the Faculty of Fine, Applied and Performing Arts, the University of Gothenburg, Sweden.

Darla's publications include a collaborative volume with Kathleen Coessens and Anne Douglas, *The Artistic Turn: A Manifesto* (Leuven, 2009) and numerous book chapters and articles. Some of the more recent of these include 'Allotropes of Advocacy: a model for categorizing persuasiveness in musical performances', co-authored with Jeremy Cox, in *Music & Practice*, Vol. 1 (1) 2013 and 'Of Arnold Schoenberg's *Klavierstück Op. 33a*, "a Game of Chess," and the Emergence of New Epistemic Things', in Experimental Systems—*Future Knowledge in Artistic Research*, ed. Michael Schwab (Leuven 2014). She is currently working on a book entitled *The Solo Piano Works of the Second Viennese School: Performance, Ethics and Understanding* (Boydell and Brewer).

Nicolas Donin

Nicolas Donin is head of the Analysis of Musical Practices team, a joint research group of Institut de Recherche et de Coordination Acoustique/Musique, Université Pierre et Marie Curie and Centre National de la Recherche Scientifique in Paris. His work focuses on contemporary musical practices, particularly composition and performance, using both musicological and ethnographic/cognitive approaches. From 2009 to 2011 he led the MuTeC project (a series of case studies in compositional processes from the 1930s to the present) funded by the Agence nationale de la recherche, and established the biennial conference TCPM (Tracking the Creative Process in Music). He co-edited with Rémy Campos *L'analyse musicale, une pratique et son histoire* (Droz/Conservatoire de Genève, 2009), and with Laurent Feneyrou *Théories de la composition musicale au XXe siècle (Symétrie).* His recent work has been published in *Contemporary Music Review, Genesis: Revue Internationale de Critique Génétique, Musicae Scientiae*, as well as in various edited collections in French and English. He also co-authored several documentary films on the creative process of composers Georges Aperghis, Luca Francesconi, Philipp Maintz, Roque Rivas and Marco Stroppa.

Thomas Eder

Thomas Eder is a lecturer at the University of Vienna. He focuses on literary theory and avant-garde studies. He completed his Ph.D. by a study on the Austrian post-war avant-garde and has authored several articles as well edited books on this topic. Currently he is working on his habilitation thesis on *Cognitive Poetics. A Critical Re-evaluation* in which he critically examines theoretical issues of applying Cognitive Studies to the understanding of literary texts. Metaphor theory and synesthesia are the two focuses of his research which relate to the cooperation with Clemens Gadenstätter. Additionally Thomas Eder works in the Office of Publications for the Federal Chancellor of Austria and is the head of the unit for Corporate Design strategies and publications.

Harald Fripertinger

As a flutist with completed studies in concert class and pedagogy he is teaching at music school of Köflach, Austria. Habilitated in mathematics he is teaching at University of Graz and Graz University of Technology. His key aspects of activity are combinatorics with group operations applied in coding theory and mathematical music theory as well as functional equations and iteration theory.

Daniel Mayer

Born 1967, Daniel Mayer completed degrees in pure mathematics and philosophy at the University of Graz and music composition with Prof. Gerd Kühr at the University of Music and Performing Arts Graz, Austria. 2001/02 he continued his studies at the Electronic Studio of the Music Academy of Basel, Switzerland, with Hanspeter Kyburz. He was a guest composer at the Center for Art and Media Karlsruhe (2003/04) and IEM Graz (2005) and is using generative computer algorithms in electronic and instrumental music.

Guerino Mazzola

Born 1947, Guerino Mazzola qualified as a professor in mathematics (1980) and in computational science (2003) at the University of Zürich. Visiting professor at the Ecole Normale Supérieure in Paris in 2005. Since 2007 he is professor at the School of Music, University of Minnesota. He developed a Mathematical Music Theory and software presto and Rubato. Since 2007 he is the president of the Society for Mathematics and Computation in Music. He has published 24 books and 120 papers, 24 jazz CDs, and a classical sonata.

Gerhard Nierhaus

Gerhard Nierhaus studied composition with Peter Michael Hamel, Gerd Kühr and Beat Furrer. Working within both traditional and contemporary digital and intermedial formats, his compositional output includes numerous works of acoustic and electronic music and includes visual media. Gerhard Nierhaus currently is a researcher at the Institute of Electronic Music and Acoustics at the University of Music and Performing Arts Graz, Austria, and teaches Computer Music and Algorithmic Composition. His book *Algorithmic Composition: Paradigms of Automated Music Composition* was published by Springer Wien/NewYork in 2009.

Hanns Holger Rutz

Hanns Holger Rutz (b. 1977 in Germany) studied computer music and audio engineering at the Technical University Berlin, and from 2004–2009 worked as assistant professor at the Studio for electroacoustic Music (SeaM) Weimar. In 2014 he received a Ph.D. from the Interdisciplinary Centre for Computer Music Research (ICCMR) in Plymouth (UK). His compositions include tape music, works with video, as well as collaborative works with theatre and dance. His recent focus is on sound installation, electronic live improvisation and the development of novel algorithmic systems. His work has been presented in international festivals, exhibitions and concerts, including Germany, Austria, Romania, Latvia, Denmark, England, Spain, Slovenia, and Venezuela. He currently lives in Graz and works at the Institute for Electronic Music and Acoustics (IEM).